Methods in Molecular Biology

Series Editor
John M. Walker
School of Life and Medical Sciences
University of Hertfordshire
Hatfield, Hertfordshire, UK

For further volumes:
http://www.springer.com/series/7651

For over 35 years, biological scientists have come to rely on the research protocols and methodologies in the critically acclaimed *Methods in Molecular Biology series.* The series was the first to introduce the step-by-step protocols approach that has become the standard in all biomedical protocol publishing. Each protocol is provided in readily-reproducible step-by step fashion, opening with an introductory overview, a list of the materials and reagents needed to complete the experiment, and followed by a detailed procedure that is supported with a helpful notes section offering tips and tricks of the trade as well as troubleshooting advice. These hallmark features were introduced by series editor Dr. John Walker and constitute the key ingredient in each and every volume of the *Methods in Molecular Biology series.* Tested and trusted, comprehensive and reliable, all protocols from the series are indexed in PubMed.

Cerebral Cortex Development

Methods and Protocols

Edited by

Koh-ichi Nagata

Institute for Developmental Research, Aichi Developmental Disability Center, Aichi, Japan

Humana Press

Editor
Koh-ichi Nagata
Institute for Developmental Research
Aichi Developmental Disability Center
Aichi, Japan

ISSN 1064-3745 ISSN 1940-6029 (electronic)
Methods in Molecular Biology
ISBN 978-1-0716-3809-5 ISBN 978-1-0716-3810-1 (eBook)
https://doi.org/10.1007/978-1-0716-3810-1

Cover Caption: Image courtesy of Dr. Nanako Hamada.

This Humana imprint is published by the registered company Springer Science+Business Media, LLC, part of Springer Nature.
The registered company address is: 1 New York Plaza, New York, NY 10004, U.S.A.

Paper in this product is recyclable.

Preface

It would be my great pleasure to issue this volume *Cerebral Cortex Development* of the highly successful laboratory protocol book series Methods in Molecular Biology.

As the volume editor, I wish to express my appreciation to the chapter contributors whom are very active, as well as established researchers and the senior editor, Prof. Emeritus John Walker, for his guidance.

Following the tradition of the series, each chapter aims to describe a specific technique or application, presented in an easy-to-follow and step-by-step format. In this volume, contributors introduce 28 established methods that have been "tried and tested" by the research community, rather than very new methods that are yet to be used and validated. Based on the policy of this series, this book does not contain methods that are still to be fully established. This volume is suitable for beginners to experienced researchers. Readers will get useful hints and tips to easily image respective experiments, with the support of the valuable "Notes" section.

Meanwhile, genetic and in silico analyses, including database analyses, are tremendously developing and represent a relevant prerequisite to current research. Although these methods are not included in this book, it is my great pleasure to have the Overview by Prof. Emeritus Mariko Y. Momoi, who discussed the most cutting-edge methodologies on genetics and bioinformatics in the field. It is noteworthy that ample references in the Overview allows the readers to easily access updated information about new and developing methods as well as in silico techniques.

Finally, I would like to thank the late Prof. Alan Hall (Univ. College London) for his mentorship still influencing me and Junko for her long unwavering encouragement.

Aichi, Japan *Koh-ichi Nagata*

Contents

Preface . . . . *v*
Contributors . . . . *ix*

1 Overview: Research on the Genetic Architecture of the Developing Cerebral Cortex in Norms and Diseases 1
Mariko Y. Momoi

2 In Situ Hybridization of Brain Slices 13
Tomohiro Kanayama, Hiroyuki Tomita, and Akira Hara

3 Immunohistochemistry of Brain Tissues 21
Akira Hara, Toshiaki Taniguchi, Tomohiro Kanayama, and Hiroyuki Tomita

4 Two-Photon FRET/FLIM Imaging of Cerebral Neurons 33
Thomas T. Luyben, Jayant Rai, Bingyue Zhou, Hang Li, and Kenichi Okamoto

5 Pre- and Postembedding Immunoelectron Microscopy to Analyze the Central Nervous System 45
Fumino Fujiyama and Fuyuki Karube

6 Mouse Brain Tissue Preparation for Scanning Electron Microscopy 63
Kosuke Mori, Chihiro Takada, Hideshi Okada, and Hiroyuki Tomita

7 Simple Method for the Preparation of Postsynaptic Density Fraction from Mouse Brain 71
Hidenori Ito, Rika Morishita, and Koh-ichi Nagata

8 Reconstitution of Neuronal Motor Traffic on Septin-Associated Microtubules 79
Yani Suber and Elias T. Spiliotis

9 BioID Analysis of Actin-Binding Proteins 95
E. Emily Joo and Michael F. Olson

10 Proteomic Analyses of the Mouse Brain 105
Jing Li

11 Generation of iPSCs Using Sendai Virus Vectors 121
Hitomi Aoki

12 Directed Differentiation of Neurons from Human iPSCs for Modeling Neurological Disorders 141
Cheng Wang, Jonas Cerneckis, and Yanhong Shi

13 Generating Homogeneous Brain Organoids from Human iPSCs 157
Xianwei Chen and Yanhong Shi

14 Primary Culture of Dissociated Neurons from the Embryonic Cerebral Cortex 169
Shiho Ito, Megumi Kawasaki, and Takeshi Kawauchi

15 Immunocytochemistry of Primary Cultured Cerebral Cortical Neurons 177
Shiho Ito and Takeshi Kawauchi

16 Brain Sample Preparation After Performing in Utero Electroporation to Proliferating and Differentiated Cells in the Developing Mouse Neocortex 187
Minkyung Shin and Yoichi Kosodo

17 Live Imaging of Migrating Neurons and Glial Progenitors Visualized by in Utero Electroporation 201
Masashi Nishikawa, Koh-ichi Nagata, and Hidenori Tabata

18 Ca^{2+} Imaging in Immature Cortical Neurons 211
Shin-ichiro Horigane and Sayaka Takemoto-Kimura

19 In Vitro Patch-Clamp 221
Toshihiro Nomura

20 In Vivo Whole-Cell Recording from the Mouse Brain 245
Masahiro Kawatani and Takayuki Yamashita

21 In Vivo Brain Microdialysis 259
Makoto Kondo

22 CT Scan Data Analysis in Malformations of Cortical Development 271
Marcello Scala and Mariasavina Severino

23 MRI Data Analysis in Malformations of Cortical Development 281
Mariasavina Severino, Domenico Tortora, and Marcello Scala

24 Genomic Copy Number Analysis Using Droplet Digital PCR: A Simple Method with EvaGreen Single-Color Fluorescent Design 293
Takeaki Tamura, Taichi Imaizumi, Keiko Shimojima Yamamoto, and Toshiyuki Yamamoto

25 Locomotor Assays in *Drosophila* Larvae and Adult Flies 305
Leo Tsuda, Huynh Man Anh, and Hideki Yoshida

26 Analyses of Genetic Regulation of the Nervous System in the Nematode *Caenorhabditis elegans* 313
Tzu-Ting Huang and Ikue Mori

27 Analyses of Neural Circuits Governing Behavioral Plasticity in the Nematode *Caenorhabditis elegans* 321
Tzu-Ting Huang and Ikue Mori

28 Social Behavior in Animal Models of Autism Spectrum Disorder 331
Hitomi Kurahashi, Kazuo Kunisawa, and Akihiro Mouri

29 Preparation of Healthy Single Neuron or Astrocyte Suspension from Adult Mouse Brain for RNA-seq 341
Neng-Yuan Hu, Tian-Ming Gao, and Yi-Hua Chen

Index *353*

Contributors

HUYNH MAN ANH • *Kyoto Institute of Technology, Matsugasaki, Kyoto, Japan*

HITOMI AOKI • *Department of Stem Cell and Regenerative Medicine, Gifu University Graduate School of Medicine, Gifu, Japan*

JONAS CERNECKIS • *Department of Neurodegenerative Diseases, Beckman Research Institute of City of Hope, Duarte, CA, USA; Irell & Manella Graduate School of Biological Sciences, Beckman Research Institute of City of Hope, Duarte, CA, USA*

XIANWEI CHEN • *Department of Neurodegenerative Diseases, Beckman Research Institute of City of Hope, Duarte, CA, USA*

YI-HUA CHEN • *State Key Laboratory of Organ Failure Research, Key Laboratory of Mental Health of the Ministry of Education, Guangdong-Hong Kong-Macao Greater Bay Area Center for Brain Science and Brain-Inspired Intelligence, Guangdong-Hong Kong Joint Laboratory for Psychiatric Disorders, Guangdong Province Key Laboratory of Psychiatric Disorders, Guangdong Basic Research Center of Excellence for Integrated Traditional and Western Medicine for Qingzhi Diseases, Department of Neurobiology, School of Basic Medical Sciences, Southern Medical University, Guangzhou, China*

FUMINO FUJIYAMA • *Laboratory of Histology and Cytology, Faculty of Medicine and Graduate School of Medicine, Hokkaido University, Sapporo, Japan*

TIAN-MING GAO • *State Key Laboratory of Organ Failure Research, Key Laboratory of Mental Health of the Ministry of Education, Guangdong-Hong Kong-Macao Greater Bay Area Center for Brain Science and Brain-Inspired Intelligence, Guangdong-Hong Kong Joint Laboratory for Psychiatric Disorders, Guangdong Province Key Laboratory of Psychiatric Disorders, Guangdong Basic Research Center of Excellence for Integrated Traditional and Western Medicine for Qingzhi Diseases, Department of Neurobiology, School of Basic Medical Sciences, Southern Medical University, Guangzhou, China*

AKIRA HARA • *Department of Tumor Pathology, Gifu University Graduate School of Medicine, Gifu, Japan*

SHIN-ICHIRO HORIGANE • *Department of Neuroscience I, Research Institute of Environmental Medicine, Nagoya University, Nagoya, Japan; Molecular Cellular Neuroscience, Graduate School of Medicine, Nagoya University, Nagoya, Japan*

NENG-YUAN HU • *State Key Laboratory of Organ Failure Research, Key Laboratory of Mental Health of the Ministry of Education, Guangdong-Hong Kong-Macao Greater Bay Area Center for Brain Science and Brain-Inspired Intelligence, Guangdong-Hong Kong Joint Laboratory for Psychiatric Disorders, Guangdong Province Key Laboratory of Psychiatric Disorders, Guangdong Basic Research Center of Excellence for Integrated Traditional and Western Medicine for Qingzhi Diseases, Department of Neurobiology, School of Basic Medical Sciences, Southern Medical University, Guangzhou, China*

TZU-TING HUANG • *Neuroscience Institute, Graduate School of Science, Nagoya University, Nagoya, Japan; Graduate School of Life Sciences, Tohoku University, Sendai, Japan*

TAICHI IMAIZUMI • *Department of Pediatrics, St. Marianna University School of Medicine, Kawasaki, Japan*

HIDENORI ITO • *Department of Molecular Neurobiology, Institute for Developmental Research, Aichi Developmental Disability Center, Aichi, Japan*

Shiho Ito • *Department of Adaptive and Maladaptive Responses in Health and Disease, Graduate School of Medicine, Kyoto University, Sakyo-ku, Kyoto, Japan*
E. Emily Joo • *Department of Chemistry and Biology, Toronto Metropolitan University, Toronto, ON, Canada*
Tomohiro Kanayama • *Department of Tumor Pathology, Gifu University Graduate School of Medicine, Gifu, Japan*
Fuyuki Karube • *Laboratory of Histology and Cytology, Faculty of Medicine and Graduate School of Medicine, Hokkaido University, Sapporo, Japan*
Megumi Kawasaki • *Department of Adaptive and Maladaptive Responses in Health and Disease, Graduate School of Medicine, Kyoto University, Kyoto, Japan*
Masahiro Kawatani • *Department of Physiology, Fujita Health University School of Medicine, Toyoake, Japan*
Takeshi Kawauchi • *Department of Adaptive and Maladaptive Responses in Health and Disease, Graduate School of Medicine, Kyoto University, Sakyo-ku, Kyoto, Japan; Department of Physiology, Keio University School of Medicine, Shinjuku-ku, Tokyo, Japan*
Makoto Kondo • *Department of Anatomy and Neuroscience, Graduate School of Medicine, Osaka Metropolitan University, Osaka, Japan*
Yoichi Kosodo • *Korea Brain Research Institute (KBRI), Daegu, Republic of Korea; Department of Brain Sciences, Daegu Gyeongbuk Institute of Science and Technology (DGIST), Daegu, Republic of Korea*
Kazuo Kunisawa • *Department of Regulatory Science for Evaluation and Development of Pharmaceuticals and Devices, Graduate School of Health Science, Fujita Health University, Toyoake, Aichi, Japan*
Hitomi Kurahashi • *Department of Regulatory Science for Evaluation and Development of Pharmaceuticals and Devices, Graduate School of Health Science, Fujita Health University, Toyoake, Aichi, Japan*
Hang Li • *Lunenfeld-Tanenbaum Research Institute, Mount Sinai Hospital, Toronto, ON, Canada*
Jing Li • *Institute for Translational Medicine, The Affiliated Hospital of Qingdao University, Medical College, Qingdao University, Qingdao, China*
Thomas T. Luyben • *Lunenfeld-Tanenbaum Research Institute, Mount Sinai Hospital, Toronto, ON, Canada; Department of Molecular Genetics, Faculty of Medicine, University of Toronto, Toronto, ON, Canada*
Mariko Y. Momoi • *Ryomo Seishi Ryogoen Rehabilitation Hospital for Children with Disabilities, Gunma, Japan*
Ikue Mori • *Neuroscience Institute, Graduate School of Science, Nagoya University, Nagoya, Japan; Chinese Institute for Brain Research, Beijing (CIBR), Zhongguancun Life Science Park Changping District, Beijing, China*
Kosuke Mori • *Departments of Tumor Pathology, Gifu University Graduate School of Medicine, Gifu, Japan*
Rika Morishita • *Department of Molecular Neurobiology, Institute for Developmental Research, Aichi Developmental Disability Center, Aichi, Japan*
Akihiro Mouri • *Department of Regulatory Science for Evaluation and Development of Pharmaceuticals and Devices, Graduate School of Health Science, Fujita Health University, Toyoake, Aichi, Japan*
Koh-ichi Nagata • *Department of Molecular Neurobiology, Institute for Developmental Research, Aichi Developmental Disability Center, Kasugai, Aichi, Japan; Department of Neurochemistry, Nagoya University Graduate School of Medicine, Nagoya, Japan*

MASASHI NISHIKAWA • *Department of Molecular Neurobiology, Institute for Developmental Research, Aichi Developmental Disability Center, Kasugai, Aichi, Japan*
TOSHIHIRO NOMURA • *Department of Neuroscience, Feinberg School of Medicine, Northwestern University, Chicago, IL, USA*
HIDESHI OKADA • *Departments of Emergency and Disaster Medicine, Gifu University Graduate School of Medicine, Gifu, Japan*
KENICHI OKAMOTO • *Lunenfeld-Tanenbaum Research Institute, Mount Sinai Hospital, Toronto, ON, Canada; Department of Molecular Genetics, Faculty of Medicine, University of Toronto, Toronto, ON, Canada*
MICHAEL F. OLSON • *Department of Chemistry and Biology, Toronto Metropolitan University, Toronto, ON, Canada*
JAYANT RAI • *Lunenfeld-Tanenbaum Research Institute, Mount Sinai Hospital, Toronto, ON, Canada; Department of Molecular Genetics, Faculty of Medicine, University of Toronto, Toronto, ON, Canada*
MARCELLO SCALA • *UOC Genetica Medica, IRCCS Istituto Giannina Gaslini, Genoa, Italy; Department of Neurosciences, Rehabilitation, Ophthalmology, Genetics, Maternal and Child Health, University of Genoa, Genoa, Italy*
MARIASAVINA SEVERINO • *Department of Neurosciences, Rehabilitation, Ophthalmology, Genetics, Maternal and Child Health, University of Genoa, Genoa, Italy; Neuroradiology Unit, IRCCS Istituto Giannina Gaslini, Genoa, Italy*
YANHONG SHI • *Department of Neurodegenerative Diseases, Beckman Research Institute of City of Hope, Duarte, CA, USA; Irell & Manella Graduate School of Biological Sciences, Beckman Research Institute of City of Hope, Duarte, CA, USA*
KEIKO SHIMOJIMA YAMAMOTO • *Institute of Medical Genetics, Tokyo Women's Medical University, Tokyo, Japan; Department of Transfusion Medicine and Cell Processing, Tokyo Women's Medical University, Tokyo, Japan*
MINKYUNG SHIN • *Korea Brain Research Institute (KBRI), Daegu, Republic of Korea*
ELIAS T. SPILIOTIS • *Department of Biology, Drexel University, Philadelphia, PA, USA; Department of Cell Biology, University of Virginia School of Medicine, Charlottesville, VA, USA*
YANI SUBER • *Department of Biology, Drexel University, Philadelphia, PA, USA*
HIDENORI TABATA • *Department of Molecular Neurobiology, Institute for Developmental Research, Aichi Developmental Disability Center, Kasugai, Aichi, Japan*
CHIHIRO TAKADA • *Departments of Emergency and Disaster Medicine, Gifu University Graduate School of Medicine, Gifu, Japan*
SAYAKA TAKEMOTO-KIMURA • *Department of Neuroscience I, Research Institute of Environmental Medicine, Nagoya University, Nagoya, Japan; Molecular Cellular Neuroscience, Graduate School of Medicine, Nagoya University, Nagoya, Japan*
TAKEAKI TAMURA • *Department of Pediatrics and Child Health, Nihon University School of Medicine, Tokyo, Japan; Division of Gene Medicine, Graduate School of Medical Science, Tokyo Women's Medical University, Tokyo, Japan*
TOSHIAKI TANIGUCHI • *Department of Tumor Pathology, Gifu University Graduate School of Medicine, Gifu, Japan*
HIROYUKI TOMITA • *Department of Tumor Pathology, Gifu University Graduate School of Medicine, Gifu, Japan*
DOMENICO TORTORA • *UO Neuroradiologia, IRCCS Istituto Giannina Gaslini, Genoa, Italy*

LEO TSUDA • *Kansai Gakken Laboratory, Kankyo Eisei Yakuhin Co. Led, Seika-cho, Kyoto, Japan*
CHENG WANG • *Department of Neurodegenerative Diseases, Beckman Research Institute of City of Hope, Duarte, CA, USA*
TOSHIYUKI YAMAMOTO • *Division of Gene Medicine, Graduate School of Medical Science, Tokyo Women's Medical University, Tokyo, Japan; Institute of Medical Genetics, Tokyo Women's Medical University, Tokyo, Japan*
TAKAYUKI YAMASHITA • *Department of Physiology, Fujita Health University School of Medicine, Toyoake, Japan*
HIDEKI YOSHIDA • *Kyoto Institute of Technology, Matsugasaki, Kyoto, Japan*
BINGYUE ZHOU • *Lunenfeld-Tanenbaum Research Institute, Mount Sinai Hospital, Toronto, ON, Canada; Department of Molecular Genetics, Faculty of Medicine, University of Toronto, Toronto, ON, Canada*

Chapter 1

Overview: Research on the Genetic Architecture of the Developing Cerebral Cortex in Norms and Diseases

Mariko Y. Momoi

Abstract

The human brain is characterized by high cell numbers, diverse cell types with diverse functions, and intricate connectivity with an exceedingly broad surface of the cortex. Human-specific brain development was accomplished by a long timeline for maturation from the prenatal period to the third decade of life. The long timeline makes complicated architecture and circuits of human cerebral cortex possible, and it makes human brain vulnerable to intrinsic and extrinsic insults resulting in the development of variety of neuropsychiatric disorders. Unraveling the molecular and cellular processes underlying human brain development under the elaborate regulation of gene expression in a spatiotemporally specific manner, especially that of the cortex will provide a biological understanding of human cognition and behavior in health and diseases. Global research consortia and the advancing technologies in brain science including functional genomics equipped with emergent neuroinformatics such as single-cell multiomics, novel human models, and high-volume databases with high-throughput computation facilitate the biological understanding of the development of the human brain cortex. Knowing the process of interplay of the genome and the environment in cortex development will lead us to understand the human-specific cognitive function and its individual diversity. Thus, it is worthwhile to overview the recent progress in neurotechnology to foresee further understanding of the human brain and norms and diseases.

Key words Development of human cerebral cortex, Pathophysiology of neuropsychiatric disorders, Interplay of the genome and the environment, Integrated functional genomics

1 Introduction

Exploring human brain development has been a central theme of brain science. Despite progress in research, how the brain's unique cognitive function is acquired and which mechanisms make the human brain unique remain incompletely understood.

The human brain is characterized by high cell numbers, diverse cell types with different functions, and intricate connectivity with an exceedingly broad surface due to cortical gyrification. This human-specific brain complexity is accomplished by delaying development with a long childhood termed neoteny [1, 2]. The slower

Koh-ichi Nagata (ed.), *Cerebral Cortex Development: Methods and Protocols*, Methods in Molecular Biology, vol. 2794,
https://doi.org/10.1007/978-1-0716-3810-1_1,

development compared with that of nonhuman primates makes the brain's intricate architecture possible through a long timeline for maturation, extending from the prenatal to long after the postnatal period over two decades. This results in a large brain size and a highly developed neocortex that makes the human brain unique in its architecture and circuits.

In exchange for the highly developed neocortex, the long time spent in the fetal stage and immaturity at birth followed by a long childhood, adolescence, and young adulthood makes the human neocortex vulnerable to adverse genetic and environmental influences that make the human brain susceptible to a variety of neuropsychiatric disorders including developmental disorders. Of all brain regions, the prefrontal cortex plays a major role in executive functions such as regulating behavior and emotions, decision-making, and cognitive control. The prefrontal cortex is known to be a late-developing region that is susceptible to environmental influences. In fact, the region is the site of the central lesion in most neuropsychiatric disorders [3–8].

Unraveling human-specific, spatiotemporally specific molecular and cellular dynamics in the developing cortex, especially the prefrontal region, by comprehensive analyses of genetic and environmental impacts on their formation will directly or indirectly lead us to understand the biological mechanisms of neuropsychiatric disorders.

Considering the fact that most neuropsychiatric disorders present high prevalence and seriously adversely influence the well-being of the affected individuals and society, targeted therapeutic strategies based on the biological basis of the specific pathophysiology of each disorder would give affected individuals and society an enormous advantage.

Unraveling the molecular and cellular processes underlying the complex architecture and function of the human brain constructed under the elaborate regulation of gene expression in a spatiotemporally specific manner will provide deep insights into human cognition and behavior in health and disease. Growing evidence indicates the prenatal origin of most neuropsychiatric disorders [9, 10]. Thus, understanding human brain development, especially the development of the neocortex which serves human-specific cognitive and behavioral functions, is indispensable to unraveling normal brain function and the pathophysiology of neuropsychiatric disorders, where the derangement of cognitive function is the major phenotype. While the prenatal period and early childhood are vulnerable times in brain development, the prefrontal cortex, the site of high executive function, is a vulnerable spatial locus. Consequently, the elucidation of developmental processes of the human brain, specifically the cortex, is the major theme in understanding the development of normal human brain individuality and the mechanisms underlying the diverse phenotypes exhibited by various neuropsychiatric disorders.

The growing scientific and social interests in neuropsychiatric disorders have led to large projects or consortia such as the US BRAIN Initiative [11], the Japanese Brain/MINDS [12], and the European Human Brain Project [13] for innovating neurotechnology. Along with these national or international projects, the evolving capacity of methods such as whole-genome sequencing (WGS) and RNA sequencing has accelerated the accumulation of functional genomic data. National and international consortia have constructed catalogs of these genomic, transcriptomic, epigenomic, and proteosomic data (together named multiomics) from cells, postmortem human brain specimens, and model organisms to provide open resources for further analyses by research groups worldwide.

The advancing technologies in brain science in the twenty-first century, including functional genomics with multiomic analysis, new models of the human brain, and high-volume datasets equipped with high-throughput computation, have accelerated the unraveling of the biological mystery of the developing neocortex. Thus, it is worthwhile to review recent methodological progress to foresee the future development of our understanding of human cortical formation and the biological mechanisms of neuropsychiatric disorders.

2 From a Single Gene to Integrated Functional Genomic Analysis

Unraveling the developmental process of the human brain, specifically the neocortex, will directly and indirectly elucidate the biological mechanisms of neuropsychiatric disorders, in which the major phenotypes in most are cognitive and behavioral dysfunction.

Detecting causative disease-related alleles in genes affecting neuropsychiatric disorders was the major theme of clinical neuroscience in the last decades of the twentieth century. This was expected to reveal disease mechanisms and subsequently elucidate candidate molecular targets for specific treatments.

The discovery of the abnormally hypermethylated noncoding region in the *FMR1* gene in patients with fragile X syndrome that results in the loss of function of the FMR1 protein (FMRP) was one of the most intriguing findings in the molecular and cellular pathology of mental disability and autism spectrum disorder, both of which are central clinical features of the disease [14–16]. At around the same time, de novo mutations were reported in the *MECP2* gene in patients with Rett syndrome, which shares the major clinical phenotypes of intellectual disability and autism spectrum disorder [17]. These findings increased the research community's expectation of a complete understanding of human intelligence and the biological mechanisms of intellectual disability. However, the revelation of the functions of causative genes was the beginning of another mystery. FMRP has diverse functions that serve a central role in brain development. It selectively binds to mRNAs to regulate translation involving synaptic formation and plasticity. In addition, FMRP affects RNA stability, editing, and transport to regulate spatially and temporally specific translation [18–22].

A large proportion of mutations in *MECP2* are in the CpG region where regulatory proteins bind methylated DNA to suppress its transcription. MeCP2 serves a major role in chromosome maintenance and transcriptional silencing by regulating the dynamic condensation of heterochromatin [23]. The mutation in *MECP2* induces the disruption of constitutive heterochromatin condensation, which perturbs the orchestrated silencing of genes functioning in the developing brain.

Knowledge of FMRP and MeCP2 control of spatiotemporally specific gene expression in the developing cortex that governs the dynamic cellular and tissue architecture and neuronal circuit integration indicated that comprehensive functional genomic analysis is a requisite for understanding the biology of these disorders in which intellectual dysfunction is the major phenotype.

Besides syndromic intellectual disability, risk alleles of most of the nonsyndromic neuropsychiatric disorders presenting with heritability and heterogeneous clinical phenotypes including cognitive and behavioral dysfunction were detected in the noncoding regions [24–27]. This led to the implementation of comprehensive multiomics analyses equipped with next-generation sequencing for further elucidation of biological mechanisms of cortical formation and neuropsychiatric disorders.

Most of the progress in identifying the risk loci of neuropsychiatric disorders, including developmental disorders, has been the product of genome-wide association studies (GWASs), which have identified many risk genes for each disorder. Still, understanding the biological mechanisms of nonsyndromic neuropsychiatric disorders is challenging, mainly because of their clinical and genetic heterogeneity, despite their evident clinical heritability.

Rare mutations in a substantial number of genes and large de novo copy number variants (CNVs) in risk loci have been reported to be associated with idiopathic autism disorders, schizophrenia, and other disorders. Large cohort studies have revealed that a certain portion of CNVs at identical loci are shared among neuropsychiatric disorders, including intellectual disability, epilepsy, bipolar disorder, attention deficit hyperactivity disorder, and specific learning disorder, some of which are comorbid with each other. Single-nucleotide variants (SNVs) that damage gene or protein functions have also been identified at the same risk loci across multiple phenotypes.

The pervasive nature of pleiotropy and cross-genetics observed among the major neuropsychiatric disorders is a key to understanding their biomechanisms and the heterogeneity of each disorder under the current diagnostic manuals [28].

Pleiotropy makes the neuropsychiatric disorders distinct from other mobidities. Therefore, the investigation of the pleiotropy will reveal the shared modulation of cortical development underlying neurodevelopmental disorders. Conversely, pleiotropic genetic processes are the fundamental mechanisms in the developing cortex that govern human cognitive ability. While the disorders share the pleiotropic genetic processes, additional sets of genetic variations and environmental impacts operating during the long developmental timeline create each distinct phenotype and the heterogeneous features of each disorder.

Together, information on the pleiotropic genetic architecture and additional gene–environment interplay executing in a spatiotemporally specific manner in the developing brain are essential to interpret the mechanisms of these multifactorial disorders. RNA sequencing data provide evidence for the prenatal origin of brain-related disorders. Together, GWAS studies that have detected common noncoding genetic variants and human brain transcriptome profiling by RNA sequencing have revealed expression quantitative trait loci (eQTLs) that increase the risk of neuropsychiatric disorders [29–32].

Considering the long timeline of human cortical development from the prenatal period to the third decade of life, exposure to a variety of supportive or adverse impacts, both endogenous and exogenous, on development is inevitable. The balance of adverse and supportive genomic changes that disrupt or sustain the robustness of neurogenesis and the gene–environment interplay at a given time and in a given region needs to be investigated with functional genomics to reveal the developmental patterns of gene expression. Recent developments in RNA technology, including single-cell RNA analysis, have contributed to the integrated genomic analysis of the brain and its diverse cell types [33–36]. Considering the accumulating evidence for the critical roles of noncoding RNAs including microRNAs and long noncoding RNAs in the developing brain and brain disorders, cell-specific and spatiotemporally specific profiling of noncoding RNAs contribute to the understanding of molecular and cellular events in the developing cortex [37–45].

To analyze spatiotemporally specific gene expression and its developmental trajectories, large datasets consisting of risk alleles, quantitative trait loci (QTLs), RNA sequence data of a variety of RNA species, expression QTLs (eQTLs), splicing QTLs (sQTLs), transcriptomes, proteosomes, and epigenomes of a variety of human brain specimens and models are crucial to understanding health and disease.

To promote the investigation of the spatiotemporally specific regulatory system of gene expression of the human tissues, the NIH Roadmap Epigenomics Consortium collected a large number of human epigenomes and has started developing a map of tissue- or cell-specific regulatory elements in the human genome [46, 47].

3 From Nonhuman Models to Human Brain Organoids

The human brain is characterized by its expanded neocortex and consequent gyrification. The coordinated development of the neocortex requires spatiotemporally specific molecular and cellular regulation to form its complex architecture and neural connections during the long developmental period. Valid models representing human-specific features of brain development are needed to construct functional genomic datasets and analyses due to the limited availability of developing human brain specimens. To substitute for the human material, a variety of organisms have been used. These include *Caenorhabditis elegans*, *Drosophila melanogaster*, *Danio rerio*, rodents, and nonhuman primates. Each model has both advantages and disadvantages depending on the aim of the experiment, and whether and how completely the results can be extrapolated to the human brain is often unclear.

Neural organoids from pluripotent cells are new alternatives to the existing brain models. They not only substitute for unavailable specimens of human brains in early development but also offer three-dimensionality and lasting samples. This makes possible an integrated analysis of dynamic gene regulation, morphology, and function. The reproducibility and fidelity of organoid experimental protocols are rapidly improving, but it is still challenging to reproduce interregional cortical layers containing diverse nonneuronal cells, vessels, and gyrification to represent human-specific features of the brain [48–53].

These new organoid models make it possible to investigate the vulnerability of the developing brain to any internal and external influences. It is also possible to model the developmental processes of monogenic and polygenic neuropsychiatric disorders. A multiomics approach to organoids derived from normal and diseased individuals with known genetic changes will relate the genetic risk to processes underlying the initiation and development of malfunction. To date, human brain organoids from a variety of disorders such as cortical dysplasia, developmental disorders, and neuropsychiatric disorders have been produced and investigated [54–56]. Further development is expected with improved technologies. Together with accelerated progress in computer biology, organoids developed with improved protocols are a promising model for the developing brain and a variety of human neuropsychiatric disorders, providing the information needed for their prevention and treatment.

Human brain models such as organoids combined with functional genomic technologies shed light on the biological mechanisms of human brain development and the initiation and development of neuropsychiatric disorders. Despite the unprecedented advantages of neural organoids, which will transform brain science, reproducing longevity, fidelity, tissue maturity (e.g., gyrification), and neural connectivity of the human brain is still a challenge. Together with accelerated progress in computer biology, organoids developed with improved protocols are a promising model for the developing brain and a variety of human neuropsychiatric disorders of congenital, monogenic, and polygenic origins, providing the information needed for prevention and treatment.

4 Big Datasets and Computer-Driven Neuroscience

To address the gene expression trajectory of the developing human brain, big datasets and high-performance computer-driven analyses have emerged in brain science. Dysregulation of the gene expression trajectory in the developing human brain is the underlying fundamental pathology of neuropsychiatric disorders, regardless of the initiation age of the disorders. The ENCODE (Encyclopedia of DNA Elements) Consortium was constructed to annotate regulatory elements in the genomes of a variety of species including humans. It aims to comprehensively catalog functional elements of the human genome and has expanded to include more than 10,000 projects, mostly with data from cell lines [57]. To focus on neuropsychiatric disorders, the PsychENCODE Consortium was constructed as an open resource of multidimensional genomic data of human adult postmortem brain specimens and model systems [58–61].

In addition to ENCODE and PsychENCODE, an enormous number of open-access functional genomics data are available through resources such as the BrainSpan Consortium (Atlas of the Developing Human Brain) [62] and the NIH Roadmap Epigenomics Consortium for mining the DNA-interacting elements and chromatin modifiers of the human brain with the aid of high-performance computing technology.

Cortical function is organized as a complex network of communicating regions—a functional connectome. Connectomes are comprehensive maps of functional neural connections revealed by the use of innovative MRI technology [63–65]. The strategy visualizes the functional neural connections both within and between the regions of the brain. Comparative studies of normal and diseased brains are likely to suggest functional biomarkers of specific neurophenotypes. The technology broadens the visualized scale from macro- and meso- to microscale studies that could enable the development of functional biomarkers for healthy and diseased brains [66]. The fetal brain connectome and disease-related connectomes have been studied, and further accumulation of micro- to macroscale data from a variety of regions in the developing human brain is expected to identify functional biomarkers for identifying the diversity of healthy brain as well as the varieties of diseased brain [67, 68].

The ENIGMA (Enhancing Neuroimaging Genetics Through Meta Analysis) Consortium constructed a worldwide alliance with the aim of a large-scale genetic study of neuroimaging designed to reveal the structural and genetic processes of brain development in health and disease [69, 70].

The evolution of the infrastructure of big data and continuously improving neuroinformatics will make it possible to link genes and the pathology of their expression with and without the influence of the environment and thus promote research on the pathophysiology of neuropsychiatric disorders [71–73]. Many of the developmental disorders are heterogeneous in nature, sharing some phenotypes that lack interpretable biomarkers; this is one of the major obstacles for clinical and basic researchers. Multiomic, including connectomic, data could lead us to novel insights regarding disease subgrouping, treatment targets, prevention, and even presymptomatic pathology assessment in each individual. Whether we should acquire biomarkers for the accurate presymptomatic assessment of the risk of neuropsychiatric disorders is a new ethical issue in medicine and society.

5 Unraveling the Underlying Mechanisms of Neuropsychiatric Disorders Will Lead Us to Precision Medicine and a Biological Understanding of Neurodiversity

Comprehensive functional genomics equipped with emergent neuroinformatics and novel human models is accelerating the progress of brain science. The application of emerging strategies in neuroscience demands consideration of several ethical issues ranging from the governance of big datasets to the aims of any investigation using such resources. Projects have been expanding across national boundaries where researchers may encounter different ethical cultures and rules. It is imperative to construct a fair and responsible governance system to ensure the further progress of human brain science [74–76].

By elucidating the mechanisms of human brain development until adulthood, we will gain a biological understanding of the cognition and behavior that make humans unique. Comprehensive knowledge contributes to the evolution of AI with deep learning, which in turn facilitates a paradigm shift in brain research.

Understanding how the human brain acquires its function through the interplay of the genome and the environment will also lead us to understand the fundamental process of integration of our individuality. Knowing the developmental mechanisms of human-specific cognitive ability will lead to an understanding of individual differences in cognitive aptitude that may yield predictive indicators of cognitive levels and diversity at the early developmental stages. Whether having a tool that predicts intelligence levels early in life is beneficial and allowed will be another ethical issue for the research community and society.

Future research may also provide highly accurate presymptomatic biomarkers of neuropsychiatric disorders. These biomarkers may benefit individuals who will be affected only when a preventive or therapeutic approach is available for the identified condition.

Among neuropsychiatric disorders, developmental disorders have unique traits that differ from psychiatric disorders in which most of the symptoms are detrimental to well-being. Developmental disorders include intellectual disorder, autism spectrum disorder, attention-deficit hyperactivity disorder, and specific learning disorder. These have long been recognized and accepted as disorders, deficits, and diseases under the medical paradigm and are thus automatically considered undesirable deviations from the norms of individuals and society. According to the conventional medical paradigm, such conditions must be diagnosed and treated to restore the functional deviation.

Defining norms and deviations based on the medical model of disability may not be the most cogent way to accept developmental disorders when we understand the biological mechanisms of individuality of brain function. Instead, a paradigm shift from medical disorder to neurodiversity may be at least partly beneficial by encouraging human society to understand individuals as they are. Whether it is possible to make society to adopt that understanding seems like an open question. Treatment should not target the disorder itself, but rather the phenotype that disturbs the individual's well-being in society. Understanding the complete biology of neuropsychiatric disorders through an innovative research strategy will give us new insights into the biology of brain individuality and the raison d'être of human beings and provide us with a better way to recognize human individuality and to achieve personal well-being throughout life. This may be an ultimate goal of precision medicine in the treatment of neuropsychiatric disorders.

References

1. Li ML, Tang H, Shao Y, Wang MS, Xu HB, Wang S et al (2020) Evolution and transition of expression trajectory during human brain development. BMC Evol Biol 20:72

2. Somel M, Franz H, Yan Z, Lorenc A, Guo S, Giger T et al (2009) Transcriptional neoteny in the human brain. Proc Natl Acad Sci USA 106: 5743–5748

3. Schneider B, Koenigs M (2017) Human lesion studies of ventromedial prefrontal cortex. Neuropsychologia 107:84–93
4. Teffer K, Semendeferi K (2012) Human prefrontal cortex: evolution, development and pathology. Prog Brain Res 195:191–218
5. Hiser J, Koenigs M (2018) The multifaceted role of the ventromedial prefrontal cortex in emotion, decision making, social cognition, and psychopathology. Biol Psychiatry 83:638–647
6. Kolk SM, Rakic P (2022) Development of prefrontal cortex. Neuropsychopharmacology 47:41–57
7. Yan Z, Rein B (2022) Mechanisms of synaptic transmission dysregulation in the prefrontal cortex: Pathophysiological implications. Mol Psychiatry 27:445–465
8. Pan X, Chen A, Li Y, Xing X, Hui L (2019) Medial prefrontal cortex in neurological diseases. Physiol Genomics 51:432–442
9. Isles AR (2022) The contribution of imprinted genes to neurodevelopmental and neuropsychiatric disorders. Transl Psychiatry 12:210
10. Amgalan A, Andescavage N, Limperopoulos C (2021) Prenatal origins of neuropsychiatric disorders. Acta Paediatr 110:1741–1749
11. BRAIN Initiative—American Brain Coalition (2013). https://www.amerikanbrainkoalition.org/page/BRAINInitiative
12. Brain/MINDS. https://brainminds.jp/en/
13. Huan Brain Project. https://www.humanbrainproject.eu/en/
14. Kremer EJ, Pritchard M, Lynch M, Yu S, Holman K, Baker E et al (1991) Mapping of DNA instability at the fragile X to a trinucleotide repeat sequence p(CCG)n. Science 252:1711–1714
15. Bell MV, Hirs MC, Nakahori Y, MacKinnon RN, Roche A, Flint TJ et al (1991) Physical mapping across the fragile X: hypermethylation and clinical expression of the fragile X syndrome. Cell 64:861–866
16. Heitz D, Rousseau F, Devys D, Saccone S, Abderrahim H, Le Paslier D et al (1991) Isolation of sequences that span the fragile X and identification of a fragile X-related CpG Island. Science 251:1236–1239
17. Amir RE, Van den Veyver IB, Wan M, Tran CQ, Francke U, Zoghbi HY (1999) Rett syndrome is caused by mutations in X-linked MECP2, encoding methyl-CpG-binding protein 2. Nat Genet 23:185–188
18. Weiler IJ, Irwin SA, Klintsova AY, Spencer CM, Brazelton AD, Miyashiro K et al (1997) Fragile X mental retardation protein is translated near synapses in response to neurotransmitter activation. Proc Natl Acad Sci USA 94:5395–5400
19. Zalfa F, Adinolfi S, Napoli I, Kühn-Hölsken E, Urlaub H, Achsel T et al (2005) Fragile X mental retardation protein (FMRP) binds specifically to the brain cytoplasmic RNAs BC1/BC200 via a novel RNA-binding motif. J Biol Chem 280:33403–33410
20. Bhogal B, Jepson JE, Savva YA, Pepper AS-R, Reenan RA, Jongens TA (2011) Modulation of dADAR-dependent RNA editing by the *Drosophila* fragile X mental retardation protein. Nat Neurosci 14:1517–1524
21. Muddashetty RS, Nalavadi VC, Gross C, Yao X, Xing L, Laur O, Warren ST, Bassell GJ (2011) Reversible inhibition of PSD-95 mRNA translation by miR-125a, FMRP phosphorylation, and mGluR signaling. Mol Cell 42:673–688
22. Edbauer D, Neilson JR, Foster KA, Wang CF, Seeburg DP, Batterton MN et al (2010) Regulation of synaptic structure and function by FMRP-associated microRNAs miR-125b and miR-132. Neuron 65:373–384
23. Li CH, Coffey EL, Dall'Agnese A, Hannett NM, Tang X, Henninger JE et al (2020) MeCP2 links heterochromatin condensates and neurodevelopmental disease. Nature 586:440–444
24. Ripke S, Neale BM, Corvin A, Walters JT, Farh KH, Holmans PA et al (2014) Biological insights from 108 schizophrenia-associated genetic loci. Nature 511:421–427
25. Maurano MT, Humbert R, Rynes E, Thurman RE, Haugen E, Wang H et al (2012) Systematic localization of common disease-associated variation in regulatory DNA. Science 337:1190–1195
26. An JY, Lin K, Zhu L, Werling DM, Dong S, Brand H et al (2018) Genome-wide de novo risk score implicates promoter variation in autism spectrum disorder. Science 362. https://doi.org/10.1126/science.aat6576
27. Hall LS, Pain O, O'Brien HE, Anney R, WaltersJTR OMJ et al (2021) *Cis*-effects on gene expression in the human prenatal brain associated with genetic risk for neuropsychiatric disorders. Mol Psychiatry 26:2082–2088
28. Lee PH, Feng Y-CA, Smoller JW (2021) Pleiotropy and cross disorder genetics among psychiatric disorders. Biol Psychiatry Biol Psychiatry 89:20–31
29. Croteau-Chonka DC, Rogers AJ, Raj T, McGeachie MJ, Qiu W, Ziniti JP et al (2015) Expression quantitative trait loci information improves predictive modeling of disease

relevance of non-coding genetic variation. PLoS One 10:e0140758

30. Walker RL, Ramaswami G, Hartl C, Mancuso N, Gandal MJ, de la Torre-Ubieta L et al (2019) Genetic control of expression and splicing in developing human brain informs disease mechanisms. Cell 179:750–771.e22
31. Bryois J, Calini D, Macnair D, Foo L, Urich E, Ortmann W et al (2022) Cell-type-specific cis-eQTLs in eight human brain cell types identify novel risk genes for psychiatric and neurological disorders. Nat Neurosci 25:1104–1112
32. O'Brien HE, Hannon E, Hill MJ, Toste CC, Robertson MJ, Morgan JE et al (2018) Expression quantitative trait loci in the developing human brain and their enrichment in neuropsychiatric disorders. Genome Biol 19:194
33. Tang F, Barbacioru C, Nordman E, Li B, Xu N, Bashkirov VI et al (2020) RNA-Seq analysis to capture the transcriptome landscape of a single cell. Nat Protoc 5:516–535
34. Tang F, Barbacioru C, Wang Y, Nordman E, Lee C, Xu N et al (2009) mRNA-Seq whole-transcriptome analysis of a single cell. Nat Methods 6:377–382
35. Vandereyken K, Sifrim A, Thienpont B, Voet T (2023) Methods and applications for single-cell and spatial multi-omics. Nat Rev Genet 2: 1–22
36. Baysoy A, Bai Z, Satija R, Fan R (2023) The technological landscape and applications of single-cell multi-omics. Nat Rev Mol Cell Biol 6:1–19
37. Gibbons NA, Udawela M, Dean B (2018) Non-coding RNA as novel players in the pathophysiology of schizophrenia. Noncoding RNA 4:11
38. Liu Y, Chang X, Hahn C-G, Gur RE, Sleiman PAM, Hakonarson H (2018) Non-coding RNA dysregulation in the amygdala region of schizophrenia patients contributes to the pathogenesis of the disease. Transl Psychiatry 8:44
39. Najaf H, Naseri M, Zahiri J, Totonchi M, Sadeghizadeh M (2019) Identification of the molecular events involved in the development of prefrontal cortex through the analysis of RNA-Seq data from BrainSpan. ASN Neuro 11:1759091419854627
40. Liau W-S, Samaddar S, Banerjee S, Bredy TW (2021) On the functional relevance of spatiotemporally-specific patterns of experience-dependent long noncoding RNA expression in the brain. RNA Biol 18:1025–1036
41. Morgunova A, Flores C (2021) MicroRNA regulation of prefrontal cortex development and psychiatric risk in adolescence. Semin Cell Dev Biol 118:83–91
42. Qureshi IA, Mehler MF (2012) Emerging roles of non-coding RNAs in brain evolution, development, plasticity and disease. Nat Rev Neurosci 13:528–541
43. Tsujimura K, Shiohama T, Takahashi E (2022) microRNA biology on brain development and neuroimaging approach. Brain Sci 12:1366
44. Andersen R, Lim DA (2018) Forging our understanding of lncRNAs in the brain. Cell Tissue Res 371:55–71
45. Aliperti V, Skonieczna J, Cerase A (2021) Long non-coding RNA (lncRNA) roles in cell biology, neurodevelopment and neurological disorders. Noncoding RNA 7(2):36
46. Bernstein BE, Stamatoyannopoulos JA, Ren B, Milosavljevic A, Meissner A, Kellis M et al (2010) The NIH roadmap epigenomics mapping consortium. Nat Biotechnol 28: 1045–1048
47. Chadwick LH (2012) The NIH roadmap epigenomics program data resource. Epigenomics 4:317–324
48. Benito-Kwiecinski S, Madeline A (2020) Brain organoids: human neurodevelopment in a dish. Lancaster Cold Spring Harb Perspect Biol 12: a035709
49. Eiraku M, Watanabe K, Matsuo-Takasaki M, Kawada M, Yonemura S, Matsumura M et al (2008) Self-organized formation of polarized cortical tissues from ESCs and its active manipulation by extrinsic signals. Cell Stem Cell 3: 519–532
50. Camp JG, Badsha F, Florio M, Kanton S, Gerber T, Wilsch-Bräuninger M et al (2015) Human cerebral organoids recapitulate gene expression programs of fetal neocortex development. Proc Natl Acad Sci USA 112:15672–15677
51. Khakipoor S, Crouch EE, Mayer S (2021) Human organoids to model the developing human neocortex in health and disease. Brain Res 1742:146803
52. Cakir B, Xiang Y, Tanaka Y, Kural MH, Parent M, Kang Y-J et al (2019) Engineering of human brain organoids with a functional vascular-like system. Nat Methods 16:1169–1175
53. Qian FX, Su Y, Adam CD, Deutschmann AU, Pather SR, Goldberg EM et al (2020) Sliced human cortical organoids for modeling distinct cortical layer. Stem Cell 26:766–781
54. Avansini SH, Puppo F, Adams JW, Vieira AS, Coan AC, Rogerio F et al (2022) Junctional instability in neuroepithelium and network

hyperexcitability in a focal cortical dysplasia human model. Brain 145:1962–1977

55. Chan WK, Griffiths R, Price DJ, Mason JO (2020) Cerebral organoids as tools to identify the developmental roots of autism. Mol Autism 11:58
56. Notaras M, Lodhi A, Dündar F, Collier P, Sayles NM, Tilgner H et al (2022) Schizophrenia is defined by cell-specific neuropathology and multiple neurodevelopmental mechanisms in patient-derived cerebral organoids. Mol Psychiatry 27:1416–1434
57. ENCODE Project Consortium (2004) The ENCODE (ENCyclopedia of DNA elements) project. Science 00306:636–640
58. PsychENCODE Consortium. http://psychencode.org
59. Li M, Santpere G, Kawasawa YI, Evgrafov OV, Gulden FO, Pochareddy S et al (2018) Integrative functional genomic analysis of human brain development and neuropsychiatric risks. Science 362:eaat7615
60. Wang D, Liu S, Warrell J, Won H, Shi X, Navarro FCP et al (2018) Comprehensive functional genomic resource and integrative model for the human brain. Science 362: eaat8464
61. Jourdon A, Scuderi S, Capauto AA, Vaccarino FM (2021) PsychENCODE and beyond: transcriptomics and epigenomics of brain development and organoids. Neuropsychopharmacology 46:70–85
62. Brain Span Consortia. Brain Span Atlas of Human Brain/Atlas of Developing Human Brain/ https://brainspan.org
63. Sporns O, Tononi G, Kötter R (2005) The human connectome: a structural description of the human brain. PLoS Comput Biol 1: e42. https://doi.org/10.1371/journal.pcbi.0010042
64. Human Connectome Project (HCP) (2011). http://humanconnectome.org/consortia/ Front Neuroinform 5:4
65. Marcus DS, Harwell J, Olsen T, Hodge M, Glasser MF, Prior F et al (2011) Informatics and data mining tools and strategies for the human connectome project. Front Neuroinform 5:4
66. Huang SY, Witzel T, Keil B, Scholz A, Davids M, Dietz P et al (2021) Connectome 2.0: developing the next-generation ultra-high gradient strength human MRI scanner for bridging studies of the micro-, meso- and macro-connectome. NeuroImage 243:118530
67. Turk E, van den Heuvel MI, Benders MJ, de Heus R, Franx A, Manning JH et al (2019) Functional connectome of the fetal brain. J Neurosci 39:9716–9724
68. van den Heuvel MP, Sporns O (2019) A cross-disorder connectome landscape of brain dysconnectivity. Nat Rev Neurosci 20:435–446
69. Grasby KL, Jahanshad N, Painter JN, Colodro-Conde L, Bralten J, Hibar DP et al (2021) The genetic architecture of the human cerebral cortex. Science 367:eaay6690
70. Thompson PM, Jahanshad N, Ching CRK, Salminen LE, Thomopoulos SI, Bright J, for the ENIGMA Consortium et al (2020) ENIGMA and global neuroscience: a decade of large-scale studies of the brain in health and disease across more than 40 countries. Transl Psychiatry 10:100
71. Hoogman M, van Rooij D, Klein M, Boedhoe P, Ilioska I, Li T et al (2022) Consortium neuroscience of attention deficit/hyperactivity disorder and autism spectrum disorder: the ENIGMA adventure. Hum Brain Mapp 43: 37–55
72. Ching CRK, Hibar DP, Gurholt TP, Nunes A, Thomopoulos SI, Abé C, ENIGMA Bipolar Disorder Working Group et al (2022) What we learn about bipolar disorder from large-scale neuroimaging: findings and future directions from the ENIGMA Bipolar Disorder Working Group. Hum Brain Mapp 43:56–82
73. Sha Z, van Rooij D, Anagnostou E, Arango C, Auzias G, Behrmann M et al (2022) Subtly altered topological asymmetry of brain structural covariance networks in autism spectrum disorder across 43 datasets from the ENIGMA consortium. Mol Psychiatry 27:2114–2125
74. Fothergill BT, Knight W, Stahl BC, Ulnicane I (2019) Responsible data governance of neuroscience big data. Front Neuroinform 13:28
75. Eke DO, Bernard A, Bjaalie JG, Chavarriaga R, Hanakawa T, Hannan AJ et al (2023) International data governance for neuroscience. Neuron 110:600–612
76. Chen S, He Z, Han X, He X, Li R, Zhu H et al (2019) How big data and high-performance computing drive brain science. Genomics Proteomics Bioinformatics 17:381–392

Chapter 2

In Situ Hybridization of Brain Slices

Tomohiro Kanayama, Hiroyuki Tomita, and Akira Hara

Abstract

In situ hybridization (ISH) is an important technique for identifying gene expression at the cellular level in various organs, including brain slices. This approach hybridizes nucleic acid probes to cellular mRNA, allowing the detection of transcriptional products. Recent advances have enabled RNA preservation in formalin-fixed paraffin-embedded (FFPE) samples, making ISH applicable to brain tumor diagnosis and research. Here, we provide a concise overview of the standard application of chromogenic ISH in neuroscience research and neuropathology practice using FFPE blocks of brain slice sections.

Key words In situ hybridization, Formalin-fixed paraffin-embedded (FFPE), Brain slice, Mouse

1 Introduction

In situ hybridization (ISH) is a crucial technique for detecting gene expression at the cellular level, which enables the identification of transcriptional products and their respective locations within tissue sections of various organs, including brain slices.

ISH is based on the hybridization of cellular mRNA to a nucleic acid probe (mostly RNA) labeled with a radioisotope or a detectable molecule (e.g., digoxigenin or biotin) in frozen sections.

In the past, RNA preservation in both aged tissue blocks and stored tissue sections has been an issue, leading to biased results and inappropriate interpretations of RNA expression profiles in formalin-fixed paraffin-embedded (FFPE) samples [1]. This has been explored using multiple techniques, including RNA ISH [2]. Recently, methods have been developed to fix RNA without destroying it and to preserve it in a stable and long-term manner. Thus, ISH of sections from FFPE blocks has been developed not only in basic research but also in the clinical diagnosis of brain tumors.

In the clinical setting, genomic abnormalities in brain tumors have been found to correlate with pathogenesis and clinical

Koh-ichi Nagata (ed.), *Cerebral Cortex Development: Methods and Protocols*, Methods in Molecular Biology, vol. 2794, https://doi.org/10.1007/978-1-0716-3810-1_2,

outcomes through comprehensive genomic analysis. Therefore, the identification of genomic abnormalities has been deemed essential or desirable for the diagnosis and grading of brain tumors in the fifth edition of the WHO classification [3]. For example, the detection of 1p/19q co-deletion is facilitated by fluorescence in situ hybridization (FISH), which involves fluorescently labeled probes to detect the loss of 1p and 19q. Although array comparative genomic hybridization is preferable for the precise detection of 1p/19q codeletions, FISH can be performed using sliced sections from FFPE blocks and is an alternative cost-effective approach [4].

In this chapter, we describe the basic and standard methods of chromogenic ISH applied to FFPE blocks of brain slice sections in the daily practice of neuroscience research and neuropathology.

2 Materials

2.1 Probes, Kit, and Equipment

1. RNAscope Singleplex Target Probe [species]–[gene]: probe targeting specific RNA (commercially available).
2. RNAscope 2.5 HD Reagent Kit (BROWN) (commercially available):
   - RNAscope hydrogen peroxide (commercially available).
   - RNAscope Protease Plus (commercially available).
   - RNAscope Target Retrieval (10×) (commercially available).
   - RNAscope 2.5 HD AMP 1, 2, 3, 4, 5, 6, A, and B (commercially available).
3. RNAscope Wash Buffer (50×), distilled water (commercially available).
4. HybEZ Oven (commercially available).
5. HybEZ Humidity Control Tray (with a lid) (commercially available).

2.2 Solutions

1. Deionized and distilled water.
2. Neutral buffered formalin, 10%.
3. 100% and 70% ethanol (EtOH).
4. 0.02% ammonia water.
5. Xylene.
6. DAB substrate solution.
7. Mounting medium.

3 Methods

3.1 Preparation of Brain Slices and FFPE Blocks

1. Anesthetize animals with sodium pentobarbital.
2. Transcardially perfuse animals with saline and then with neutral buffered formalin (10%).
3. Remove the brain and cut it into suitable slices.
4. Immerse the brain slices in 10% phosphate-buffered formalin for additional fixation for 24 h.
5. Process for paraffin embedding.
6. Cut several 5-μm-thick sections from the FFPE tissue blocks using a microtome and mount on adhesion-treated slides.

3.2 Prepare Slides

1. Bake the slides in a dry oven for 1 h at 60 °C (see **Note 1**).
2. Place the slides in a slide rack in xylene for 2 × 5 min.
3. Incubate the slides in 100% EtOH for 2 × 1 min.
4. Remove the slides from the rack. Air-dry the slides for 5 min at room temperature (*see* **Note 2**).

3.3 Pretreat the Samples

1. Set the HybEZ oven to 40 °C and warm the HybEZ Humidity Control Tray for 30 min before use. Keep the tray warm during the assay.
2. Prepare 700 mL fresh 1× Target Retrieval in a beaker. Cover the beaker with foil, heat the solution to a mild boiling temperature, and maintain that temperature (*see* **Note 3**).
3. Add approximately 5–8 drops of hydrogen peroxide to each section for 10 min at room temperature.
4. Place the slides into a rack submerged in distilled water.
5. Wash the slides with distilled water by moving the rack up and down 3–5 times and repeat the process with fresh distilled water.
6. Using a pair of forceps, very slowly submerge the slide rack into the boiling 1× Target Retrieval solution for 15 min at 98–102 ° C (*see* **Note 4**).
7. Immediately transfer the hot slide rack to a staining dish containing distilled water.
8. Wash the slides with distilled water by moving the rack up and down 3–5 times and repeat the process with fresh distilled water.
9. Wash the slides using fresh 100% EtOH by moving the rack up and down 3–5 times and then air-dry.
10. Draw 2–4 times around the tissue using a pap pen. Dry completely overnight at room temperature.

11. Place the slides in the HybEZ Slide Rack and add 5 drops of Protease Plus to each section.
12. Place the HybEZ Slide Rack in the prewarmed HybEZ Humidity Control Tray.
13. Seal the tray and insert it back into the HybEZ Oven. Incubate at 40 °C for 30 min.
14. Wash the slides with distilled water by moving the rack up and down 3–5 times and repeat the process with fresh distilled water.

3.4 Prepare the Materials

1. Prepare 3 L 1× wash buffer by adding 2.94 L of distilled water and 1 bottle (60 mL) of 50× wash buffer to a large carboy. Mix well. Warm 50× wash buffer up to 40 °C for 10–20 min before preparing 1× wash buffer.
2. Prepare 50% hematoxylin and 0.02% ammonia water.
3. Prepare dehydrating reagents: 200 mL xylene in a clearing agent dish and 2 × 200 mL 100% EtOH and 200 mL 70% EtOH in staining dishes.
4. Equilibrate reagents and equipment:
   - Remove Amp 1–6 from the refrigerator.
   - Warm the probes for 10 min at 40 °C and cool at room temperature.

3.5 Run the Assay

1. To hybridize the probes, remove excess liquid from the slides, place the slides in the HybEZ Slide Rack, and add 4 drops of probe to each section.
2. Insert the sealed tray containing the HybEZ Slide Rack back into the HybEZ Oven for 2 h at 40 °C. Then, remove the slide rack (*see* **Note 5**).
3. Wash the slides with 1× Wash Buffer for 2 min at room temperature. Repeat the process with fresh 1× wash buffer.
4. To hybridize Amp 1 (30 min at 40 °C), remove excess liquid from the slides, place the slides in the HybEZ Slide Rack, and add 4 drops of Amp 1 to each section.
5. Insert the sealed tray containing the HybEZ Slide Rack into the HybEZ Oven for 30 min at 40 °C. Then, remove the slide rack.
6. Wash the slides with 1× wash buffer for 2 min at room temperature. Repeat with fresh 1× wash buffer.
7. To hybridize Amp 2 (15 min at 40 °C), remove excess liquid from the slides, place in the HybEZ Slide Rack, and add 4 drops of Amp 2 to each section.
8. Insert the sealed tray containing the HybEZ Slide Rack into the HybEZ Oven for 15 min at 40 °C. Then, remove the slide rack.

9. Wash slides with 1× wash buffer for 2 min at room temperature. Repeat with fresh buffer.
10. To hybridize Amp 3 (30 min at 40 °C), remove excess liquid from the slides, place in the HybEZ Slide Rack, and add 4 drops of Amp 3 to each section.
11. Insert the sealed tray containing the HybEZ Slide Rack into the HybEZ Oven for 30 min at 40 °C. Then, remove the slide rack.
12. Wash the slides in 1× wash buffer for 2 min at room temperature. Repeat with fresh buffer.
13. To hybridize Amp 4 (15 min at 40 °C), remove excess liquid from the slides, place in the HybEZ Slide Rack, and add 4 drops of Amp 4 to each section.
14. Insert the sealed tray containing the HybEZ Slide Rack into the HybEZ Oven for 15 min at 40 °C. Remove the slide rack but do not place the tray back into the oven.
15. Wash the slides with 1× Wash Buffer for 2 min at room temperature. Repeat with fresh buffer.
16. To hybridize Amp 5 (30 min at room temperature), remove excess liquid from the slides, place in the HybEZ Slide Rack, and add 4 drops of Amp 5 to each section.
17. Incubate the sealed tray containing the HybEZ Slide Rack for 30 min at room temperature.
18. Wash the slides in 1× Wash Buffer for 2 min at room temperature. Repeat with fresh buffer.
19. To hybridize Amp 6 (15 min at room temperature), remove excess liquid from the slides, place in the HybEZ Slide Rack, and add 4 drops of Amp 6 to each section.
20. Incubate the sealed tray containing the HybEZ Slide Rack for 15 min at room temperature.
21. Wash the slides with 1× wash buffer for 2 min at room temperature. Repeat with fresh buffer.
22. To detect the signal (10 min at room temperature), mix equal volumes of BROWN-A and BROWN-B in the kit.
23. Remove excess liquid from the slides, place in the HybEZ Slide Rack, and pipette 120 μL DAB onto each tissue section.
24. Incubate the sealed tray containing the HybEZ Slide Rack for 10 min at room temperature.
25. Remove DAB from the slides and wash 3–5 times with distilled water.
26. To counterstain the slides (2 min at room temperature), place them in 50% hematoxylin for 2 min at room temperature

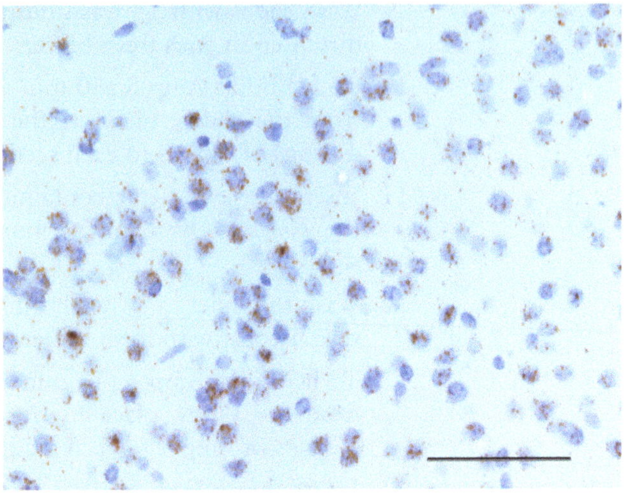

Fig. 1 Microphotograph of RNA ISH using the *Mus musculus* peptidylprolyl isomerase B (Ppib)-specific probe in an FFPE section of murine cerebral cortex tissue. Brown dots indicate positive expression signals. Scale bars, 50 μm

(*see* **Note 6**). Wash the slides 3–5 times with distilled water and repeat with fresh distilled water.

27. Wash the slides for 10 s with 0.02% ammonia water and then wash them 3–5 times with distilled water.
28. To mount the slides, incubate them in 70% EtOH for 2 min with occasional agitation.
29. Incubate the slides in 95% EtOH for 2 min with occasional agitation. Repeat with fresh EtOH.
30. Incubate the slides in xylene for 5 min with occasional agitation.
31. Add adequate mounting medium and place a coverslip over the section and air-dry (*see* **Note 7**).

3.6 Evaluate the Results

Examine the tissue sections under a standard bright-field microscope at adequate magnification (Fig. 1) (*see* **Note 8**).

4 Notes

1. Use the sectioned tissue within 1 week. Store the sections with desiccants at room temperature.
2. Air-dry overnight at room temperature (must use within 24 h) or proceed directly to the next step.
3. Do not boil for more than 30 min before use.

4. Keep the temperature above 98 °C at all times while the rack is in the retrieval solution. For mouse brains, this is done for 15 min.
5. When samples are removed from the oven, do not allow them to cool or dry out. Process them quickly. Then, quickly return them to the oven.
6. In some cases, staining with hematoxylin for a duration shorter than 2 min may be beneficial for observing the brown dots. Adjust the time accordingly.
7. After the mounting medium has properly dried, it is recommended to perform microscopic observation.
8. Since dots are as small as approximately 1–2 μm, micrographs at 40× objective or higher are recommended.

Acknowledgments

We thank all the members of our laboratories for their support. We also thank H. Nikki March, PhD, from Edanz (https://jp.edanz.com/ac) for editing a draft of this manuscript.

Funding This work was supported by JSPS KAKENHI (grant number JP20K07587 to H.T.).

References

1. Gupta S, Chen T, Destenaves B (2021) Quantitative RNA assessment and long-term stability in the FFPE tumor samples using digital spatial profiler. Immunooncol Technol 13:100069. https://doi.org/10.1016/j.iotech.2021.100069
2. Bordeaux JM, Cheng H, Welsh AW, Haffty BG, Lannin DR, Wu X, Su N, Ma XJ, Luo Y, Rimm DL (2012) Quantitative in situ measurement of estrogen receptor mRNA predicts response to tamoxifen. PLoS One 7(5):e36559. https://doi.org/10.1371/journal.pone.0036559
3. Louis DN, Perry A, Wesseling P, Brat DJ, Cree IA, Figarella-Branger D, Hawkins C, Ng HK, Pfister SM, Reifenberger G, Soffietti R, von Deimling A, Ellison DW (2021) The 2021 WHO classification of tumors of the central nervous system: a summary. Neuro-Oncology 3(8):1231–1251. https://doi.org/10.1093/neuonc/noab106
4. Pinkham MB, Telford N, Whitfield GA, Colaco RJ, O'Neill F, McBain CA (2015) FISHing tips: what every clinician should know about 1p19q analysis in gliomas using fluorescence in situ hybridisation. Clin Oncol (R Coll Radiol) 27(8):445–453. https://doi.org/10.1016/j.clon.2015.04.008

Chapter 3

Immunohistochemistry of Brain Tissues

Akira Hara, Toshiaki Taniguchi, Tomohiro Kanayama, and Hiroyuki Tomita

Abstract

Immunohistochemistry (IHC) is the basis of histological or pathological analysis and is widely used to enable the detection and characterization of proteins in various organ tissues, including brain tissues. IHC is commonly performed on formalin-fixed paraffin-embedded (FFPE) tissues because of their easy storage and versatility. IHC is a key method for providing more accurate analysis of localization and function of neurons, neuroendocrine cells, and neural stem cells in the brain and other nervous systems. The related cells such as glial cells and neurovascular units have also been analyzed by IHC. Visualization of antibody–antigen interactions can be performed primarily in one of the following ways: chromogenically stained IHC and fluorescently stained IHC. In chromogenically stained IHC, an antibody is chemically conjugated to an enzyme, such as peroxidase, that can be reacted with a suitable substrate to give a colored product. In fluorescently stained IHC, the antibodies are finally tagged with fluorescent chemicals such as fluorescein isothiocyanate (FITC) or rhodamine. Here, we describe the standard methods of IHC applied to brain slice sections. Furthermore, an automated immunostainer is presented as another option for standardized immunohistochemistry.

Key words Immunohistochemistry, Formalin-fixed paraffin-embedded (FFPE), Brain slice, Automated immunostainer

1 Introduction

Identification of immunohistochemical markers is now key in providing more accurate protein localization and function of neurons, glia cells, neuroendocrine cells, and neural stem cells in the brain as well as other nervous systems. The related cells such as mesenchymal cells and neurovascular units have also been analyzed by immunohistochemistry (IHC). IHC is performed on frozen sections [1, 2]; however, it is more commonly performed on formalin-fixed paraffin-embedded (FFPE) tissues [3, 4]. With the use of antigen retrieval methods, which has been remarkably improved in recent years, IHC has been performed more conveniently and usefully on FFPE tissues because of their easy storage.

Koh-ichi Nagata (ed.), *Cerebral Cortex Development: Methods and Protocols*, Methods in Molecular Biology, vol. 2794, https://doi.org/10.1007/978-1-0716-3810-1_3,

Furthermore, IHC is the first step of molecular analysis in molecular diagnosis based on the World Health Organization (WHO) classification of human tumors of the central nervous system [5]. IHC for detecting mutations such as IDH 1/2 in human brain tumors has been used to provide predictive and prognostic information, followed by various gene examinations including next-generation sequencing [6].

Rodent models have been used in the study of the central nervous system to understand the underlying pathophysiology and to explore potential treatments of human diseases because of their small size and moderate cost [7, 8]. For histomorphological preservation of brain samples of rodents, transcardial perfusion using formaldehyde-containing solutions is commonly applied in an immunohistochemical staining procedure [9]. It is well known, however, that formaldehyde fixation may mask the tissue antigens due to cross-linking of protein amino acid residues [10]. Furthermore, the fixative type and the time of fixation are also considered as preanalytical variables, which may influence (usually diminish) IHC stains. Usually, the optimal fixation time varies between fixatives. For the recovery of immunoreactivity of the fixed tissues, antigen retrieval techniques have often been used to unmask the hidden antigen epitopes [11].

In this section, we describe the basic and standard methods of IHC applied to brain slice sections in the daily practice of neuroscience research and neuropathology. The several essential steps in IHC using FFPE tissues can be summarized as follows: (1) antigen (epitope) retrieval, (2) selection and application of primary antibodies, (3) incubation with secondary antibodies that bind to the primary antibodies, and (4) addition of detection reagents to enhance and visualize the primary antibody. The major steps of IHC performed on frozen sections are almost the same as for IHC using FFPE tissues, except that no antigen retrieval is required. Furthermore, in addition to the conventional manual IHC, an automated immunostainer is presented as another option for standardized immunohistochemistry [12–14]. IHC involves many complex work steps (Fig. 1), but automated staining systems can process dozens of IHC slides simultaneously with reproducible and accurate staining results (Fig. 2).

2 Materials

2.1 Solutions

1. Neutral buffered formalin, 10% (*see* **Note 1**): Add 100 mL formalin to 1 L graduated cylinder or glass beaker. Weigh out 25.9 g Na_2HPO_4 and 4.4 g NaH_2PO_4 and transfer to cylinder. Add water to a volume of 1000 mL. The mixed solution is adjusted to pH 7.2.

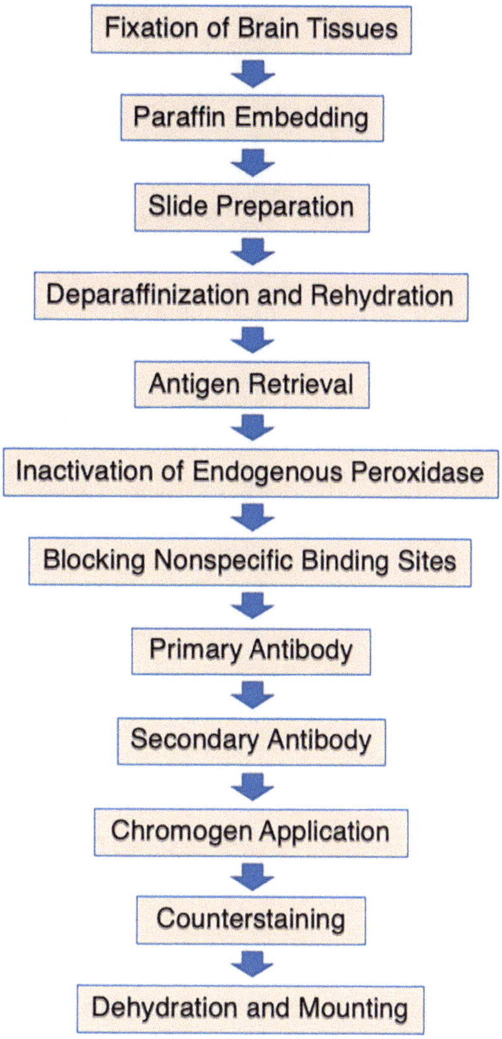

Fig. 1 Flowchart of brain tissue immunohistochemistry. The flowchart depicts the flow of procedures through the different phases of the method of immunohistochemistry in the mouse brain

2. Deionized and distilled water.
3. Xylene.
4. Ethanol, anhydrous denatured, histological grade (100, 95, 80, and 70%).
5. Hematoxylin solution according to Mayer (*see* **Note 2**).
6. Wash buffers (stock solution, 10× PBS): 0.01 M phosphate-buffered saline (PBS), pH 7.4; can be modified as an antibody following vendor's recommendations (*see* **Note 3**).

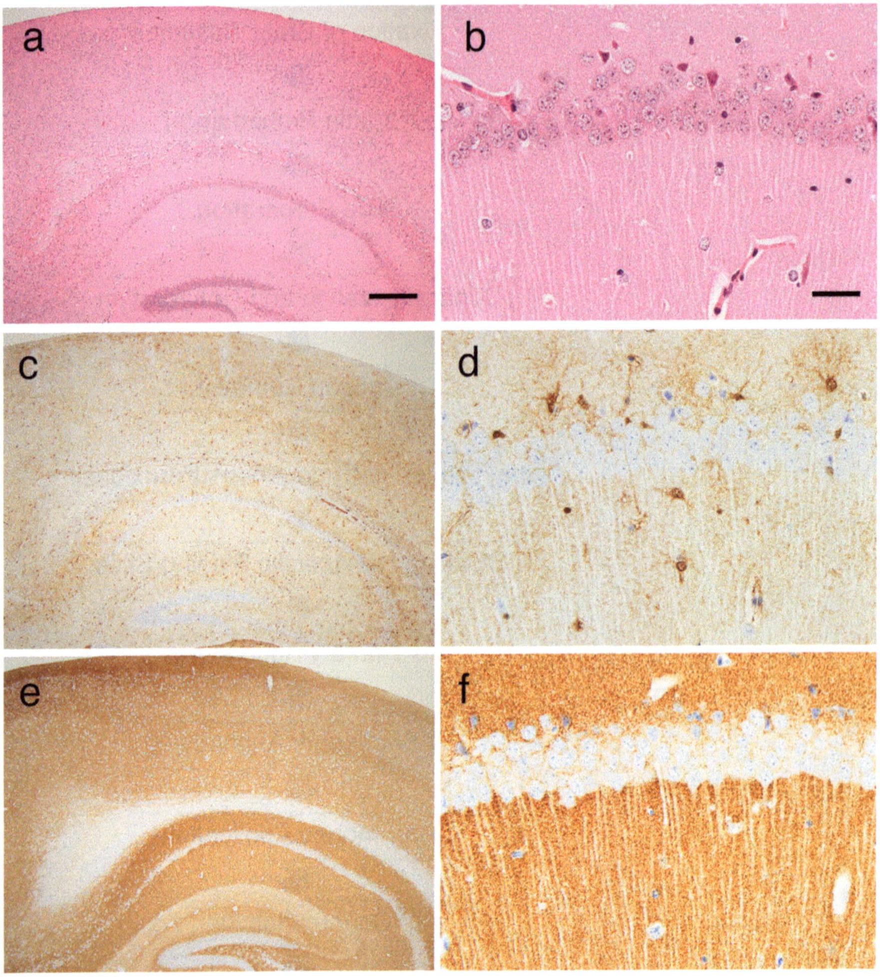

Fig. 2 Microphotographs of HE staining for histomorphological examination and immunohistochemistry for S-100 and synaptophysin. Coronal sections of the left cerebral hemisphere and hippocampus in the mouse brain are shown. The serial sections with a thickness of 3 μm are used. Low-power field (**a**, **c**, and **e**) and high-power field (**b**, **d**, and **f**) are shown for each staining. Photographs of high-power field indicate hippocampal neurons in each staining. The immunohistochemical staining was performed using an automated immunostainer. Scale bars in (**a**) and (**b**) of HE staining are 300 and 30 μm, respectively

7. Wash buffers: 0.01 M phosphate-buffered saline (PBS), pH 7.4, 0.2% Tween®20; can be modified as an antibody following vendor's recommendations (*see* **Note 4**).
8. Antigen retrieval buffer: depends on each antigen retrieval protocol. Representative antigen retrieval buffers are as follows:

   10× citrate buffer, pH 6.0 or pH 7.0, which can be diluted to a 1× solution (*see* **Note 5**).

   10× Tris–EDTA buffer, which can be diluted to a 1× solution containing 10 mM Tris and 1 mM EDTA, pH 9.0.

9. 3% hydrogen peroxide methanolic solution: Add 30 mL of 3% hydrogen peroxide to 270 mL of methanol (*see* **Note 6**).
10. Blocking solution: 0.01 M phosphate-buffered saline (PBS), pH 7.4, containing 2% bovine serum albumin (BSA) (*see* **Note 7**).

2.2 Heat-Induced Antigen Retrieval (HIAR)

Several heating devices are available for HIAR. The heating devices, specifically designed for HIAR, monitor and record temperature, pressure, and time to optimize and standardize immunostaining procedures (*see* **Note 8**).

2.3 Antibodies

1. Primary antibodies: immunoglobulins that bind to a specific antigen of interest and are typically unconjugated (*see* **Note 9**).
2. Secondary antibodies: immunoglobulins labeled with an enzyme (e.g., horseradish peroxidase), biotin (*see* **Note 10**), or polymer (*see* **Note 11**).

3 Methods

3.1 Perfusion Fixation and Process for Paraffin Embedding

1. Anesthetize animals with sodium pentobarbital (*see* **Note 12**).
2. Perfuse animals transcardially with saline and then with neutral buffered formalin (*see* **Note 13**).
3. Remove the brain and cut it into desirable slices.
4. Immerse the brain slices in neutral buffered formalin for additional fixation for 24 h (*see* **Note 14**).
5. Process them for paraffin embedding.

3.2 Slide Preparation, Deparaffinization, and Rehydration

1. Cut several sections at a thickness of 3–5 μm (*see* **Note 15**) from the FFPE tissue blocks using a microtome and mount on adhesion-treated slides.
2. Place the freshly cut paraffin sections to dry overnight in a 37 ° C oven or in a 55 °C oven for at least 2 h. The temperature of the incubation oven may be modified according to the melting temperature of paraffin (*see* **Note 16**).
3. Immerse the slides (approximately 15 dips each for at least 5 min) in 3 washes of xylene.
4. Immerse the slides (approximately 15 dips each for at least 1 min) in graded alcohols sequentially from 100%, 100%, 90%, and 80% to 70%.
5. Immerse in two changes of deionized water.

3.3 HIAR (See Note 17)

1. Place the deparaffinized and rehydrated slides in a heat-resistant plastic jar containing suitable antigen retrieval buffer (*see* **Note 18**). Slides must be fully covered with antigen retrieval buffer solution.
2. Operate the HIAR device according to the specifications of the primary antibody, for example, at 110 °C for 10 min as standard.
3. Let them cool slowly (*see* **Note 19**) at room temperature for at least 20 min.
4. After cooling the slides, transfer the slides to 0.01 M PBS, pH 7.4, and let them sit in PBS for 5 min.
5. Immerse and wash them in additional 2 changes of 0.01 M PBS, pH 7.4.

3.4 Inactivation of Endogenous Peroxidase

1. Immerse the slides in 3% hydrogen peroxide methanolic solution in a glass staining jar for 10 min.
2. Immerse and wash them in 3 changes of 0.01 M PBS, pH 7.4.

3.5 Blocking Nonspecific Binding Sites

1. Prepare a humidity chamber for immunohistochemistry and put wet tissue paper on the bottom of the chamber.
2. Wipe off the water around the tissue in each slide using a Kimwipe and encircle the tissue with a hydrophobic pen (e.g., Dako PAP Pen) (*see* **Note 20**).
3. Immediately apply several drops of blocking solution just enough to cover the tissues inside the circle drawn by PAP pen.
4. Then, put the slide in a humidity chamber.
5. Cover the chamber with the lid and incubate for 40 min at room temperature.

3.6 Primary and Secondary Antibody Application

1. Carefully shake off excess blocking solution (*see* **Note 21**) and add an appropriate volume (enough to cover the tissue section completely) of primary antibody diluted in PBS.
2. Add the same isotype serum as the primary antibody at the same dilution in another section as a negative control.
3. Then, incubate the treated tissue sections in the humidity chamber under dark conditions overnight at 4 °C.
4. After incubation, remove the primary antibody solution and wash the slides three times with PBS for at least 5 min each with shaking at room temperature (*see* **Note 22**).
5. After the final wash, rinse off excess washing solution (*see* **Note 23**) and add an appropriate volume of the secondary antibody for which a polymer detection system is recommended.

6. Incubate the slides in the humidity chamber for 30 min at room temperature.
7. After incubation, remove the secondary antibody solution and wash the slides three times with PBS for at least 5 min each with shaking at room temperature.

3.7 Chromogen Application

1. Prepare chromogenic substrate solution depending on a detection system according to the manufacturer's specifications.
2. Carefully shake off excess washing solution and apply appropriate drops of freshly prepared substrate mixture to cover the tissue section.
3. Incubate for a few min to a maximum of 10 min until a desired color reaction can be observed under low-power field of the microscope.
4. Remove the substrate solution, wash the slides with PBS with shaking, and reconfirm the chromogenic staining on the slides under a microscope (*see* **Note 24**).
5. Then, wash the slides with tap water after the completion of the staining reaction.

3.8 Counterstaining, Dehydration, and Mounting

1. Place the slides in hematoxylin solution (*see* **Note 25**) for about 30 s for nuclear counterstain (*see* **Note 26**).
2. Rinse in multiple changes of tap water until clear and then let sit in tap water for at least 5 min.
3. Dehydrate the slides by washing each with graded ethanol (70, 90, 95%, and 100% ethanol) for about 5 min.
4. Wash the slides three times with xylene (*see* **Note 27**), with each wash being for about 5 min.
5. Mount the sections using a nonaqueous mounting medium and coverslips.

3.9 Automated Immunostainer as another Option for IHC (See Note 28)

1. Enter case details to the software.
2. Enter slide details to the software.
3. Print slide labels and attach them to the slides.
4. Set sliced paraffin sections before deparaffinization in a slide tray.
5. Place the detection system and the marker containers for primary antibodies onto the processing module of an automated immunostainer.
6. Start the processing software according to the manufacturer's manual.

4 Notes

1. Formalin is composed of formaldehyde (37–40% (weight/volume)), where 10% neutral buffered formalin is basically a 4% (volume/volume) formaldehyde solution diluted using neutral phosphate buffer. 10% neutral buffered formalin is commonly used for general purpose histological examination including immunohistochemistry [15, 16]. 10% neutral buffered formalin can also be used for transcardial perfusion following euthanasia in rodent models.
2. IHC signals will usually be seen in brown with the nuclei counterstained blue with Mayer's hematoxylin.
3. PBS is commonly used in biological research since it is isotonic and nontoxic to cells. Thus, PBS is frequently used as a wash and dilution buffer in IHC procedures. Tris-buffered saline (TBS), which is an optimal formulation of salts and pH stabilizers, is usually used for western blot and ELISA procedures without disruption of antibody–antigen binding interactions. TBS can be also used as a wash and dilution buffer similar to PBS in IHC procedures.
4. Wash buffers for frozen sections usually contain 0.2% Tween®20 to avoid nonspecific background staining.
5. Preparation of sodium citrate buffer:

   Add about three-fourth of the final volume of distilled water to a glass beaker.

   Add 24.41 g of sodium citrate dihydrate to the solution.

   Add 3.99 g of citric acid to the solution.

   Adjust the solution to desired pH (pH 6.0 or pH 7.0) using 0.1 N HCl or 10 N NaOH and then add distilled water until the volume is 1 L.
6. A ready-to-use peroxidase blocking solution available from various manufacturers can also be used.
7. The best results would be obtained in animal tissue samples when the normal serum of the same species as the host of the secondary antibody is used in the blocking solution.
8. The HIAR step is usually performed for antigens stained weakly or not stained according to the primary antibody specifications. The proper use of heat and pressure in conjunction with the appropriate buffer solutions is important for consistent IHC staining.
9. Unlabeled primary antibodies are commonly used for most IHC techniques using FFPE samples. Fluorochrome-labeled primary antibodies are sometimes used for fluorescent IHC staining of frozen tissue sections. Primary antibodies can be

developed as either polyclonal antibodies, which bind to multiple different epitopes on the antigen, or monoclonal antibodies, which bind to one specific epitope on the antigen.

10. Biotin-labeled secondary antibodies are commonly used in various methods such as western blotting, ELISA and IHC. Secondary antibodies are usually labeled with multiple biotin molecules. Secondary antibodies can be used with horseradish peroxidase-conjugated streptavidin- or avidin-based techniques allowing for signal amplification of lowly expressed antigen. Polymer-based detection is more sensitive than biotin-based systems.
11. Polymer-based reagents are a more recent introduction into IHC detection methodology than streptavidin and biotin conjugates. IHC polymer detection systems offer distinct advantages over the traditional methods particularly for applications such as multiple antigen labeling (multiplexing) on the same tissue section. IHC polymer detection systems essentially consist of an integrated polymer of active enzyme and secondary antibody that binds to a primary antibody target. This integrated format introduces significantly more enzyme at the site of localization, thereby generating a greater reaction with the subsequent chromogen than with a secondary antibody directly conjugated with the enzyme. Additionally, the use of a one-step polymer method shortens the IHC procedure by avoiding the two-step biotinylated secondary antibody and ABC reagent that are required for standard avidin-biotin systems (from manufacturer's instructions; Vector Laboratories).
12. Mice are usually anesthetized by intraperitoneal injection of 50 mg/kg sodium pentobarbital [17].
13. The purpose of fixation is to preserve brain tissues in a life-like state as rapidly and uniformly as possible [17]. The advantage of transcardial perfusion fixation is that the fixatives can quickly reach every brain tissue by using the structural vascular network. Formalin and paraformaldehyde (PFA) are commonly used as fixatives of brain slices in experimental animals for application of transcardial perfusion. While perfusion fixation is not commonly performed during the examination of brain tissues of human postmortem autopsy [18], it is the gold standard for processing brain tissues in animals.
14. An additional fixation time of formalin of 6–48 h is usually recommended according to the size of tissue samples. However, it is reported that prolonged formalin fixation time of up to 7 weeks has minimal effects on antigen detection for most commonly used antibodies for IHC in diagnostic pathology [19].

15. The thickness of the sections depends on the type of tissue or staining method used. For example, brain tissue is usually cut thicker than other tissues. Cold paraffin on ice allows thinner sections to be obtained by cutting easily. The paraffin sections usually stick to the regular slides and are not detached from the slide during the standard staining process. The HIAR technique (explained in the next section) in IHC can cause the paraffin sections to detach from the slide. The adhesion-treated slides can be self-made using coatings such as protein glycerol (albumine) or gelatine. Commercially available adhesive slides are used and recommended, especially in HIAR technique. Alternatively, a convenient method for attaching paraffin sections to the regular slides is described as follows: place the stored paraffin sections in a 65 °C oven for 45 min. The melting temperature of paraffin is usually 56~58 °C. Therefore, the paraffin sections in a 65 °C oven melt transiently and then solidify and stick strongly to the slides when removed from the oven and allowed to cool at room temperature. It is important to note that a prolonged exposure to a 65 °C oven will degrade the quality of the specimens.
16. The melting temperature of paraffin is usually 56~58 °C.
17. A variety of heating devices including microwave ovens, pressure cookers, vegetable steamers, and autoclaves have been used for HIAR technique in IHC. The exact mechanism by which HIAR works is unknown; however, recent studies indicate that the essential mechanism of HIAR is based on the cleavage of stable protein–protein crosslinks induced by formalin fixation [16].
18. Suitable buffer among the stocks of several antigen retrieval buffers is selected and used for achieving the best primary antibody reaction.
19. Rapid cooling appears to cause faint background staining.
20. The PAP Pen is a special marking pen that provides a thin film-like hydrophobic barrier when a circle is drawn gently around a specimen on a slide. The water-repellent barrier keeps staining solution localized on the tissue sections and allows the use of less reagents per section. The PAP Pen markings, insoluble in alcohol, can be optionally removed with xylene after the staining procedure is completed (from manufacturer's instructions).
21. Shake the slides gently to remove excess blocking solution, and a water-repellent barrier drawn by a PAP Pen appears around the specimen.
22. It is recommended to separate the washing procedures for primary antibodies and negative controls to avoid contamination of primary antibodies with negative controls.

23. Shake the slides gently to remove excess PBS washing solution, and a water-repellent barrier drawn by a PAP Pen appears again around the specimens.
24. One more treatment with chromogenic substrate solution can be done when the staining reaction is insufficient. However, excessive staining time will increase background staining.
25. To avoid too much staining of the nucleus, you can use the hematoxylin solution diluted 2× with water, especially in case if the immunoreaction is mainly observed in the nuclei.
26. Counterstain provides a good contrast to IHC chromogens for the discrimination of the IHC chromogenic signal. In addition, it allows researchers to identify the cell type and exact localization of the target of IHC. Hematoxylin is the most commonly used counterstain in IHC. Hematoxylin can stain the nuclei dark blue/purple by binding to lysine residues on nuclear histones. Methylene blue is another widely used counterstain. Methylene blue binds to tissue anions such as carboxylic acid, sulfuric acid and phosphoric acid groups and will give a quick and simple nuclear stain. Methylene blue stains both the cytoplasm and the nucleus in contrast to hematoxylin that stains mainly the nucleus. More strong stain is recognized in the nucleus than in the cytoplasm when using methylene blue; however, the staining intensity of the nucleus is not so strong as hematoxylin. Thus, methylene blue may be preferable in IHC when the target is nucleus.
27. PAP Pen markings on the slides are removed by xylene in the dehydration procedure.
28. Advantage of automated immunohistochemistry system is that the quality of staining is standardized and controlled because all reagents are prepared by the manufacturer and pretested to assure optimal performance. IHC involves many complex work steps (Fig. 1), but automated staining systems can process dozens of IHC slides simultaneously with reproducible and accurate staining results (Fig. 2).

References

1. Cao S, Li J, Yuan J et al (2020) Fast localization and sectioning of mouse locus coeruleus. Biomed Res Int 2020:2–6. https://doi.org/10.1155/2020/4860735
2. Ryu B, Sekine H, Kawamata T et al (2019) Sticker method for preparation of frozen section using adhesive film. J Neurosci Methods 328:108436. https://doi.org/10.1016/j.jneumeth.2019.108436
3. Lin F, Chen Z (2014) Standardization of diagnostic immunohistochemistry: literature review and geisinger experience. Arch Pathol Lab Med 138:1564–1577. https://doi.org/10.5858/arpa.2014-0074-RA
4. Dennis CV, Suh LS, Rodriguez ML et al (2016) Human adult neurogenesis across the ages: an immunohistochemical study. Neuropathol Appl Neurobiol 42:621–638. https://doi.org/10.1111/nan.12337
5. Louis DN, Perry A, Wesseling P et al (2021) The 2021 WHO classification of tumors of the central nervous system: a summary. Neuro

Oncol 23:1231–1251. https://doi.org/10.1093/neuonc/noab106

6. Meredith DM (2020) Advances in diagnostic immunohistochemistry for primary tumors of the central nervous system. Adv Anat Pathol 27:206–219. https://doi.org/10.1097/PAP.0000000000000225
7. Aoki H, Hara A, Kunisada T (2017) Induced haploinsufficiency of kit receptor tyrosine kinase impairs brain development. JCI Insight 2:e94385. https://doi.org/10.1172/jci.insight.94385
8. Hara A, Iwai T, Niwa M et al (1996) Immunohistochemical detection of Bax and Bcl-2 proteins in gerbil hippocampus following transient forebrain ischemia. Brain Res 711:249–253. https://doi.org/10.1016/0006-8993(95)01436-5
9. Hara A, Niwa M, Nakashima M et al (1998) Protective effect of apoptosis-inhibitory agent, N-tosyl-L-phenylalanyl chloromethyl ketone against ischemia-induced hippocampal neuronal damage. J Cereb Blood Flow Metab 18: 819–823. https://doi.org/10.1097/00004647-199808000-00001
10. Fritschy JM (2008) Is my antibody-staining specific? How to deal with pitfalls of immunohistochemistry. Eur J Neurosci 28:2365–2370. https://doi.org/10.1111/j.1460-9568.2008.06552.x
11. Shi SR, Shi Y, Taylor CR (2011) Antigen retrieval immunohistochemistry: review and future prospects in research and diagnosis over two decades. J Histochem Cytochem 59: 13–32. https://doi.org/10.1369/jhc.2010.957191
12. Kim SW, Roh J, Park CS (2016) Immunohistochemistry for pathologists: protocols, pitfalls, and tips. J Pathol Transl Med 50:411–418. https://doi.org/10.4132/jptm.2016.08.08
13. Ilie M, Khambata-Ford S, Copie-Bergman C et al (2017) Use of the 22C3 anti-PD-L1 antibody to determine PD-L1 expression in multiple automated immunohistochemistry platforms. PLoS One 12:e0183023. https://doi.org/10.1371/journal.pone.0183023
14. Koppel C, Schwellenbach H, Zielinski D et al (2018) Optimization and validation of PD-L1 immunohistochemistry staining protocols using the antibody clone 28-8 on different staining platforms. Mod Pathol 31:1630–1644. https://doi.org/10.1038/s41379-018-0071-1
15. Grizzle WE (2009) Special symposium: fixation and tissue processing models. Biotech Histochem 84:185–193. https://doi.org/10.3109/10520290903039052
16. Yamashita S (2007) Heat-induced antigen retrieval: mechanisms and application to histochemistry. Prog Histochem Cytochem 41: 141–200. https://doi.org/10.1016/j.proghi.2006.09.001
17. Hara A, Niwa M, Kumada M et al (2004) Fragmented DNA transport in dendrites of retinal neurons during apoptotic cell death. Brain Res 1007:183–187. https://doi.org/10.1016/j.brainres.2004.01.075
18. McFadden WC, Walsh H, Richter F et al (2019) Perfusion fixation in brain banking: a systematic review. Acta Neuropathol Commun 7:146. https://doi.org/10.1186/s40478-019-0799-y
19. Webster JD, Miller MA, Dusold D et al (2009) Effects of prolonged formalin fixation on diagnostic immunohistochemistry in domestic animals. J Histochem Cytochem 57:753–761. https://doi.org/10.1369/jhc.2009.953877

Chapter 4

Two-Photon FRET/FLIM Imaging of Cerebral Neurons

Thomas T. Luyben, Jayant Rai, Bingyue Zhou, Hang Li, and Kenichi Okamoto

Abstract

Two-photon FRET (Förster resonance energy transfer) and FLIM (fluorescence lifetime imaging microscopy) enable the detection of FRET changes of fluorescence reporters in deep brain tissues, which provide a valuable approach for monitoring target molecular dynamics and functions. Here, we describe two-photon FRET and FLIM imaging techniques that allow us to visualize endogenous and optogenetically induced cAMP dynamics in living neurons with genetically engineered FRET-based cAMP reporters.

Key words Two-photon microscopy, FRET, FLIM, cAMP, Fluorescence reporters, Synaptic plasticity, Dendritic spine, Optogenetics, Photoactivatable adenylyl cyclase (PAC)

1 Introduction

cAMP (adenosine 3′,5′-cyclic monophosphate) is a ubiquitous second messenger that has a variety of cellular functions [1]. In the postsynaptic site of excitatory neurons called dendritic spines, the cAMP signaling pathways are thought to play important roles in synaptic plasticity underlying learning and memory [2–5]. However, the spatiotemporal dynamics and immediate actions of postsynaptic cAMP in synaptic plasticity are still elusive.

While pharmacological or genetic approaches have been used for the cAMP studies, these approaches remain limited in their spatial and temporal specificity for interrogation of cAMP function/signaling. Recent optogenetic techniques utilizing photoactivation of adenylyl cyclase [6, 7] have demonstrated the timing-dependent manipulation of cAMP by light in the target neurons [8, 9], suggesting the ability to precisely monitor and manipulate cAMP dynamics by the combination of optogenetics and live imaging approaches. To visualize synaptic molecules in deep brain tissues, we have used two-photon fluorescence microscopy [10, 11]. Specifically, using two-photon FRET (Förster resonance

Koh-ichi Nagata (ed.), *Cerebral Cortex Development: Methods and Protocols*, Methods in Molecular Biology, vol. 2794, https://doi.org/10.1007/978-1-0716-3810-1_4,

energy transfer) and FLIM (fluorescence lifetime imaging microscopy) live imaging techniques with genetically encoded FRET (Förster-resonance energy transfer) reporters, we have previously reported postsynaptic actin and CaMKIIβ dynamics and their interaction during synaptic plasticity [12–16].

Genetically encoded cAMP FRET fluorescence reporters utilizing a pair of fluorescent proteins, such as CFP (cyan fluorescent protein) and YFP (yellow fluorescent protein), separated by a cAMP-binding domain have been developed to observe cAMP dynamics in cells [17, 18].

We describe a detailed protocol for two-photon live cAMP FRET and FLIM imaging that utilizes cAMP FRET reporters to monitor endogenous and optogenetically induced cAMP dynamics in living neurons of cultured hippocampal brain slices.

2 Materials

2.1 Animals

The organotypic slice cultures of the hippocampus were prepared from postnatal day 6–7 rats as previously described [15]. All animal procedures were performed following a protocol approved by the University Health Network Animal Care Committee in accordance with their guidelines for care and use of animals.

2.2 Cell Culture and Transfection

1. HEK293T cells (*see* **Note 1**).
2. 6-well tissue culture plates.
3. Dulbecco's Modified Eagle's Medium (DMEM) with 4.5 g/L glucose and L-glutamine.
4. Fetal bovine serum (FBS), heat inactivated.
5. Phosphate-buffered saline (PBS) solution: 137 mM NaCl, 10 mM Na_2HPO_4, and 1.8 mM KH_2PO_4.
6. CO_2 incubator (37 °C, 5% CO_2).
7. Transfection reagent.
8. Fluorescence dissection microscope.
9. Transfection medium.
10. Expression vectors (cAMP FRET reporters, Figs. 1a, 2a) (*see* **Note 2**).

2.3 Cell Lysate

1. Phosphate-buffered saline (PBS) solution: 137 mM NaCl, 10 mM Na_2HPO_4, and 1.8 mM KH_2PO_4.
2. Assay solution (40 mM HEPES/Na, pH 8.0, 0.1 mM EGTA, 5 mM magnesium acetate, 1 mM DTT, and 0.01% Tween-20).
3. Cell scrapers.

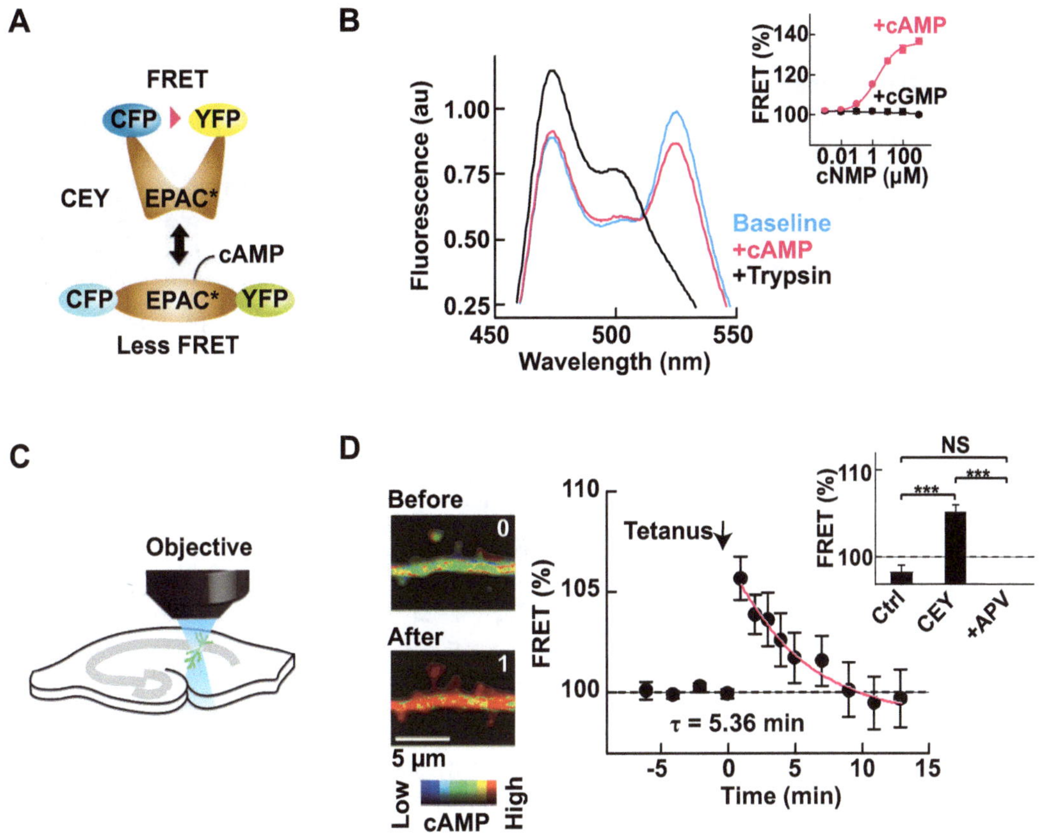

Fig. 1 Detection of cAMP dynamics in living neurons. (**a**) Detection of cAMP with FRET (CEY). (**b**) Emission spectra of HEK293 cell lysate expressing CEY before (Baseline) and after application of 100 μM 8-Br-cAMP (+cAMP) or trypsin digest (+Trypsin). Inset: Dose-dependent FRET changes (CFP/YFP ratio) to 8-Br-cAMP (+cAMP; $n = 6$) or 8-Br-cGMP (+cGMP; $n = 6$). EC_{50} of CEY to cAMP is 1.68 μM (Hill fit). (**c**) Schematic drawing of two-photon FRET imaging in the hippocampal slices. A cAMP FRET reporter (CEY) or a control reporter (Ctrl) was biolistically transfected in CA1 pyramidal neurons of the cultured hippocampal slices. (**d**) Tetanus induces transient cAMP increase in living neurons. Left: cAMP FRET (CEY) pseudocolor images of a dendrite before and after tetanus. Right: Time course of the postsynaptic FRET change after tetanus (decay time $\tau = 5.36$ min and $n = 10$; exponential fit). Black arrow indicates tetanus (100 Hz, 1 s × 3). Inset: Quantification of cAMP FRET changes by CEY immediately after tetanus. A control reporter was also separately expressed in neurons (Ctrl: fused only CFP and YFP directly). D-APV (50 μM) was bath applied for inhibition of NMDA receptors (+APV). Ctrl ($n = 13$), CEY ($n = 10$), +APV ($n = 11$). Steel–Dwass test. All data are presented as mean ± SEM. ***, $p < 0.001$

4. Sonicator.
5. Small centrifuge.

2.4 Fluorospectrometric Assay

1. Fluorescence spectrophotometer.
2. Cuvette (*see* **Note 3**).
3. Adenosine 3′,5′-cyclic monophosphate (cAMP) and Guanosine 3′,5′-cyclic monophosphate (cGMP) (*see* **Note 4**).

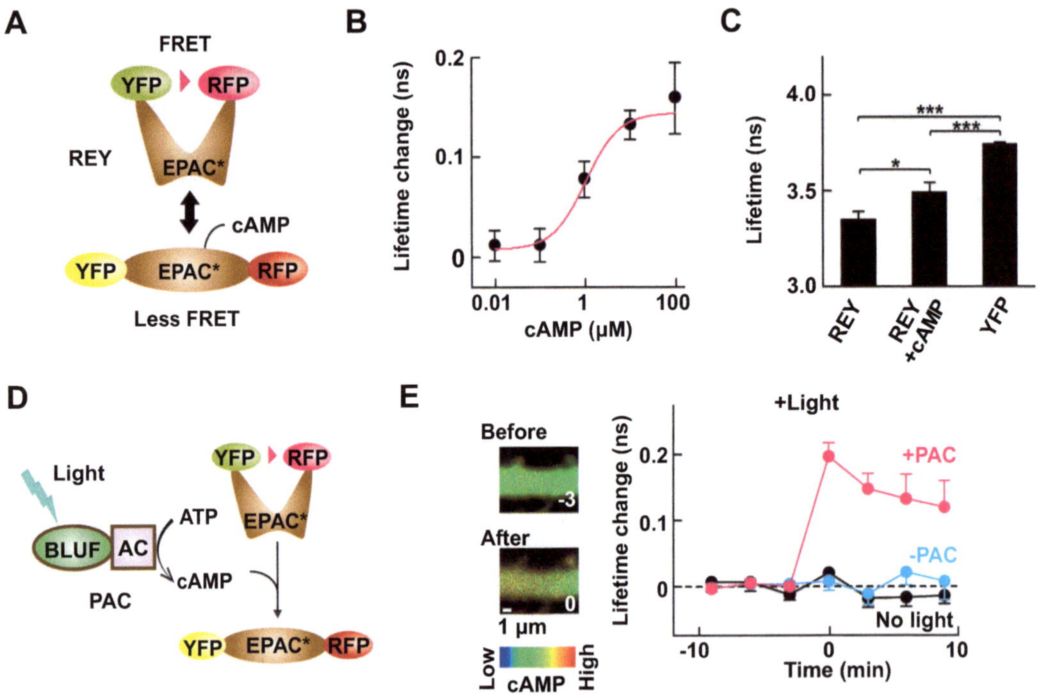

Fig. 2 Detection of PAC photoactivation. (**a**) A FLIM reporter (REY) for cAMP detection. (**b**) Dose-dependent changes of the fluorescence lifetime of REY with $EC_{50} = 1.81$ μM ($n = 9$; Hill fit). (**c**) Quantification of the lifetime before and after 100 μM 8-Br-cAMP and YFP (donor) alone. REY ($n = 9$), REY + cAMP ($n = 9$), and YFP ($n = 10$). Tukey's test. (**d**) Scheme for PAC photoactivation and its synthesized cAMP detection by REY. (**e**) Detection of increased cAMP level by PAC photoactivation in living neurons. Left: The cAMP pseudocolor REY FLIM images of a dendrite before and after photoactivation (414 ± 21 nm mercury arc lamp, 11 mW, 60 s). Right: Time course of the fluorescence lifetime change in neurons expressing REY either with PAC or without PAC (+/-PAC). No light: No PAC photoactivation during FLIM imaging in neurons expressing REY and PAC. Black bar represents PAC photoactivation (+Light). +PAC ($n = 11$), -PAC ($n = 10$), No light ($n = 15$). All data are presented as mean ± SEM. ***, $p < 0.001$; *, $p < 0.05$

4. Trypsin.
5. APV (NMDA receptor antagonist).

2.5 Organotypic Slice Culture and Transfection

1. 6-well plates.
2. Filter membrane inserts.
3. Slice culture media: Minimum Essential Medium (MEM), 26.6 mM D-glucose, 5.8 mM $NaHCO_3$, 30 mM HEPES, 2 mM $CaCl_2$, 2 mM $MgSO_4$, 0.48% (v/v) ascorbic acid, 1 uM insulin, and 20% horse serum, pH 7.2.
4. Cutting solution: 10 mM D-glucose, 4 mM KCl, 26 mM $NaHCO_3$, 248 mM sucrose, 5 mM $MgCl_2$, and 1 mM $CaCl_2$.
5. 5% CO_2/95% O_2 gas.
6. Spatulas, forceps, scissors, and pipettes.

7. Dissection microscope.
8. Tissue chopper.
9. Aspirator.
10. CO_2 incubator (35 °C, 5% CO_2).
11. Biolistic transfection equipment.
12. Expression vectors (cAMP FRET reporters, PAC: photoactivatable adenylyl cyclase) (*see* **Note 5**).

2.6 Two-Photon FRET Imaging

1. ACSF (artificial cerebrospinal fluid) solution containing 119 mM NaCl, 2.5 mM KCl, 4 mM $CaCl_2$, 4 mM $MgCl_2$, 26.2 mM $NaHCO_3$, 1 mM NaH_2PO_4, and 11 mM glucose at 30 °C equilibrated with 5% CO_2/95% O_2 (*see* **Note 6**).
2. Glass electrode filled with 0.5 M NaCl to deliver local tetanic electrical stimulation (100 Hz, 1 s, 3 trains at 10 s intervals) (*see* **Note 7**).
3. Two-photon FRET microscope system: For two-photon FRET imaging of neurons, CFP and YFP images (800 nm excitation, 460–500 nm for CFP emission, and 520–542 nm for YFP emission) were used for FRET ratiometric assay in a custom-made two-photon microscope equipped with a 60× objective lens (NA 1.0) and analyzed using imaging software [14, 15].

2.7 Two-Photon FLIM Imaging by Photoactivation

1. ACSF (artificial cerebrospinal fluid) solution for FLIM imaging of living neurons.
2. Assay solution for in vitro FLIM imaging of cell lysate.
3. Two-photon FLIM microscope with a time-correlated single-photon counting (TCSPC) system: The donor (YFP) fluorescence lifetime of the cAMP FRET reporter (REY) (900 nm two-photon excitation and 520–542 nm emission for YFP) was used for FLIM imaging in a custom-made two-photon microscope equipped with a 60x objective lens, NA 1.0 for 60 s (in vitro FLIM assay), or 30 s (neuron FLIM imaging) using single-plane imaging.
4. Photoactivation equipment with focal light from a mercury arc lamp (414 ± 21 nm, 11 mW, 60 s) under 60× objective lens (NA 1.0) under two-photon FLIM system.

3 Methods

3.1 HEK293T Cell Culture and Transfection

1. Place 5×10^5 HEK293 cells in 2 mL of (DMEM +10% FBS) per well of 6-well plates and culture the cells overnight in a CO_2 incubator (37 °C, 5% CO_2).

2. Prepare the plasmid DNAs for transfection following the vendor's instructions (mix 2.5–5.0 μg plasmid DNAs, 5 μL transfection reagent, and 125 μL transfection medium at room temperature and wait for 15 min) (*see* **Note 8**).
3. After 70–90% confluency of HEK293T cells in each well, add the plasmid DNAs with transfection reagent solution to the well for transfection (*see* **Note 9**).
4. Observe the expression of the transfected fluorescence reporters in the HEK293T cells by fluorescence dissection microscope after 36–48 h following transfection.

3.2 HEK293T Cell Lysate

1. Wash the culture medium in the well with 2 mL PBS solution for three times (*see* **Note 10**).
2. Replace with 0.5 mL assay solution for harvest.
3. Scrape the HEK293T cells in the well using a cell scraper and transfer the cells into tubes.
4. Homogenize the HEK293T cells using a sonicator (*see* **Note 11**) and centrifuge at 16,000 g for 15 min to clear large cell debris. The supernatant was then isolated and used for further analysis.

3.3 In Vitro Fluorospectrometric Assays for FRET

1. Using a fluorescence spectrophotometer, excite the cell lysate in the cuvette at 433 nm and measure basal FRET as a ratio of emission at 478 nm (CFP) to 525 nm (YFP) (Fig. 1b, Baseline). Add cAMP or a cyclic nucleotide analog such as 8-Br-cAMP to detect the FRET change (Fig. 1b, +cAMP). To validate whether the change is due to FRET, cleave the linker between the CFP and YFP in the FRET reporter by 2.5% trypsin application (Fig. 1b, +Trypsin).
2. Measure the dose-dependent FRET change by serially adding cAMP or 8-Br-cAMP (cGMP or 8-Br-cGMP as a control) (Fig. 1b inset) (*see* **Note 12**).

3.4 Organotypic Slice Culture and Transfection

1. Place a filter membrane to each well in the 6-well plates and add 0.9 mL of slice culture media. Transfer into a CO_2 incubator (37 °C, 5% CO_2).
2. Keep 40 mL cutting solution in 100 mL bottle with 5% CO_2/95% O_2 on ice.
3. Sterilize all dissection tools under UV in a hood at least for 1 h.
4. Sterilize a rat pup with 70% ethanol.
5. Remove the brain and transfer it into the cutting solution on ice for 1 min. Separate the cerebral hemispheres on a dissection microscope and place one of the hemispheres in the cutting solution.

6. Dissect the hippocampus and place it on a tissue chopper (*see* **Note 13**).
7. Prepare the hippocampal slices (400 μm thickness) by tissue chopper.
8. Transfer the slices from the tissue chopper to a dish with the cutting solution by a pipette and separate the slices by forceps under a dissection microscope.
9. Place 3–4 slices on a membrane of each well by a pipette and remove the cutting solution around the slices with an aspirator (*see* **Note 14**).
10. Maintain the 6-well plates in a CO_2 incubator (37 °C, 5% CO_2) (*see* **Note 15**).
11. Biolistically transfect after 5 days of slice preparation. In the coexpression experiments using photoactivatable adenylyl cyclase and a FRET reporter (PAC and REY), use 1:1 ratio of plasmids.
12. Perform live imaging 3–5 days after transfection.

3.5 Two-Photon FRET Neuron Imaging

1. Place the slice in a chamber with continuous perfusion of ACSF solution at 30 °C equilibrated with 5% CO_2/95% O_2.
2. Image the positive CA1 pyramidal neurons using a two-photon FRET imaging system (15–20 sections taken at 0.5 μm intervals of z-stack images of the dendrites) (Fig. 1c).
3. Deliver local tetanic electrical stimulation (100 Hz, 1 s, 3 trains at 10 s intervals) by a glass electrode filled with 0.5 M NaCl and fluorescent beads (1 μm diameter) (*see* **Note 16**) placed within 5–15 μm of the target dendritic spines (Fig. 1d).
4. To validate the activity-dependent FRET change, bath apply 50 μM APV (a potent NMDA receptor antagonist) (Fig. 1d inset, +APV). Additionally, a control FRET reporter lacking the cAMP (or cGMP) -binding domain by direct fusion of CFP and YFP may also serve as a negative control for FRET imaging (Fig. 1d inset, Ctrl).
5. Prepare pseudocolor FRET images using imaging software.

3.6 In Vitro Two-Photon FLIM Imaging

1. Turn on a two-photon FLIM system with time-correlated single-photon counting (TCSPC) (*see* **Note 17**).
2. Place the cell lysate in a chamber and measure the YFP fluorescence lifetime of the cAMP reporter (REY) (Fig. 2a). Optimize two-photon excitation laser power to detect the sufficient fluorescence lifetime and fit the lifetime decay curve by a double exponential decay model and tail-fitting analysis found in the FLIM imaging software.

3. Measure the dose-dependent FLIM change by serially adding cAMP or 8-Br-cAMP and compare with a control cell lysate which expresses the YFP alone (Fig. 2b, c).

3.7 Two-Photon FLIM Neuron Imaging

1. Place the slice in a chamber with continuous perfusion of ACSF solution at 30 °C equilibrated with 5% CO_2/95% O_2 (*see* **Note 18**).
2. Image the positive CA1 pyramidal neuron dendrites using a two-photon FLIM system (*see* **Note 19**) and measure the YFP fluorescence lifetime of the cAMP reporter (REY) (*see* **Note 20**) (Fig. 2d).
3. Detect the stable basal fluorescence lifetime during the time course imaging (single section images of the dendrites) before application of photoactivation light (Fig. 2e, no light).
4. Apply light using a photoactivation equipment (414 ± 21 nm mercury arc lamp, 11 mW, 60 s) for photoactivation of coexpressed PAC in neurons under the two-photon FLIM system and compare the time course change of YFP fluorescence lifetime of the cAMP reporter (REY) with or without PAC (Fig. 2e, +PAC, -PAC).
5. Prepare the fluorescence lifetime pseudocolor image of neurons using the pixel-by-pixel fitting function in the FLIM analysis software.

4 Notes

1. Only use cells with over 90% viability (healthy cells).
2. The cAMP FRET reporter CEY (mouse codon-optimized) was constructed following ICUE2 [19] and showed soluble distribution in rodent neurons. To enhance the brightness and FRET efficiency, CEY utilizes a C-terminal truncated bright CFP (truncated mTurquoise2) [20, 21] as a FRET donor in the N-terminus which is fused to the human Epac1-derived cAMP-binding domain (AF103905: residue 149-881) [4, 10]. For the FRET acceptor, we fused it with a YFP (Citrine) at the C-terminus and subcloned the cAMP FRET reporter into a pCAG plasmid vector [22]. The control FRET reporter (Control) lacks the cAMP-binding domain and was fused by only CFP and YFP directly. To construct REY, we selected tdTomato as the acceptor and replaced with CFP in the CEY by PCR without the linker between tdTomato and Epac1. tdTomato was selected as other red fluorescent proteins demonstrated aggregation of the resulting reporters in slice cultured hippocampal slices.
3. It is recommended to use a glass or quartz cuvette.

4. It is recommended to use 8-Bromoadenosine 3′,5′-cyclic monophosphate (8-Br-cAMP) and 8-Bromoguanosine 3′,5′--cyclic monophosphate (8-Br-cGMP) to prevent the degradation of cAMP/cGMP by endogenous phosphodiesterases in the cell lysate.
5. Photoactivatable adenylyl cyclase (PAC) with S27A mutation (by PCR of the PAC sequence using the forward primer 5′--atcggcaaggtggccatcaagaacaac-3′ and the reverse primer 5′--tgttcttgatggccaccttgccgatc-3′ to decrease dark activity) [7, 23] was fused with a 6 histidine tag at the N-terminal and subcloned into the pCAG plasmid vector (human codon-optimized).
6. Prepare 10 times concentrated the stock solution without $CaCl_2$ and $MgCl_2$. To prepare ACSF solution, dilute the stock solution and add $CaCl_2$ and $MgCl_2$ gassed with 5% CO_2/95% O_2 in advance for imaging experiments.
7. Add fluorescent beads to the electrode tip to aid with placement within the slices.
8. Use high-quality DNAs, OD 260/280 between 1.7 and 1.8.
9. Gently shake the 6-well plates after adding plasmid DNA solution to mix well without HEK293T cell detachment.
10. Slowly replace the solution to avoid detachment of HEK293T cells on the well.
11. Sonicate the cells on ice (around 40 times).
12. Administer a minimal volume of cAMP/cGMP solution to avoid the dilution of cell lysate. Use the same amount of buffer solution without cAMP/cGMP to measure the dilution effect and subtract it from the cAMP/cGMP results.
13. Place on a disposable Teflon sheet and remove cutting solution by filter paper.
14. Make sure slice culture media is attached to the bottom of the membrane insert, but do not cover the top of the membrane.
15. Replace the slice culture medium every 2–3 days.
16. To aid placement within the sample.
17. Turn on at least 30 min prior to the two-photon FLIM imaging.
18. Maintain the slices for at least 15 min to stabilize the slice position for single-plane FLIM imaging.
19. Target CA1 pyramidal neurons located near the surface of the slices for the FLIM imaging in order to adequately photoactivate PAC by single-photon light due to low penetration of the focal light into hippocampal tissue in the brain slices.

(a) The primary apical dendrites (~3 μm) or secondary apical dendrites (nearby apical dendrite bifurcations) are ideal targets for FLIM imaging.

(b) Single section: It is recommended to select a focal plane in the middle of the z-axis of the dendrite for FLIM imaging.

20. Use YFP-RFP pair of FRET reporters in combination with PAC photoactivation due to overlap in the excitation spectra of CFP-YFP FRET reporters and PAC photoactivation light (~500 nm wavelength).

References

1. Beavo JA, Brunton LL (2002) Cyclic nucleotide research—still expanding after half a century. Nat Rev Mol Cell Biol 3(9):710–718. https://doi.org/10.1038/nrm911
2. Chetkovich DM, Gray R, Johnston D, Sweatt JD (1991) N-methyl-D-aspartate receptor activation increases cAMP levels and voltage-gated Ca2+ channel activity in area CA1 of hippocampus. Proc Natl Acad Sci USA 88(15):6467–6471
3. Chetkovich DM, Sweatt JD (1993) nMDA receptor activation increases cyclic AMP in area CA1 of the hippocampus via calcium/calmodulin stimulation of adenylyl cyclase. J Neurochem 61(5):1933–1942
4. Greengard P, Jen J, Nairn AC, Stevens CF (1991) Enhancement of the glutamate response by camp-dependent protein-kinase in hippocampal-neurons. Science 253(5024): 1135–1138. https://doi.org/10.1126/science.1716001
5. Seeman P (1980) Brain dopamine-receptors. Pharmacol Rev 32(3):229–313
6. Iseki M, Matsunaga S, Murakami A, Ohno K, Shiga K, Yoshida K, Sugai M, Takahashi T, Hori T, Watanabe M (2002) A blue-light-activated adenylyl cyclase mediates photoavoidance in Euglena gracilis. Nature 415(6875): 1047–1051. https://doi.org/10.1038/4151047a
7. Stierl M, Stumpf P, Udwari D, Gueta R, Hagedorn R, Losi A, Gartner W, Petereit L, Efetova M, Schwarzel M, Oertner TG, Nagel G, Hegemann P (2011) Light modulation of cellular cAMP by a small bacterial photoactivated adenylyl cyclase, bPAC, of the soil bacterium Beggiatoa. J Biol Chem 286(2): 1181–1188. https://doi.org/10.1074/jbc.M110.185496
8. Luyben TT, Rai J, Li H, Georgiou J, Avila A, Zhen M, Collingridge GL, Tominaga T, Okamoto K (2020) Optogenetic manipulation of postsynaptic cAMP using a novel transgenic mouse line enables synaptic plasticity and enhances depolarization following tetanic stimulation in the hippocampal dentate gyrus. Front Neural Circ 14:24. https://doi.org/10.3389/fncir.2020.00024
9. Zhang SX, Lutas A, Yang S, Diaz A, Fluhr H, Nagel G, Gao S, Andermann ML (2021) Hypothalamic dopamine neurons motivate mating through persistent cAMP signalling. Nature 597(7875):245–249. https://doi.org/10.1038/s41586-021-03845-0
10. Khan M, Goldsmith CR, Huang Z, Georgiou J, Luyben TT, Roder JC, Lippard SJ, Okamoto K (2014) Two-photon imaging of Zn2+ dynamics in mossy fiber boutons of adult hippocampal slices. Proc Natl Acad Sci USA 111(18):6786–6791. https://doi.org/10.1073/pnas.1405154111
11. Ochiishi T, Futai K, Okamoto K, Kameyama K, Kosik KS (2008) Regulation of AMPA receptor trafficking by delta-catenin. Mol Cell Neurosci 39(4):499–507. https://doi.org/10.1016/j.mcn.2008.06.002
12. Kim K, Lakhanpal G, Lu HE, Khan M, Suzuki A, Hayashi MK, Narayanan R, Luyben TT, Matsuda T, Nagai T, Blanpied TA, Hayashi Y, Okamoto K (2015) A temporary gating of actin remodeling during synaptic plasticity consists of the interplay between the kinase and structural functions of CaMKII. Neuron 87(4):813–826. https://doi.org/10.1016/j.neuron.2015.07.023
13. Mower AF, Kwok S, Yu H, Majewska AK, Okamoto K, Hayashi Y, Sur M (2011) Experience-dependent regulation of CaMKII activity within single visual cortex synapses in vivo. Proc Natl Acad Sci USA 108(52): 21241–21246. https://doi.org/10.1073/pnas.1108261109
14. Okamoto K, Hayashi Y (2006) Visualization of F-actin and G-actin equilibrium using

fluorescence resonance energy transfer (FRET) in cultured cells and neurons in slices. Nat Protoc 1(2):911–919. https://doi.org/10.1038/nprot.2006.122

15. Okamoto K, Nagai T, Miyawaki A, Hayashi Y (2004) Rapid and persistent modulation of actin dynamics regulates postsynaptic reorganization underlying bidirectional plasticity. Nat Neurosci 7(10):1104–1112
16. Takao K, Okamoto K, Nakagawa T, Neve RL, Nagai T, Miyawaki A, Hashikawa T, Kobayashi S, Hayashi Y (2005) Visualization of synaptic Ca2+ /calmodulin-dependent protein kinase II activity in living neurons. J Neurosci 25(12):3107–3112. https://doi.org/10.1523/JNEUROSCI.0085-05.2005
17. Gorshkov K, Zhang J (2014) Visualization of cyclic nucleotide dynamics in neurons. Front Cell Neurosci 8:395. https://doi.org/10.3389/fncel.2014.00395
18. Sprenger JU, Nikolaev VO (2013) Biophysical techniques for detection of cAMP and cGMP in living cells. Int J Mol Sci 14(4):8025–8046. https://doi.org/10.3390/ijms14048025
19. Violin JD, DiPilato LM, Yildirim N, Elston TC, Zhang J, Lefkowitz RJ (2008) beta2-adrenergic receptor signaling and desensitization elucidated by quantitative modeling of real time cAMP dynamics. J Biol Chem 283(5):2949–2961. https://doi.org/10.1074/jbc.M707009200
20. Goedhart J, von Stetten D, Noirclerc-Savoye-M, Lelimousin M, Joosen L, Hink MA, van Weeren L, Gadella TW Jr, Royant A (2012) Structure-guided evolution of cyan fluorescent proteins towards a quantum yield of 93%. Nat Commun 3:751. https://doi.org/10.1038/ncomms1738
21. Klarenbeek JB, Goedhart J, Hink MA, Gadella TW, Jalink K (2011) A mTurquoise-based cAMP sensor for both FLIM and ratiometric read-out has improved dynamic range. PLoS One 6(4):e19170. https://doi.org/10.1371/journal.pone.0019170
22. Niwa H, Yamamura K, Miyazaki J (1991) Efficient selection for high-expression transfectants with a novel eukaryotic vector. Gene 108(2):193–199
23. Stierl M, Penzkofer A, Kennis JT, Hegemann P, Mathes T (2014) Key residues for the light regulation of the blue light-activated adenylyl cyclase from Beggiatoa sp. Biochemistry 53(31):5121–5130. https://doi.org/10.1021/bi500479v

Chapter 5

Pre- and Postembedding Immunoelectron Microscopy to Analyze the Central Nervous System

Fumino Fujiyama and Fuyuki Karube

Abstract

Immunocytochemistry, a method of delineating the subcellular localization of target proteins, was developed from immunohistochemistry. In principle, proteins are labeled using an antigen–antibody reaction. In order to observe under an electron microscope, the reaction product must scatter the electron beam with sufficient contrast while it is necessary to have an amplifying label that can withstand the observation. We have some detailed tips on making electron microscope samples to achieve this objective, and we would be happy to help you.

Key words Immunoelectron microscopy, Preembedding method, Postembedding method, Freeze-substitution

1 Introduction

Light microscopic (LM) immunostaining is a method of visualizing the localization of a substance (antigen) in tissues and cells by immunohistochemistry using specific antibodies against that antigen. When performing immunoelectron microscopy, attention must be paid to technical problems such as (1) preservation of the cell structure, (2) retention of antigenicity, and (3) penetration of antibodies. Certainly, LM immunostaining also faces these problems; however, samples for electron microscopy (EM) should be fixed stronger than LM samples to maintain a fine ultrastructure, which makes the problems more severe. Therefore, even when you use the same antibody, which works well in LM samples, there is no guarantee of success in EM samples. Particularly, the choice of fixative and setting staining conditions are important. Current immunoelectron microscopy methods are broadly divided into preembedding methods, in which sections are embedded in resin after the antigen–antibody reaction, and postembedding methods, in which the antigen–antibody reaction is performed on ultrathin

Koh-ichi Nagata (ed.), *Cerebral Cortex Development: Methods and Protocols*, Methods in Molecular Biology, vol. 2794,
https://doi.org/10.1007/978-1-0716-3810-1_5,

sections after tissue is embedded. The preembedding method is relatively similar to immunostaining for LM; however, as noted above, special care must be taken. For example, no (or a minimum concentration of) detergent is used for EM specimens to preserve the ultrastructure, which reduces penetration of reagents inside tissues. You can find general methods of fixation and immunohistochemistry for LM in other chapters; therefore, here we only briefly discuss them, except for some important tips on EM samples. Based on the authors' experience, this chapter details common preembedding methods and freeze-substituted postembedding method.

2 Materials

All solutions must be prepared using pure water (e.g., double-distilled water (DDW) or filtered DW) and high-quality grade (e.g., EM grade, etc.) reagents.

2.1 Fixation

1. Prefixatives: Ringer solution; 0.1 M phosphate-buffered saline (PBS) and others. Generally, pH is adjusted to 7.4.
2. Fixatives: For immunoelectron microscopy, paraformaldehyde (PFA) (electron microscopy grade), which produces effects similar to formalin used in LM, (however, commercially available formalin contains methanol and is not suitable for EM) is used to preserve antigenicity of some proteins and to immobilize biomolecules. PFA is a monomer of aldehyde with a small molecular weight and thus easily penetrates into tissues but its efficiency of fixation is lower than that of glutaraldehyde. In many cases, the combination with 0.1–2.5% glutaraldehyde is used (*see* **Note 1**). In addition, picric acid may be useful for fixation of polypeptides and can be added at the final step after pH adjustment (2 g picric acid powder per 1 L of PFA/glutaraldehyde solution, or common Zamboni's fixative). A fixative should be made at the last minute (up to 2 or 3 days before at most) or stored in a freezer soon after preparation (~ a month). For glycol proteins, a PLP fixative is known to be useful. In the following sections, we write PFA with/without picric acid (*see* **Note 2**).
3. Recently, fixatives without aldehyde, for example, glyoxal, which show better sensitivity for an immunoreaction, are also used for LM and EM [1].

2.2 Instruments for Perfusion Fixation (In Case of Mouse)

You could find this part elsewhere in this book:

1. A set of anatomical tools: Anatomy table, scissors, tweezers, and nose scissors (optional: forceps and blood vessel clamp).

2. Peristatic pumps: Tubes, needle (20–23 G for mouse; 18–20 G for rat); beakers (for PFA, prefixative, DDW).
3. Anesthetics: Pentobarbital (intraperitoneal injection of solution of at least 100 mg/kg of animal body weight).
4. Fume hood: Hazardous substances including formaldehyde must be treated with care according to the international/national and your institute guidelines. Waste substances and solution must be treated according to the guidelines. You must also take care of yourself with protection items.

2.3 Immunoelectron Microscopy

Many of the materials used for EM are hazardous. Ware appropriate protection goods and treat them in a hood. Also check the guidelines for treating them, including their waste. For all reactions, use clean experimental water, especially after you have obtained ultrathin sections.

1. Osmium tetroxide treatment: 1% OsO_4 in 0.1 M PB. The powder is hard to be dissolved, and it would be helpful to use an ample amount of commercially available premelted solution and dilute it with 0.1 M PB. OsO_4 solution is easy to elaborate, even when stored in a special vial with double caps for osmium solution storage, and the gas is also hazardous. Thus, it is recommended to use one freezer to store OsO_4.
2. Uranyl acetate: 1% (w/v) uranyl acetate in 70% EtOH (*see* **Note 3**). The compound is highly toxic and should be properly treated according to the guidelines.
3. Lead citrate: Boil 50 mL of DDW, completely dissolve 0.2 g of NaOH granules in water from which CO_2 has been removed, and filter through filter paper. This is followed by complete dissolution of 0.2 g of lead citrate. Filter again with filter paper before use. The final concentration is 0.4% lead citrate/0.4% sodium hydroxide solution (*see* **Note 4**). This compound is highly toxic and should be properly treated according to the guidelines.
4. Cylindrical stand: Made by pouring resin left over from embedding into capsules.
5. Incubation solution: 50 mM Tris–HCl buffer containing 0.9% NaCl, namely TBS (pH 7.4). This blocking solution can also be used for incubation. Basically, no detergent is recommended to use for the preservation of the fine ultrastructure; however, when you cannot achieve enough immunostaining, a low concentration of a detergent (e.g., 0.01–0.04% Triton X; TBST) might be helpful, with only a slight damage to the ultrastructure.
6. Blocking solution: 50 mM TBS (pH 7.4) containing 10% normal goat (or donkey; depending on animal species in

which the antibodies are raised) and 2% bovine serum albumin (BSA). You can find the description on pH for commercially available BSA, and use the one with pH 7. TBST (50 mM) containing 2% human serum albumin (HSA).

7. Antibodies: Use antibodies which work well for light microscopy. Due to differences in fixation and incubation conditions, longer incubation (~1 week) at 4 °C may be helpful to obtain enough signals. It will be sometimes helpful to use antibodies with smaller molecular weight (Fab, Fab', or F(ab')2) instead of IgG. Recently, antibodies derived from the *Camelidae* species, called nanobodies, are applied in EM.
8. Avidin-biotin-peroxidase complex (ABC): Before use, dilute at 1:100 to 1:200 with 50 mM TBS (or TBST) and leave on a shaker for 30 min.
9. DAB (3,3′-diaminobenzidine) (or nickel-DAB) solution: DAB is a potential carcinogen, and take care according to the guidelines. Prepare just before use. Alternatively, store at −25 °C and filter before use. Add 2 mg of DAB to 10 mL of TB (without NaCl) to prepare a 0.02% solution. For nickel-DAB, add 500 uL of 200 mM nickel ammonium sulfate to the DAB solution.
10. H_2O_2 solution: For stock, prepare 0.1–1% solution in 50 mM TB.
11. Epoxy resin: Prepare resin according to the manufacturer's instructions using a plastic disposable cup and pipettes. In most cases, it takes time to mix, so make it early. Some components are toxic and need care to treat them. Tools used for making resin must be polymerized and then discarded. After mixing, air bubbles can be removed using a small acrylic vacuum chamber.
12. Graded concentration of sucrose in 0.1 M PB: Prepare 0.5, 1.0, and 2.0 M sucrose solution in 0.1 M PB for freeze substation.
13. Lowycril HM20 or LR white resin: Prepare them according to the manufacturer's guidelines.
14. Dry absolute ethanol: Put molecular sieve 3A into commercial 99% EtOH and store it until use.
15. Sodium ethanolate.
16. Polyethylene glycol.
17. 2% glutaraldehyde in TBS.

2.4 Instruments for Preembedding Immunoreaction

1. Vibrating microtome.
2. Bioshaker.

3. Well plates: Select the size of the well depending on the section size and number. For all reactions, the sections must be well soaked in solution.

2.5 Semithin and Ultrathin Sectioning

1. Toluidine blue solution: Add 0.5 g of sodium tetraborate to 100 mL of DDW and stir until dissolved, and then add 0.5 g toluidine blue.
2. Single-slit grids with a formvar support film or mesh grids: Place 1% formvar/chloroform solution in a film-making device and make a uniform thickness film on a slide glass. Shortly after drying, cut all edges of the film on the slide. Then, gently put one edge of the slide into DDW in a large petri dish, and when you find the edge starts peeling off the slide and float on the water surface, slowly go ahead the slide to get a complete film of formvar. Put the precleaned grids on the floating film carefully using tweezers. Finally, put the parafilm, which must be slightly larger than the formvar film, above the grid and pick up the parafilm. Now the grids are caught between the parafilm and the formvar film. Using tweezers, gently push the formvar film at the edge of each grid to excise them. The formvar grids can be stored in a grid box until use. Take care to distinguish the formvar coated side.
3. Eyelash tool: Glue a eyelash on the tip of the thin bar using Instantbond.

3 Methods

3.1 Fixation (In Case of Perfusion Fixation of Mice)

It would be desirable to have a fixative that retains antigenicity of proteins while preserving the tissue structure; however, these two events are contradictory and therefore difficult. The purpose of tissue fixation is to deactivate degrading enzymes in tissues using PFA and glutaraldehyde, to prevent further denaturation, and to maintain morphology by cross-linking peptide chains, to stabilize cells and tissues in their natural state, and to make tissue hard enough to be sectioned. The best condition of fixation may vary depending on antigens and antibodies. Here, we present a basic method, which must be adjusted for your own case. Particularly, the concentration of glutaraldehyde largely affects the goodness of the immunoreaction and the preservation of the ultrastructure.

1. Mice are given an anesthetic solution intraperitoneally (100 mg/kg of their body weight), and if they no longer respond to pain stimuli, they are fixed supine on the dissection table.
2. After laparotomy, the diaphragm is incised and the ribs are incised in a cranial direction.

3. The xiphoid is pinched with forceps and flipped in the direction of the head, stopping it with a needle to expose the heart.
4. The right atrial appendage is cut with scissors.
5. A needle must be attached to the tube and filled with a prefixative before starting surgery. The needle is inserted into the left ventricle and the prefixative is allowed to flow (*see* **Note 5**).
6. The pump (keeping the bubbles out of the tube) is stopped and the end of the tube is replaced with the fixative from the prefixative (*see* **Note 6**). The needle inside the heart is grasped with tweezers or a blood vessel clamp. Soon after fixative flowing, it is confirmed whether both sides of the body start to be fixed. If not, the needle position is slightly shifted (forward or backward) to prevent uneven flowing.
7. Brain is removed and soaked in the same fixative for 3–12 h as postfixation. Enough fixation is necessary for EM samples; however, longer fixation may result in the loss of antigenicity of some proteins.

3.2 Preembedding Method

This method involves immunostaining prior to resin embedding. Peroxidase labeling is common. The immunostaining method is almost the same as the ordinary LM level, with indirect immunostaining or streptavidin use (Fig. 1) (*see* **Note 7**).

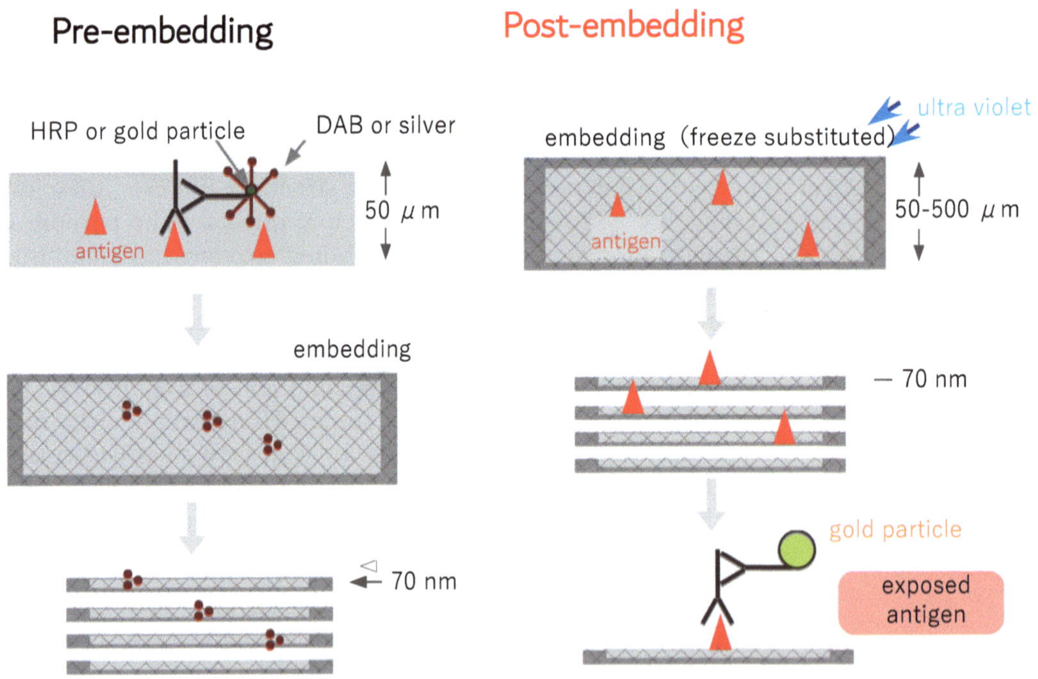

Fig. 1 Preembedding and postembedding methods. The preembedding method involves immunostaining prior to resin embedding. Peroxidase labeling is common. The postembedding method involves embedding a sample in a resin or other materials and performing immunostaining on the ultrathin section

3.2.1 Sectioning with Vibrating Microtome

We usually make a 50 μm section with a vibratome to fit the target area and rinse the sections with PB (*see* **Note 8**). All reactions are conducted in a well under floating conditions.

3.2.2 Immuno-histochemical Procedures (In Case of a Peroxidase Reaction)

1. Block the nonspecific reactions by incubation of the sections with blocking solution for 30–60 min on a bioshaker (*see* **Note 9**).
2. Dilute the primary antibody to an appropriate concentration and incubate for several hours to several days on a bioshaker (at room temperature or 4 °C).
3. After washing with incubation buffer (20 min, 3 times), incubate with peroxidase-labeled secondary antibodies. At this step, you may use gold particle–conjugated secondary antibodies. To intensify signals, biotin-conjugated secondary antibodies can be used, followed by the ABC reaction. Care should be taken to avoid the penetration of reagents since ABC is a large molecule.
4. After washing with incubation buffer (20 min, 3 times), perform the 3,3′-diaminobenzidine (DAB) reaction: After conducting the avidin-biotin complex (ABC) reaction (1–2 h, room temperature), wash with TBS. If you use gold particle–conjugated secondary antibodies, silver enhancement should be performed according to the manufacturer's protocol. Because dense DAB-osmium products fill the intracellular space and sometimes it is hard to observe the ultrastructure of intracellular organs (e.g., synaptic vesicles and spine apparatus) clearly during EM observation. Instead of DAB, nickel-DAB can be useful to solve the problem partially. Generally, DAB deposition is diffusible and thus the localization of the products at an ultrastructural level might not be reliable. Immunogold labeling, as discussed in the postembedding method section, can also be applied to the preembedding method.

3.2.3 Postfixation in Osmium Tetroxide

1. After washing, osmium tetroxide treatment (1% OsO_4 in 0.1 M PB, 25 min) is applied to form "osmium black" by reacting with DAB. It intensifies contrast for observation by transmission EM (*see* **Note 10**) or even in LM. Recently, the reduced-osmium tetroxide-thiocarbohydrazide-osmium tetroxide (rOTO) method is used to intensify electron density of the reaction products. After osmium incubation, the sections become black and rigid, which are easy to crack.
2. After washing with DDW (10 min, twice), counterstain with 1% (w/v) uranyl acetate in 70% EtOH for 1 h in the dark.

3.2.4 Dehydration and Embedding in Resin

1. Perform dehydration with graded series of volume for volume dilutions of ethanol in DDW (enough volume of 50, 70, 90, 95, and 100% EtOH and absolute ethanol 10 min each) (*see* **Note 11**). It is better to prepare three series of 100% EtOH to completely remove water in the sections.
2. Perform substitution with propylene oxide, 10 min, twice. Propylene oxide is an organic solvent and used to remove ethanol before embedding with epoxy resins (*see* **Note 12**). Insufficient dehydration and substitution with propylene oxide can lead to incomplete or nonuniform polymerization of resin.
3. Pour the resin into a container made of aluminum foil and let the sections infiltrated overnight at room temperature. Alternatively, sections are soaked into a mixture of propylene oxide and resin (1:1) for a few hours and then transferred to pure resin.
4. Flat embedding (*see* **Note 13**): Excess resin can be removed by absorbing it on to filter paper. Polymerization is performed for 24–48 h at 60 °C.

3.2.5 LM Observation and Excising of Region of Interest for EM

Observe the sections using a light microscope and determine the region of interest for LM imaging and EM observation. Due to osmification, intensive transmitting light may be necessary to observe the samples in detail. One of the advantages of the pre-embedding method is correlation of both LM and EM observations. You should check the quality of immunosignals: The degree of false positive or false negative signals and how deep inside sections you can observe signals. If they were unlikely to be reliable, you should change the conditions for fixation and immunoreaction. Once you get good specimens, proceed to the next step.

1. Remove the cover glass with a blade carefully.
2. Excise a target area under a stereomicroscope. To trim and ultrathin-section a sample embedded in a flat plate, the sample must be cut out once from the flat plate using a sharp blade. When trimming the sample, you must consider the direction of ultrathin sectioning. Then pick up the sample carefully with tweezers, and put it on instant glue placed on a cylindrical platform after considering the direction. To prevent the remaining sections from being damaged, put the coverslip again and fix it with tape. The cylindrical platform can be stored in a tube or box of appropriate size to avoid dust and moisture.

3.2.6 Trimming the Sections for Ultramicrotome

1. Using a stereomicroscope, observe the sample on a resin platform. To make orientation of ultrathin sections obvious, one or two corners of the sample are cut into a clean trapezoid about 0.5 mm by 0.5 mm above the surface (*see* **Note 14**). Use a new blade to keep the edges of the trimmed section clean.

3.2.7 Semithin Sectioning

1. After trimming, the sample block is mounted on an ultramicrotome (*see* **Note 15**). In general, a glass knife is equipped with water receptacles called boats, which are filled with DDW, and the cut sections are expanded above the water surface for collection. But you do not necessarily need this boat to make rough or confirmatory semithin sections.

   Glass knives are made by cutting a commercially available glass plate for knives at a knife manufacturer. A glass knife maker is commercially available and you can make a knife by yourself. To get a good knife, some proficiency will be necessary.
2. Picking up semithin sections: Before sectioning, it is necessary to adjust the angles between a glass knife and the section surfaces. Cut a 1 μm- thick slice and dry without a water boat attached to the glass knife to float the slice (*see* **Note 16**). Put the semithin section on the slide glass placed on a hot plate at 80 °C to evaporate water and secure the sections on the glass.
3. Observation of semithin sections: Place a 0.5% toluidine blue solution on the semithin section on the glass slide, heat for 20 s, rinse the dye under running water, dry again, and then examine it under a light microscope to check the approximate orientation in the tissue of the area where you are going to make the section. Because the surface of the section could be covered with resin, sometimes you will need to cut a couple of sections until the tissue appears.

3.2.8 Ultrathin Sectioning

1. Attach the sample block to a microtome and attach a diamond knife with a boat filled with DDW and make ultrathin sections. The volume of DDW should be adjusted to get a flat or slightly concave meniscus. You can cut successive ultrathin sections using a good glass knife too. The speed of sectioning is usually guided by the knife manufacturer, generally ~1 mm/s. In the case of a relatively soft sample, decrease the speed of cut. Precise adjustment of the block and knife surfaces is crucial to obtain uniform and nondistorted ultrathin sections (*see* **Note 17**). The thickness of ultrathin sections is 60–90 nm, which can be estimated with the interference color on water of the cut ultrathin sections. The sections with 60–80 nm looks silver gray. Thinner sections become pale gray and thicker ones gold. If the color largely varies from ultrathin sections to sections, you may have a potential problem on Epon embedding or circumference of your cutting room (temperature, wind, etc.).
2. Collecting sections on the grid: The ultrathin sections floating on the surface of water are gathered with a tool called an eyelash (or baby hair) tool to make a set of approximately two to ten sections. A mesh grid or slot grid with a formvar support film is placed over the series of sections to adsorb the sections. To store the grids, it is better to use a moisture-proof grid box.

3.2.9 Staining of Ultrathin Sections

Double staining is performed with uranium acetate and lead citrate to intensify contrast to the electron beam. Uranium acetate staining is performed for 5 min, and a grid with ultrathin sections is washed with 50% ethanol, then washed with DDW, and transferred to lead stain. Lead citrate staining takes 1 min. After dyeing, it is rinsed by dipping in beakers filled with DDW one after another. After washing, it is dried and examined under an electron microscope.

3.3 Postembedding Method (In Case of Freeze-Substituted with Lowicryl)

The method involves embedding a sample in a resin or other materials (Lowicryl, LR White, etc.), and performing immunostaining on the ultrathin section. We usually use the colloidal gold labeling method to visualize signals (Fig. 1).

Advantages

It is an excellent method for antibodies to access the antigens (Fig. 1), and multiple staining is also possible, for example, by using antibodies conjugated with colloidal gold nanoparticles of different sizes. Usually, it is common to label gold particles directly to the secondary antibody or to label immunogold particles, which have the property of binding specifically to the primary antibody, and observe the presence of these gold particles. Since the number of gold particles and the density of antigens show some correlation, the distribution of antigens can be quantified under certain conditions.

For example, ion channel receptors such as the $GABA_A$ receptors are difficult to detect at the synaptic site by the preembedding method because of the poor penetration of antibodies into the synaptic structure. Therefore, observation by the postembedding method is often performed. The localization and quantitative analysis of the $GABA_A$ receptor subunits have been demonstrated in the striatum [2] and substantia nigra [3] by the postembedding immunogold method with freeze substitution [4].

Figure 2 shows an example of postembedding labeling, which shows triple labeling for $GABA_A$ receptor subunits (Fig. 2, [4]).

Figure 3 shows an example of postembedding labeling to quantify $GABA_A$ receptors and AMPA receptors (Fig. 3, [3]).

Disadvantages

The problem with postembedding immunoelectron microscopy is that in the process up to the preparation of ultrathin sections, antigenic inactivation may be significant due to fixation (especially with glutaraldehyde or osmium tetroxide) or resin embedding operations, resulting in the so-called false negative reactions that fail to respond where they should. For example, in peptide-producing cells, immune responses are common in secretory granules, which are thought to have a high density of antigens but not in the Golgi apparatus or the lumen of the rough endoplasmic

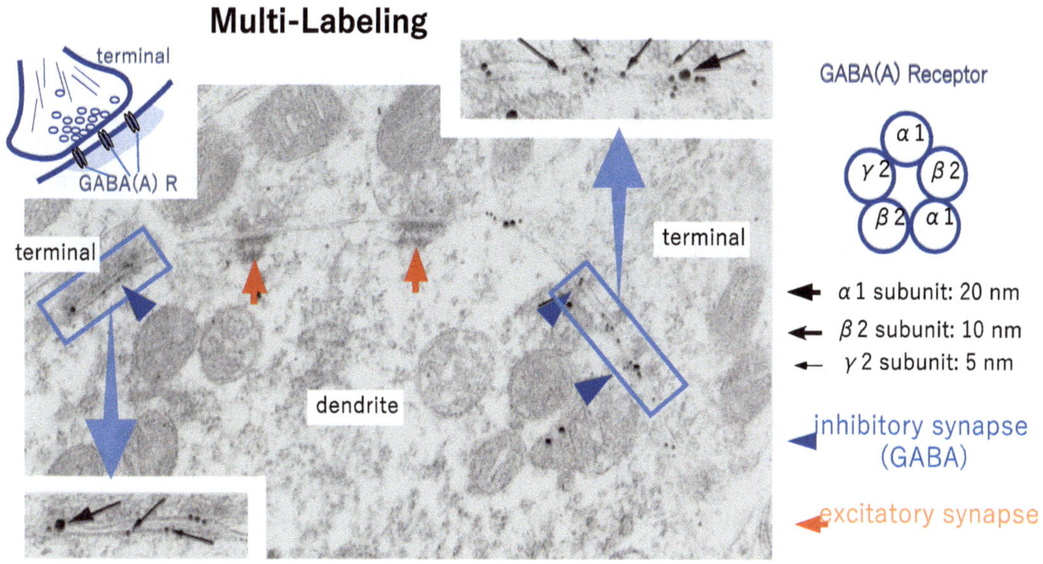

Fig. 2 Triple labeling for $GABA_A$ receptor subunits. The α1 subunit of the $GABA_A$ receptor is labeled with 20 nm, the β2 subunit is labeled with 10 nm, and the γ2 subunit is labeled with 5 nm immunocolloidal gold. (Modified from Fujiyama et al., The Basal Ganglia, 2003)

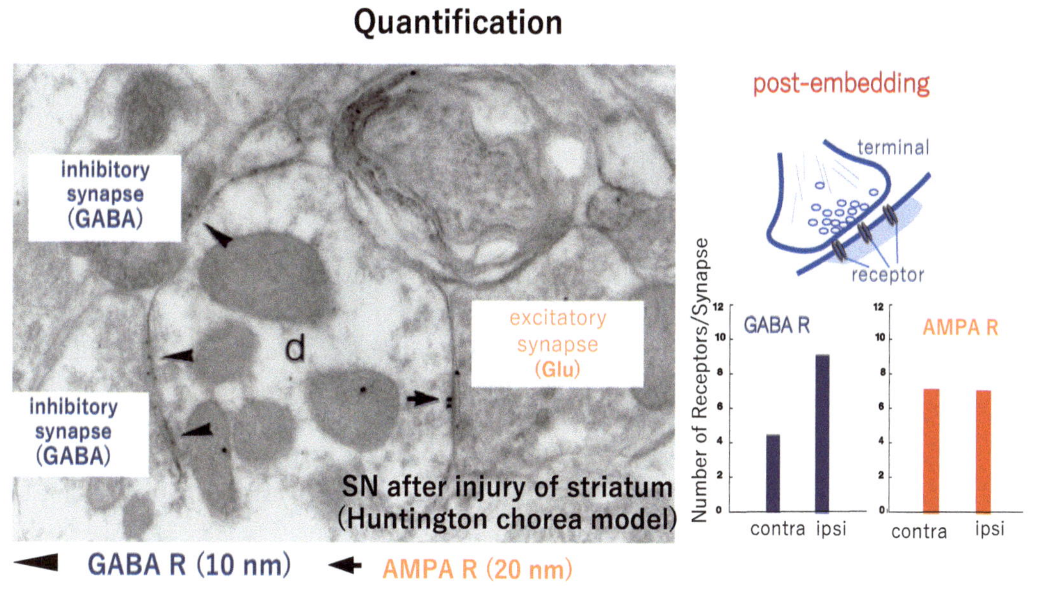

Fig. 3 Quantification of $GABA_A$ receptors and AMPA receptors. Comparison of the number of excitatory and inhibitory receptors at synapses in the substantia nigra of Huntington's chorea model and control rats. The number of $GABA_A$ receptors in substantia nigra ipsilateral to degenerating side of the striatum was up-regulated

reticulum where the antigen should be present. Another disadvantage is the difficulty of infiltration of antibodies inside the resin. Thus, antigen should be placed just on the surface of ultrathin sections. One way to solve this problem is to use low-temperature polymerized resins such as Lowicryl or LR white. However, since osmium tetroxide cannot be used as a fixative when ultraviolet polymerization is performed, the contrast of the electron microscopy images is particularly poor and tissue preservation is not ideal using low-temperature polymerized resins in general (Figs. 2 and 3). In addition, when polymerization is incomplete or not uniform, the Lowicryl block becomes hard to cut precisely into ultrathin sections. It will be better to remove residual humidity or water in tissue during the preparation and polymerization of Lowicryl, even though it is hydrophilic.

3.3.1 Sectioning with Vibrating Microtome

We usually make 500 μm sections to fit the target area with a vibrating microtome [5].

3.3.2 Cryoprotection

Immerse the pieces of section in increasing concentrations of sucrose in 0.1 M PB (0.5, 1.0, and 2.0 M; 30 min in each) and store overnight at 4 °C in 2.0 M sucrose in 0.1 M PB until the tissue sinks.

3.3.3 Trimming of 500 μm Cubic Sections of Tissue

Trim the cryoprotected block in 2.0 M sucrose-PB with a blade.

3.3.4 Plunge-Freezing in Liquid Nitrogen (Metal Contact Method, MM80)

The sections are slammed on a polished copper block cooled with liquid nitrogen (Reichert MM80E).

3.3.5 Freeze-Substitution (Automatic Freeze Substitution System) [6]

1. 0.5% uranyl acetate in dry absolute methanol, −80 °C, 24 h.
2. Gradually increase the temperature (4 °C/h) until −45 °C is reached.
3. Wash 3 times every 3 h with precooled dry absolute methanol to remove excess uranyl acetate.
4. Infiltrate with increasing concentrations of the low-temperature resin Lowicryl HM20 in methanol (1:1 and 2:1, 2 h in each), followed by pure resin (2 h) and fresh pure Lowicryl overnight at −45 °C.
5. Place the plastic capsules filled with fresh pure Lowicryl.
6. Polymerize Lowicryl at −50 °C with UV light for 48 h.
7. Gradually increase the temperature (10 °C/h) until 20 °C and maintain 20 °C for 24 h.
8. Remove the samples from the freeze substitution apparatus and expose to natural light in a fume cupboard for 3 days at RT until completely hardened.

3.3.6 Ultrathin Sectioning

See Subheading 3.3.8. Ultrathin Sectioning (*see* **Note 18**).

*In many cases, the ultrathin section has a thickness of 100 nm or less.

3.3.7 Immunogold Labeling [7] (See Note 19)

1. Perform the etching process for 10 min with 3% H_2O_2 or saturated solution of sodium ethanolate (2–3 s).
2. Wash with DDW for about 10 s, dry slightly, and float in distilled water on the parafilm.
3. Wash 2 times with 50 mM TBS for 5 min.
4. Block the nonspecific reaction by reacting with 2% HSA in TBS containing 0.01% TritonX (TBST) for 30 min.
5. Place a drop of primary antibody diluted with 2% HSA-TBST in a parafilm and suspend a grid over it and allow to react in a wet box at room temperature overnight (*see* **Note 20**).
6. Wash 3 times with TBS for 10 min.
7. Block the nonspecific reaction by reacting with 2% HSA-TBST for 10 min.
8. Perform the conjugation of secondary antibodies with colloidal gold (5–20 nm diameter) in 2% HSA in TBST supplemented with 5 mg/mL polyethylene glycol for about 2 h.
9. Wash 4 times with TBS for 30 min.
10. Incubate with 2% glutaraldehyde in TBS for 2 min.
11. Wash 4 times with DDW for 2 min and air-dry (improve contrast).

3.3.8 Negative Staining

1. Stain in aqueous 1% uranyl acetate (15–20 min), wash thoroughly in DDW for 35 min in the dark, and air-dry (improve contrast).
2. Stain on lead citrate for 1 min, wash thoroughly in DDW as before, and then air-dry.
3. For the control experiment, it is necessary to control the specificity of the secondary antibodies, especially for double or triple labeling. The sections are incubated with a mixture of primary antibodies from different species, at appropriate dilutions, and then with secondary gold-conjugated antibodies directed against immunoglobulins of the other species. In the same immunolabeling run, positive controls for each antibody are included. In each case, the inappropriate gold-conjugated secondary antibody failed to produce specific labeling (namely the resulting labels can be considered as false positives), although the positive controls revealed specific labeling (e.g., symmetrical synapses and membranes in the case of antibodies against $GABA_A$ receptor subunits or intracellular labeling of cellular elements in the case of antibodies against intracellular antigens).

3.4 Observation

The optical design of an electron microscope is basically the same as that of a light microscope, which uses electrons (electron beams) instead of light (rays). But by turning a beam of light into an electron beam, the operating environment is markedly different. Electrons not only are unrecognizable by the naked eye as colors but also have electric charges and weak permeability and annihilate by interaction with other particles only in a vacuum so that they cannot be extracted as electron beams. Therefore, it must be placed in a vacuum. Moreover, the electron beam cannot be refracted using a glass lens and so a magnetic lens (electron lens) is used.

The final magnified image is observed by converting electrons into visible light with a fluorescent plate. Because of these circumstances, the sample must be thin enough to be maintained in a vacuum and to transmit electron beams (*see* **Note 21**).

4 Notes

1. Formaldehyde is a cross-linking agent that can inactivate, stabilize, or immobilize proteins. Glutaraldehyde reacts and produces irreversible cross-linking networks throughout the cytoplasm in seconds to minutes.
2. Since PFA does not dissolve at room temperature, warm it between 60 and 70 °C and add NaOH solution little by little until the powdered PFA is dissolved and then stop adding NaOH (if you add too much, the pH becomes too alkaline).
3. Uranium acetate is dissolved in 70% ethanol to a concentration of 1%. This concentration is so close to saturation that it takes time to dissolve completely. They are denatured by ultraviolet rays, so they should be protected from light and stored in a refrigerator.
4. Lead citrate should be blocked from the air and stored in a refrigerator using a disposable syringe with a plug.
5. Insertion of the needle into the aorta often results in good fixation. Ringer's blood removal is judged by the color of the liver and the color of the effluent from the right atrial appendage. When perfusion solution enters the lung, you will see the expansion of the lung and a leak of the fixative from the nose. It prevents efficient flowing of solution to the brain; thus, in that case, clamp the lung with forceps to shut down the flow to the lung. Stop the pump when the effluent becomes clear. Since red blood cells contain peroxidase, the remaining blood results in high-background signals. If blood remains in the brain, incubate the sections with 1–2% H_2O_2 in PB for 30 min to deplete peroxidase. Then, wash the sections thrice for 20 min.
6. Make sure that the whole body is stiff.

7. Positive immunostaining sites are stained at most 20–30 μm from the surface of the section, so it is necessary to select sites with ultrastructural preservation and good staining properties.
8. Compared with cryostat sections, sectioning with a vibratome is a less invasive sectioning method that avoids the destruction of microstructures caused by freezing.
9. Blocking is important to absorb nonspecific antigens.
10. With the osmium tetroxide reaction, the sections are easy to bend, so handle the sections carefully so that they stick to the bottom. Osmium tetroxide reacts with unsaturated acyl chains of membrane lipids and nucleophiles like amino and sulfhydryl groups and is used for contrasting as it increases the retention of lipids in the tissue.
11. When you want to make it flat in the dehydration step, insert a slice between a microscope glass and a cover glass at the stage of 70% EtOH and dehydrate it by dropping alcohol from the surroundings. 100% EtOH can be derived by adding molecular sieve 3A to 99% EtOH.
12. When you use propylene oxide, use glass vials because propylene oxide dissolves plastic. In addition, propylene oxide is toxic and very easy to ignite.
13. For flat embedding, gently warm a container made of aluminum foil and the resin on a hot plate and transfer the sections on the microscope glass and cover with silicon-coated cover glass. To remove excess resin, putting small weights on the coverslips is helpful.
14. The angle between the top and sides of the carved pyramid of a resin block is about 120°, and it is better to be sharp without rounding.
15. Instead of cutting an ultrathin section abruptly, first flatten the trimmed surface, cut a semithin section about 1 μ thick, and observe it under a light microscope to see if the tissue cells of interest are exposed on the surface. Ultrathin sections are then made. Glass knives are usually used to make the semithin sections for surfacing and sample texture confirmation.
16. The sections curl a bit and stick to the knife edge, so pinch them with pointed tweezers or a platinum loop and drop them onto the droplets on the glass slide.
17. Move the specimen surface closer to the cutting edge to about 2–3 mm by eye measurement. Slowly raise and lower the sample block manually and check the position of the sample with the knife again.

    When filling a diamond knife boat with water, use a needle syringe of about 5 mL. Firstly, pour enough DDW to build up surface tension so that the DDW settles down to the cutting edge. After about 30 s, the water is now sucked in with a syringe, making a slight dent in the surface.

While looking at the binocular of the microtome, use your left hand to turn the knife knob little by little until the knife is about 0.1–2 mm in front of the sample block.

At this time, the right hand is placed on the rotating handle that moves the sample block up and down, and the handle is turned as necessary to move the block up and down and check the position of the sample surface and the knife.

When the distance between the knife and the specimen surface approaches about 0.1 mm, the shadow of the black knife can be clearly seen on the specimen surface. If the cutting edge is parallel to the sample plane, this shadow or bright band will be rectangular.

In addition, when the sample block is moved up and down, if the width of the shadow or bright band of this cutting edge is the same, it indicates that the distance from the cutting edge is constant over the entire surface of the sample.

While moving the sample block up and down, move the knife forward with a slight motion to start sectioning. Sectioning speed is usually 1 mm/s. The thickness of the section is indicated by the reflected interference color. The silver slice is 70 nm, the gold slice is 90 nm, and the gray slice is 50 nm. When the water level is adequate, section one after another across the surface of the water to form ribbons.

18. It is recommended that the grid on which the ultrathin sections are placed be made of nickel rather than copper. We usually use gold or nickel single-slot grids coated with pioloform or adhesive-coated gold or nickel mesh grids.

19. For immunogold labeling in the postembedding method, spread the parafilm in a wet plastic dish, arrange the following solutions 1 drop at a time, and float a grid on it with the side of the ultrathin section down. (Sinking the grid into the solution increases nonspecific reactions.)

20. Dilution of antibody in the postembedding method is 10–20 times more concentrated than in the preembedding method.

21. Since the size of the immunoglobulin is 8 nm and if we use secondary antibody with 15 nm colloidal gold, the center of the gold particle may be at most about 24 nm away from the antigenic site (Fig. 4). Given that the plasma membrane is 5 nm wide and the synaptic cleft is 10–20 nm, we must be cautious in identifying the location of the receptor at the synapse. Therefore, the receptor sites should be analyzed by quantitative analysis.

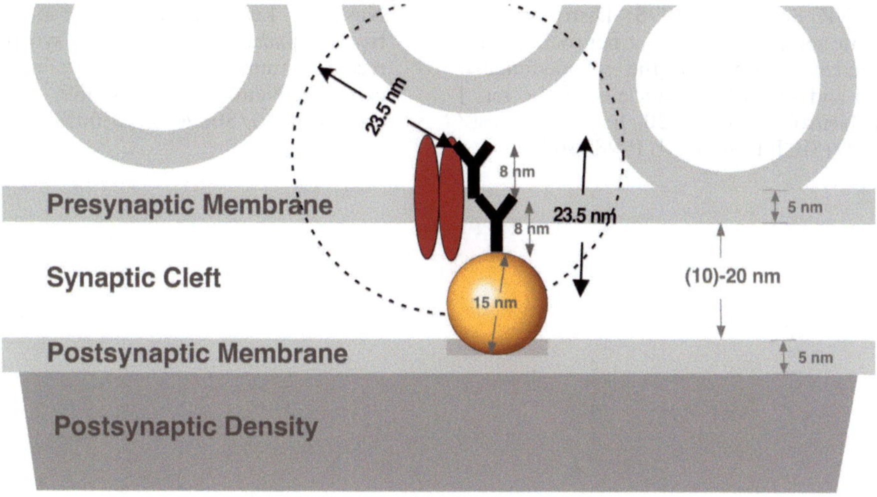

Fig. 4 Cautious in identifying the location of the receptor at the synapse. Since the size of the immunoglobulin is 8 nm and if we use the secondary antibody with 15 nm colloidal gold, the center of the gold particle may be at most about 24 nm away from the antigenic site. Given that the plasma membrane is 5 nm wide and the synaptic cleft is 10–20 nm, we must be cautious in identifying the location of the receptor at the synapse

Acknowledgments

This study was funded by Grants-in-Aid from the Ministry of Education, Culture, Sports, Science, and Technology (MEXT) and Japan Society for the Promotion of Science (JSPS) for Scientific Research (20H03549 to FF) for Exploratory Research (20 K20671 to FF) and for Scientific Researches on Innovative Areas "Hyper-Adaptability" (20H05484 to FF), and for Scientific Researches on Innovative Areas "Adaptation Circuit Census" (21H05241 to FF and to F. Karube), and HFSP Research Grant.

References

1. Richter KN, Revelo NH, Seitz KJ et al (2018) Glyoxal as an alternative fixative to formaldehyde in immunostaining and super-resolution microscopy. EMBO J 37:139–159. https://doi.org/10.15252/embj.201695709
2. Fujiyama F, Fritschy JM, Stephenson FA, Bolam JP (1999) Synaptic localization of GABA (A) receptor subunits in the striatum of the rat. J Comp Neurol 416:158–172. https://doi.org/10.1002/(SICI)1096-9861(20000110)
3. Fujiyama F, Stephenson FA, Bolam JP (2002) Synaptic localization of GABAA receptor subunits in the substantia nigra of the rat: effects of quinolinic acid lesions of the striatum. Eur J Neurosci 15:1961–1975. https://doi.org/10.1046/j.1460-9568.2002.02017.x
4. Fujiyama F, Fritschy J-M, Stephenson FABJ (2003) Synaptic localization of GABAA receptor subunits in the basal ganglia of the rat., The Basal. Kluwer Academic/Plenum Publishers, New York
5. Baude A, Nusser Z, Roberts JDB et al (1993) The metabotropic glutamate receptor (mGluRlα) is concentrated at perisynaptic membrane of neuronal subpopulations as detected by immunogold reaction. Neuron 11:771–787. https://doi.org/10.1016/0896-6273(93)90086-7

6. Clarke NP, Bolam JP (1998) Distribution of glutamate receptor subunits at neurochemically characterized synapses in the entopeduncular nucleus and subthalamic nucleus of the rat. J Comp Neurol 397:403–420. https://doi.org/10.1002/(SICI)1096-9861(19980803)

7. Nusser Z, Lujan R, Laube G et al (1998) Cell type and pathway dependence of synaptic AMPA receptor number and variability in the hippocampus. Neuron 21:545–559. https://doi.org/10.1016/S0896-6273(00)80565-6

Chapter 6

Mouse Brain Tissue Preparation for Scanning Electron Microscopy

Kosuke Mori, Chihiro Takada, Hideshi Okada, and Hiroyuki Tomita

Abstract

Scanning electron microscopy (SEM) is used to observe the surface structure of an object by irradiating an electron beam onto the sample and detecting the reflected and emitted electrons. Because of its large depth of focus, SEM can provide the three-dimensional structure of small surfaces that cannot be observed using an optical microscope. Furthermore, the cross-sectional structure of the tissue can be observed by freeze-cracking. Observing the ultrastructure of organisms that contain large amounts of water in their bodies while maintaining high resolution is challenging; however, this has recently become possible. Here, we explain the fixation and freeze-cracking method for mouse brain samples.

Key words Scanning electron microscope, Perfusion fixation, Freeze-cracking, Freeze-drying

1 Introduction

Scanning electron microscopy (SEM) allows observation of the surface structure of an object by irradiating an electron beam onto the sample and detecting the reflected and emitted electrons (Fig. 1) [1, 2]. In SEM, an electron beam is used to magnify small objects for observation. Optical microscopes shine light on an object and observe the reflected or transmitted light as an image, whereas SEM uses electrons instead of light. Although these techniques seem to be similar in principle, they have some differences (Fig. 2). One of the main advantages of SEM is its high resolution. Electrons have considerably a shorter wavelength than light; therefore, smaller objects can be isolated and viewed. For example, cellular organelles, such as mitochondria and viruses, which are too small to be observed in detail using an optical microscope, can be observed using SEM. SEM also has a large depth of focus, which provides images of deep specimens with minimal blurring. Optical microscopes use glass lenses to form images, whereas electron microscopes use electron lenses that employ a magnetic field to

Koh-ichi Nagata (ed.), *Cerebral Cortex Development: Methods and Protocols*, Methods in Molecular Biology, vol. 2794,
https://doi.org/10.1007/978-1-0716-3810-1_6,

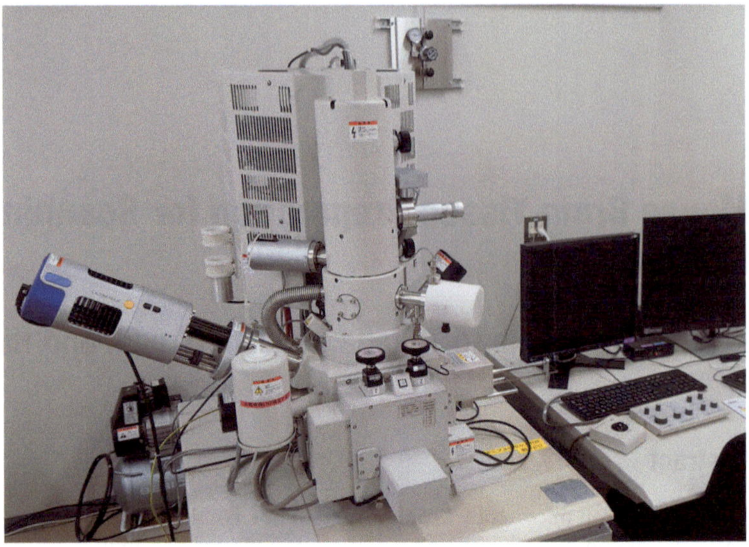

Fig. 1 Schematic of structural differences between the optical and scanning electron microscopes

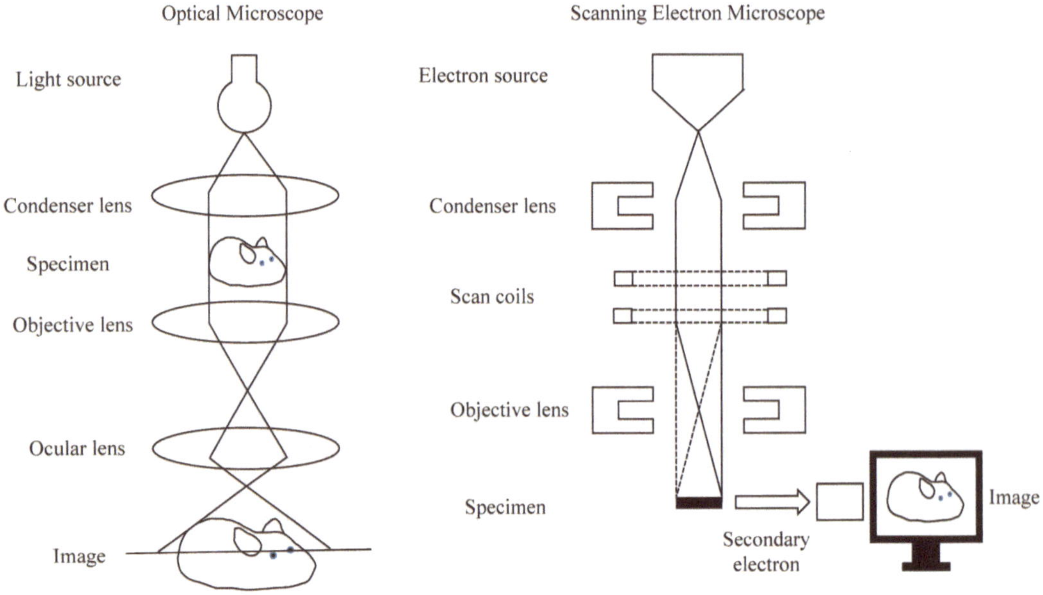

Fig. 2 Exterior view of a scanning electron microscope. (Made in Japan, Hitachi S-4800)

focus the electron beam. Furthermore, unlike light, which can penetrate air, a sample must be observed in vacuum for electron beam penetration. As electron beams cannot be seen by the naked eye, an electron detector and a camera are required for their visualization. SEM can thus be used to observe a sample at magnifications of 100,000 to 1 million times [1, 2].

Recently, SEM imaging of mouse and human tissues has become possible partly because of the development of methods for fixation and surface exposure (in this case, the freeze-cracking method) [3–5]. Attaining high-resolution ultrastructural observations of organisms with substantial water content in their bodies poses a significant challenge. In this chapter, we focus on how to process specimens for SEM, particularly mouse brain tissues.

2 Materials

2.1 Solutions

1. 0.1 M Phosphate buffer: Dissolve 3.3 g of $NaH_2PO_4{\cdot}2H_2$ and 28.7 g of $Na_2HPO_4{\cdot}12H_2O$ in distilled water and make up to 1 L with distilled water.
2. 2.5% Glutaraldehyde: Mix 5 mL of 25% glutaraldehyde and 0.1 M phosphate buffer. Shade from light with aluminum foil.

3 Methods

Carry out all the procedures at room temperature unless otherwise specified.

3.1 Perfusion Fixation

Anesthetize the mouse with sevoflurane (or isoflurane) and then fix it on its back on the dissecting table. After opening the abdomen, incise the diaphragm with the right and left ribs toward the head. Pinch the xiphoid process and invert it toward the head, hold it, and expose the heart. Perfuse fixation solution from the left ventricle of the mouse (*see* **Note 1**). Cut the right atrium of the heart ear to debleed. Use a tape or other means as a stopper to prevent excessive penetration of the left ventricle using the 22G butterfly needle after its insertion (Fig. 3). Insert the butterfly needle into the left ventricle and pass the fixation solution, 2.5% glutaraldehyde, for 4 min (2 mL/min) (*see* **Note 2**). Harvest the organ to be observed as the sample. Dice the sample into ten cubes of 5 mm size. Soak them in 2.5% glutaraldehyde and store them at 4 °C overnight. Wash the samples twice with phosphate buffer for 5 min each (Fig. 4).

3.2 Dehydration

Dehydrate the samples with 80% ethanol (EtOH) in distilled water (DW) under ice cold conditions as follows (*see* **Note 3**): First, immerse the samples twice in 50% EtOH for 5 min each, twice in 70% EtOH for 5 min each, and finally twice in 80% EtOH for 5 min each. Allow the samples to remain in 80% EtOH overnight. Finally, immerse the samples twice in 95% EtOH for 5 min each and then twice in 99.5% (100%) EtOH for 5 min each. Place the dehydrated samples in a Petri dish.

Fig. 3 Modification of the butterfly needle. A tape is wrapped around the needle to regulate it so that it does not penetrate deeply into the mouse heart

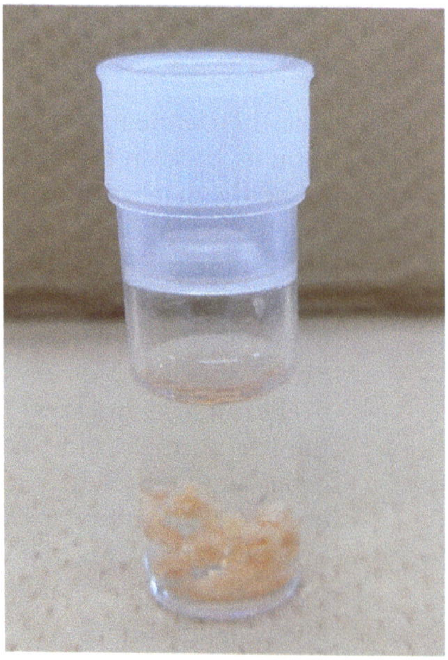

Fig. 4 Small pieces of brain samples in glutaraldehyde solution

3.3 Freezing

Cool a metal ingot sufficiently with liquid nitrogen. Place the sample on the metal ingot (*see* **Note 4**) and then pour 99.5% (100%) EtOH over the sample. Wait until the sample freezes (Fig. 5).

Fig. 5 Samples covered with ethanol on the metal ingot

Fig. 6 Sample divided into two pieces

3.4 Cracking

To avoid hitting the sample, apply a carving knife or another blade to the frozen area and break it with an impact (Fig. 6). After cracking, place the sample in a bottle filled with 99.5% (100%) EtOH. Melt butyl alcohol in a pan that previously held hot water. Replace ethanol with butyl alcohol and soak the samples twice in butyl alcohol for 5 min each.

3.5 Freeze-Drying

Place the samples at 4 °C and freeze with butyl alcohol. If freezing does not occur, repeat the substitution with more butyl alcohol. Dry the samples in a freeze-dryer (Fig. 7). After drying, store them in a desiccator.

Fig. 7 Samples in a freeze-dryer

3.6 Fixation

When examining the split cross section under a microscope, fix the sample to the specimen base using craft glue or other means. After conductive treatment with an osmium plasma coater, observe the sample using an SEM (Fig. 8).

4 Notes

1. Work as quickly as possible after the chest of the mouse is opened (diaphragm and ribs are cut open), as breathing will not be possible.
2. Check the liver and kidneys for whitish debridement as they are perfused.
3. Always use ethanol from bottles and not from metal cans to prevent adulteration.
4. When cracking, the engraving knife, tray, and metal ingot should be cooled in advance using liquid nitrogen.

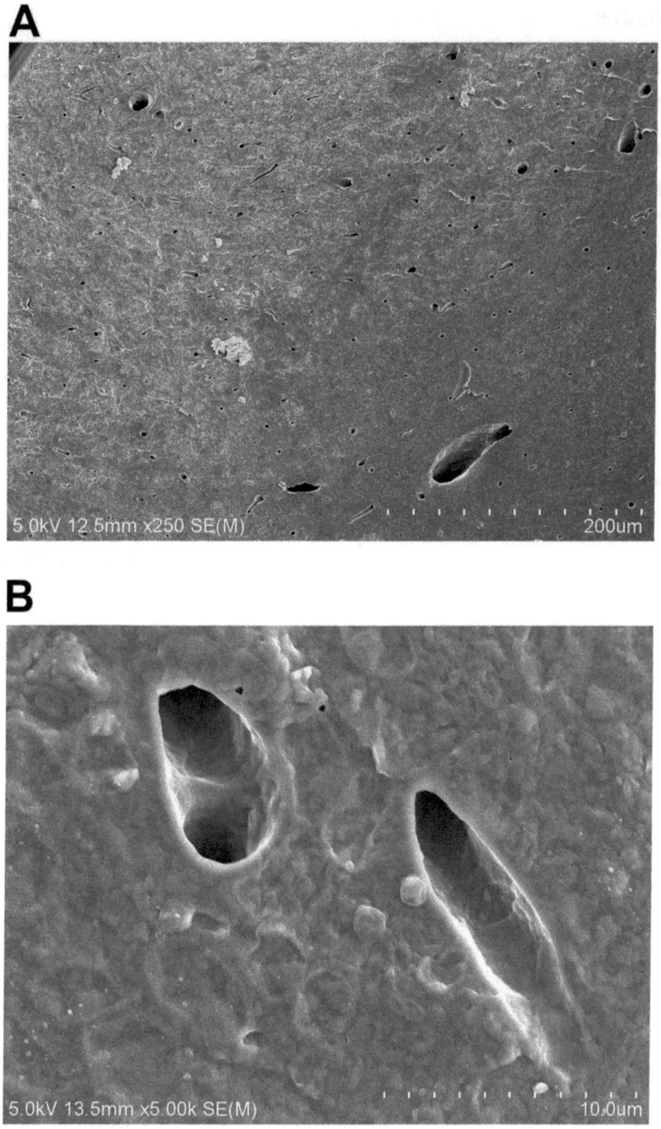

Fig. 8 Representative scanning electron microscopy image of the mouse brain: (**a**) split surface of the brain; (**b**) high magnification of capillary vessels

Acknowledgments

We thank all the members of our laboratories for their support. We would like to thank Editage (www.editage.com) for English language editing.

Funding This work was supported by JSPS KAKENHI (grant number JP20K07587) (H.T.).

References

1. Fischer ER, Hansen BT, Nair V, Hoyt FH, Dorward DW (2012) Scanning electron microscopy. Curr Protoc Microbiol. https://doi.org/10.1002/9780471729259.mc02b02s25
2. Nguyen JNT, Harbison AM (2017) Scanning electron microscopy sample preparation and imaging. Methods Mol Biol 1606:71–84. https://doi.org/10.1007/978-1-4939-6990-6_5
3. Ando Y, Okada H, Takemura G, Suzuki K, Takada C, Tomita H, Zaikokuji R, Hotta Y, Miyazaki N, Yano H, Muraki I, Kuroda A, Fukuda H, Kawasaki Y, Okamoto H, Kawaguchi T, Watanabe T, Doi T, Yoshida T, Ushikoshi H, Yoshida S, Ogura S (2018) Brain-specific ultrastructure of capillary endothelial glycocalyx and its possible contribution for blood brain barrier. Sci Rep 8(1):17523. https://doi.org/10.1038/s41598-018-35976-2
4. Tachi M, Okada H, Matsuhashi N, Takemura G, Suzuki K, Fukuda H, Niwa A, Tanaka T, Mori H, Hara A, Yoshida K, Ogura S, Tomita H (2019) Human colorectal cancer infrastructure constructed by the glycocalyx. J Clin Med 8(9). https://doi.org/10.3390/jcm8091270
5. Miyai M, Kanayama T, Hyodo F, Kinoshita T, Ishihara T, Okada H, Suzuki H, Takashima S, Wu Z, Hatano Y, Egashira Y, Enomoto Y, Nakayama N, Soeda A, Yano H, Hirata A, Niwa M, Sugie S, Mori T, Maekawa Y, Iwama T, Matsuo M, Hara A, Tomita H (2021) Glucose transporter Glut1 controls diffuse invasion phenotype with perineuronal satellitosis in diffuse glioma microenvironment. Neurooncol Adv 3(1):vdaa150. https://doi.org/10.1093/noajnl/vdaa150

Chapter 7

Simple Method for the Preparation of Postsynaptic Density Fraction from Mouse Brain

Hidenori Ito, Rika Morishita, and Koh-ichi Nagata

Abstract

Postsynaptic density (PSD) is a morphologically and functionally specialized postsynaptic membrane structure of excitatory synapses. It contains hundreds of proteins such as neurotransmitter receptors, adhesion molecules, cytoskeletal proteins, and signaling enzymes. The study of the molecular architecture of the PSD is one of the most intriguing issues in neuroscience research. The isolation of the PSD from the brain of an animal is necessary for subsequent biochemical and morphological analyses. Many laboratories have developed methods to isolate PSD from the animal brain. In this chapter, we present a simple method to isolate PSD from the mouse brain using sucrose density gradient–based purification of synaptosomes followed by detergent extraction.

Key words Cerebrum, Ultracentrifugation, Subcellular fractionation, Synapse, Postsynaptic density

1 Introduction

Synaptic transmission between neurons in the central nervous system is mediated by the release of neurotransmitter from presynaptic neurons and their reception by the specific receptors located on the membrane surface of postsynaptic neurons. Electron microscopy revealed the existence of a dense band of material beneath the postsynaptic membrane, called the postsynaptic thickening [1], the postsynaptic web [2], or the postsynaptic density (PSD) [3, 4]. Typical PSDs are found at type 1 glutamatergic synapses [5]. The PSD is composed of proteins with diverse functions, including scaffolding molecules, adhesion proteins, G proteins, and their effectors, kinases, and phosphatases [5–7]. PSD95 is perhaps the most abundant membrane-associated guanylate kinase (MAGUK) in the PSD and a key scaffolding protein. For example, PSD95 interacts with the cell adhesion molecules neuroligins [8], voltage-gated K^+ channel [9], *N*-methyl-D-aspartic acid (NMDA)

Koh-ichi Nagata (ed.), *Cerebral Cortex Development: Methods and Protocols*, Methods in Molecular Biology, vol. 2794, https://doi.org/10.1007/978-1-0716-3810-1_7,

receptor subunit [10], and neuronal nitric oxide synthase [11] as well as a disintegrin and metalloprotease (ADAM) 22 [12].

The isolation of a pure fraction of the PSD is necessary to analyze the chemical and enzymatic properties of individual components. The history of the development of the methods for the purification of synaptic junctions and their components has been succinctly described [13]. Several methods using different detergents such as Triton X-100 [14–16], deoxycholate [17], and *N*-lauroyl sarcosinate [18] have been investigated for the isolation of PSD. Among these, the Triton X-100-extracted PSD resembles the postsynaptic junctional structures seen in intact synaptosomal membranes [19]. This method reveals the presence of major components such as actin [19], calmodulin [20], and a calmodulin-activated protein kinase [21] in the PSD.

In this chapter, we present a simple method for isolating purified PSD from the mouse brain using Triton X-100. This method is based on the protocol described by Kim et al. [22].

2 Materials

1. 1 M HEPES-KOH (pH 7.4): Dissolve 119.2 g of HEPES (2-[4-(2-hydroxyethyl)-1-piperazinyl]ethanesulfonic acid) in dH_2O, adjust pH to 7.4 with KOH, and make up to 500 mL with dH_2O.
2. 1 M $MgCl_2$: Dissolve 101.7 g of $MgCl_2{\cdot}6H_2O$ in 500 mL of dH_2O.
3. 0.5 M $CaCl_2$: Dissolve 7.35 g of $CaCl_2{\cdot}2H_2O$ in 100 mL of dH_2O.
4. Buffer A [5 mM HEPES-KOH (pH 7.4), 1 mM $MgCl_2$, and 0.5 mM $CaCl_2$]: Add 0.5 mL of 1 M HEPES-KOH (pH 7.4), 0.1 mL of 1 M $MgCl_2$, and 0.1 mL of 0.5 M $CaCl_2$, in dH_2O, and make up to 100 mL (*see* **Note 1**).
5. 1 M Tris–HCl (pH 8.0): Dissolve 60.57 g of Tris (hydroxymethyl)aminomethane (Tris) in dH_2O, adjust the pH to 8.0 with HCl, and make up to 500 mL with dH_2O.
6. 6 mM Tris–HCl (pH 8.0): Dilute 3 mL of 1 M Tris–HCl (pH 8.0) with dH_2O and make up to 500 mL with dH_2O.
7. 1.5 M sucrose in 6 mM Tris–HCl (pH 8.0): Dissolve 51.3 g of sucrose in dH_2O. Add 0.6 mL of 1 M Tris–HCl (pH 8.0) and make up to 100 mL with dH_2O.
8. 1.15 M sucrose in 6 mM Tris–HCl (pH 8.0): Add 34.5 mL of 1.5 M sucrose to 6 mM Tris–HCl (pH 8.0) and 10.5 mL of 6 mM Tris–HCl (pH 8.0).

9. 1 M sucrose in 6 mM Tris–HCl (pH 8.0): Add 30 mL of 1.5 M sucrose to 6 mM Tris–HCl (pH 8.0) and 15 mL of 6 mM Tris–HCl (pH 8.0).
10. 0.85 M sucrose in 6 mM Tris–HCl (pH 8.0): Add 25.5 mL of 1.5 M sucrose to 19.5 mL of 6 mM Tris–HCl (pH 8.0).
11. Buffer B [(0.32 M sucrose in 6 mM Tris–HCl (pH 8.0)]: Add 9.6 mL of 1.5 M sucrose to 6 mM Tris–HCl (pH 8.0) and 35.4 mL of 6 mM Tris–HCl (pH 8.0) (*see* **Note 1**).
12. 10 w/v% Triton X-100: Dissolve 10 g of Triton X-100 in 100 mL of dH_2O.

3 Methods

All steps are performed at 4 °C unless otherwise stated. The procedure is outlined in Fig. 1:

1. Collect the cerebrums from 5 to 7 mice, place them in a beaker on ice, and weigh the collected brains (~about 2 g) (*see* **Note 2**).
2. Mince the brains with scissors and transfer them to the glassware of a Teflon homogenizer.
3. Add 18 mL (9 vol.) of buffer A and homogenize it with the Teflon homogenizer (12 strokes).
4. Centrifuge the resultant extract (homogenate) at 1,400 × *g* for 10 min (RA15*, 3400 rpm) (*see* **Note 3**).
5. Collect the supernatant (S1). Add 0.5 mL of dH_2O to the resultant pellet and suspend with a pipette (P1). Centrifuge S1 for 10 min at 13,800 × *g* (RPR20-2**, 10,600 rpm) (*see* **Note 3**).
6. Remove the supernatant (S2). Add 1 mL of buffer B with the protease inhibitor to the pellet and suspend with a pipette (P2). The volume of the resultant mixture will be ~3.8 mL. Add 0.77 mL of 1.5 M sucrose to 6 mM Tris–HCl (pH 8.0) to make up the final sucrose concentration to 0.32 M (*see* **Note 4**).
7. Homogenize the mixture with the Teflon homogenizer (5 strokes).
8. Layer the suspended P2 on the discontinuous sucrose gradient consisting of 8.5 mL each of 1.15 M, 1.0 M, and 0.85 M sucrose in 6 mM Tris–HCl (pH 8.0) (*see* **Note 5**).
9. Centrifuge at 82,500 × *g* for 2 h (SRP28SA**, 25,200 rpm) (*see* **Notes 3** and **6**).
10. Collect the bands in the interface between 1 M and 1.15 M sucrose using an 18G needle (*see* **Note 7**).

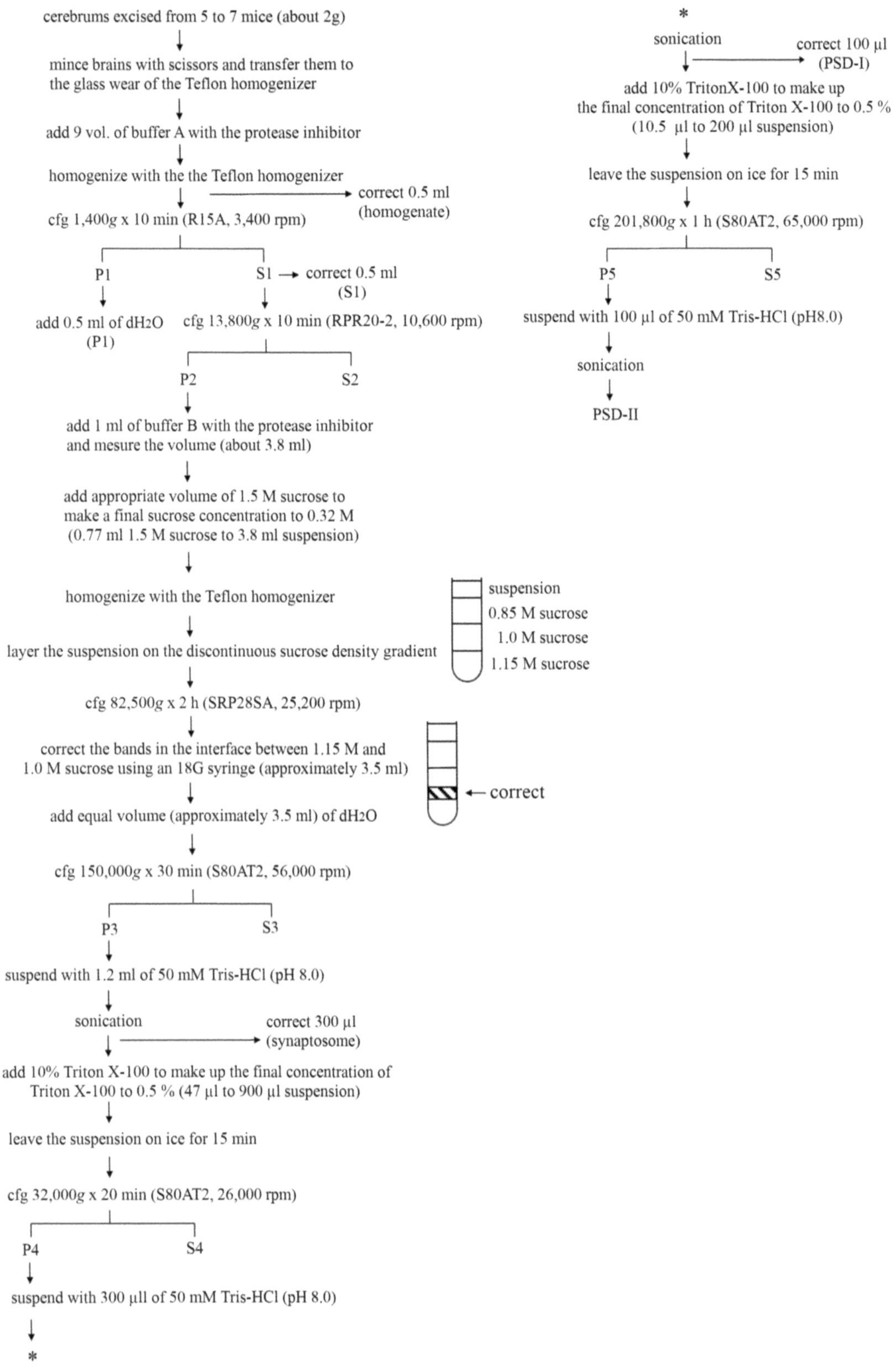

Fig. 1 Scheme of the purification of PSD by subcellular fractionation using sucrose density gradient centrifugation. Examples of rotors and centrifugation speed that we usually use are shown

11. Add an equal volume of dH_2O (approximately 3.5 mL) to the collected mixture and suspend with a pipette. Add 500 μL to each centrifugal tube and centrifuge at 150,000 × *g* for 30 min (S80AT2***, 56,000 rpm) (*see* **Notes 3** and **8**).
12. Discard the supernatant. Suspend with 1.2 mL of 50 mM Tris–HCl (pH 8.0) (*see* **Note 9**).
13. Sonicate for 10 ms × 6 pulses and collect 300 μL of the resultant suspension (synaptosome).
14. Add 47 μL of 10% Triton X-100 to the remaining 900 μL suspension to make up the final concentration of Triton X-100 to 0.5% (*see* **Note 10**).
15. Leave the suspension on ice for 15 min and then centrifuge at 32,000 × *g* for 20 min (S80AT2***, 26,000 rpm).
16. Discard the supernatant. Suspend with 300 μL of 50 mM Tris–HCl (pH 8.0) (*see* **Note 11**). Sonicate for 10 ms × 6 pulses, and collect 100 μL of the suspension (PSD-I).
17. Add 10.5 μL of Triton X-100 to the remaining 200 μL of suspension to make up the final concentration of Triton X-100 to 0.5% (*see* **Note 10**).
18. Leave the suspension on ice for 15 min and then centrifuge at 201,800 × *g* for 1 h (S80AT2***, 65,000 rpm) (*see* **Note 3**).
19. Discard the supernatant. Add 100 μL of 50 mM Tris–HCl (pH 8.0) to the resultant pellet and sonicate with 10 ms × 6 pulses (PSD-II).
20. Subject the fractions containing 5 μg of protein to SDS-PAGE followed by western blot analysis using antibodies against synaptic proteins such as PSD95 or synaptophysin to confirm successful fractionation (Fig. 2).

4 Notes

1. Just before use, add an appropriate volume of a commercially available protease inhibitor cocktail. The protease inhibitor cocktail that we usually use consists of 100 mM 4-(2-aminoethyl)benzenesulfonyl fluoride hydrochloride, 80 μM aprotinin, 1.5 mM E-64, 2 mM leupeptin hemisulfate monohydrate, 5 mM bestatin, and 1 mM pepstatin.
2. It is preferable to use a freshly isolated brain. Some proteins such as Ca^{2+}/calmodulin-dependent kinase II (CaMKII) [23] and tubulin [24] accumulate in the PSD after decapitation. In contrast, cold treatment causes the exit of tubulin, CaMKII, cytoplasmic dynein heavy chain and microtuble-associated protein 2 from dendritic spines of cultured hippocampal neurons [25]

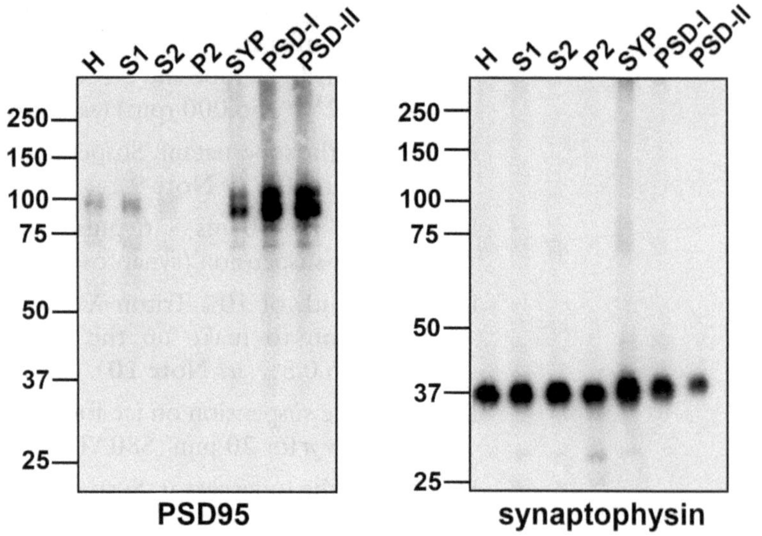

Fig. 2 Example of PSD fractionation. Fractions containing 5 μg of protein were subjected to SDS-PAGE followed by western blot analysis using anti-PSD95 (left panel) and anti-synaptophysin (right panel). The molecular markers are shown on the left

3. Rotors marked with *, **, and *** are those for centrifuges of the himac CR21N, himac CP65 ß, and himac CS100FNX, respectively. These centrifuges need not be used. Equivalent machines may be used.
4. The volume of the solution is fixed at about 24 mL for 10 g of starting tissue. We usually start the fractionation with about 2 g of the cerebrum and adjust the volume to about 4.8 mL.
5. Carefully overlay the materials so as not to disturb the interface.
6. Acceleration and deceleration of the centrifuge should be done slowly to avoid disruption of the interface. The setting of ACCEL and DECEL of the centrifuge is "0."
7. Discard the solution above the bands, take care not to disturb the band, and collect the fraction using the disposable syringe with an 18G needle.
8. The concentration of the synaptosomal protein is usually low in the resulting solution and this step is necessary to concentrate the protein.
9. Suspend the pellet as follows: Add 800 μL of 50 mM Tris–HCl (pH 8.0) to the tube and suspend by pipetting. Transfer the suspension to the next tube and suspend by pipetting. Suspend all tubes in the same way and transfer to a new tube. Add 400 μL of 50 mM Tris–HCl (pH 8.0) and suspend all tubes. After collecting all the residual pellets, combine the suspension with the previously collected suspension.

10. If Triton X-100 is added at the beginning, it will foam and prevent ultrasonic breaking, so add it later.
11. Suspend the pellet as follows: Add 200 μL of 50 mM Tris–HCl (pH 8.0) to the tube and suspend by pipetting. Transfer the suspension to the next tube and suspend by pipetting. Suspend all tubes in the same way and transfer to a new tube. Add 300 μL of 50 mM Tris–HCl (pH 8.0) and suspend all tubes. After collecting all the residual pellets, combine the suspension with the previously collected suspension.

References

1. Gray EG (1959) Axo-somatic and axo-dendritic synapses of the cerebral cortex: an electron microscope study. J Anat 93:420–433
2. De Robertis EDP (1964) Histophysiology of synapses and neurosecretion: international series of monographs on pure and applied biology: modern trends in physiological sciences. Pergamon Press, Oxford
3. Bloom FE, Aghajanian GK (1966) Cytochemistry of synapses: selective staining for electron microscopy. Science 154:1575–1577. https://doi.org/10.1126/science.154.3756.1575
4. Akert K, Moor H, Pfenninger K, Sandri C (1969) Contributions of new impregnation methods and freeze etching to the problems of synaptic fine structure. Prog Brain Res 31:223–240. https://doi.org/10.1016/S0079-6123(08)63241-0
5. Ziff EB (1997) Enlightening the postsynaptic density. Neuron 19:1163–1174. https://doi.org/10.1016/s0896-6273(00)80409-2
6. Kennedy MB (2000) Signal-processing machines at the postsynaptic density. Science 290:750–754. https://doi.org/10.1126/science.290.5492.750
7. Sheng M, Kim E (2011) The postsynaptic organization of synapses. Cold Spring Harb Perspect Biol 3. https://doi.org/10.1101/cshperspect.a005678
8. Irie M, Hata Y, Takeuchi M, Ichtchenko K, Toyoda A, Hirao K, Takai Y, Rosahl TW, Südhof TC (1997) Binding of neuroligins to PSD-95. Science 277:1511–1515. https://doi.org/10.1126/science.277.5331.1511
9. Kim E, Niethammer M, Rothschild A, Nung Jan Y, Sheng M (1995) Clustering of shaker-type K+ channels by interaction with a family of membrane-associated guanylate kinases. Nature 378:85–88. https://doi.org/10.1038/378085a0
10. Kornau H-C, Schenker LT, Kennedy MB, Seeburg PH (1995) Domain interaction between NMDA receptor subunits and the postsynaptic density protein PSD-95. Science 269:1737–1740. https://doi.org/10.1126/science.7569905
11. Christopherson KS, Hillier BJ, Lim WA, Bredt DS (1999) PSD-95 assembles a ternary complex with the N-methyl-D-aspartic acid receptor and a bivalent neuronal NO synthase PDZ domain. J Biol Chem 274:27467–27473. https://doi.org/10.1074/jbc.274.39.27467
12. Fukata Y, Adesnik H, Iwanaga T, Bredt DS, Nicoll RA, Fukata M (2006) Epilepsy-related ligand/receptor complex LGI1 and ADAM22 regulate synaptic transmission. Science 313:1792–1795
13. Somerville RA, Merz PA, Carp RI (1984) The effects of detergents on the composition of postsynaptic densities. J Neurochem 43:184–191. https://doi.org/10.1111/j.1471-4159.1984.tb06695.x
14. Carlin RK, Grab DJ, Cohen RS, Siekevitz P (1980) Isolation and characterization of postsynaptic densities from various brain regions: enrichment of different types of postsynaptic densities. J Cell Biol 86:831–843. https://doi.org/10.1083/jcb.86.3.831
15. Wu K, Carlin R, Siekevitz P (1986) Binding of L-[3H]glutamate to fresh or frozen synaptic membrane and postsynaptic density fractions isolated from cerebral cortex and cerebellum of fresh or frozen canine brain. J Neurochem 46:831–841. https://doi.org/10.1111/j.1471-4159.1986.tb13047.x
16. Cohen RS, Blomberg F, Berzins K, Siekevitz P (1977) The structure of postsynaptic densities isolated from dog cerebral cortex. I. Overall morphology and protein composition. J Cell Biol 74:181–203. https://doi.org/10.1083/jcb.74.1.181
17. Matus AI, Walters BB (1975) Ultrastructure of the synaptic junctional lattice isolated from

mammalian brain. J Neurocytol 4:369–375. https://doi.org/10.1007/BF01102119/METRICS

18. Cotman CW, Banker G, Churchill L, Taylor D (1974) Isolation of postsynaptic densities from rat brain. J Cell Biol 63:441–455. https://doi.org/10.1083/jcb.63.2.441
19. Matus AI, Taff-Jones DH (1978) Morphology and molecular composition of isolated postsynaptic junctional structures. Proc R Soc Lond B Biol Sci 203:135–151. https://doi.org/10.1098/rspb.1978.0097
20. Grab DJ, Berzins K, Cohen RS, Siekevitz P (1979) Presence of calmodulin in postsynaptic densities isolated from canine cerebral cortex. J Biol Chem 254:8690–8696
21. Grab DJ, Carlin RK, Siekevitz P (1981) Function of calmodulin in postsynaptic densities. II. Presence of a calmodulin- activatable protein kinase activity. J Cell Biol 89:440–448. https://doi.org/10.1083/JCB.89.3.440
22. Kim MJ, Futai K, Jo J, Hayashi Y, Cho K, Sheng M (2007) Synaptic accumulation of PSD-95 and synaptic function regulated by phosphorylation of serine-295 of PSD-95. Neuron 56:488–502. https://doi.org/10.1016/j.neuron.2007.09.007
23. Suzuki T, Okumura-Noji K, Tanaka R, Tada T (1994) Rapid translocation of cytosolic Ca2+/Calmodulin-dependent protein kinase II into postsynaptic density after decapitation. J Neurochem 63:1529–1537. https://doi.org/10.1046/J.1471-4159.1994.63041529.X
24. Carlin RK, Grab DJ, Siekevitz P (1982) Postmortem accumulation of tubulin in postsynaptic density preparations. J Neurochem 38:94–100. https://doi.org/10.1111/J.1471-4159.1982.TB10858.X
25. Cheng HH, Huang ZH, Lin WH, Chow WY, Chang YC (2009) Cold-induced exodus of postsynaptic proteins from dendritic spines. J Neurosci Res 87:460–469. https://doi.org/10.1002/JNR.21852

Chapter 8

Reconstitution of Neuronal Motor Traffic on Septin-Associated Microtubules

Yani Suber and Elias T. Spiliotis

Abstract

Reconstitution of intracellular transport in cell-free in vitro assays enables the understanding and dissection of the molecular mechanisms that underlie membrane traffic. Using total internal reflection fluorescence (TIRF) microscopy and microtubules, which are immobilized to a functionalized glass surface, the kinetic properties of single kinesin molecules can be imaged and analyzed in the presence or absence of microtubule-associated proteins. Here, we describe methods for the in vitro reconstitution of the motility of the neuronal kinesin motor KIF1A on microtubules associated with heteromeric septin (SEPT2/6/7) complexes. This method can be adapted for various neuronal septin complexes and kinesin motors, leading to new insights into the spatial regulation of neuronal membrane traffic by microtubule-associated septins.

Key words Septins, Microtubules, Kinesins, Membrane traffic, Microtubule-associated septins, In vitro reconstitution, In vitro single-molecule motility assays, Total internal reflection fluorescence (TIRF) microscopy

1 Introduction

In vitro cell-free microscopy-based assays have been historically pivotal in the studies of microtubule-dependent transport, enabling the discovery of kinesin and dynein-activating adaptors [1–5]. In the advent of total internal reflection fluorescence (TIRF) microscopy, which limits detection to ~100 nm above a glass substrate, and camera technology of improved sensitivity, in vitro motility assays evolved to capture the dynamics of single motor molecules [6–8]. Owing to unprecedented signal-to-noise ratios, the kinetic properties of kinesin and dynein motors can be analyzed on single microtubules in the absence or presence of regulatory cofactors and microtubule-associated proteins (MAPs). In these assays, intracellular transport is reconstituted by its minimal essential components, which enables a deeper understanding of the mechanisms that

Koh-ichi Nagata (ed.), *Cerebral Cortex Development: Methods and Protocols*, Methods in Molecular Biology, vol. 2794, https://doi.org/10.1007/978-1-0716-3810-1_8,

underlie membrane traffic and defects caused by disease-linked mutations in motors and MAPs.

In our recent work, we used in vitro motility assays to explore how microtubule-associated septins, a poorly understood family of heteromeric GTP-binding proteins, affect membrane traffic [9, 10]. We reconstituted septin-associated microtubule tracks by decorating glass-bound microtubules with recombinant fluorescent (mCherry-tagged) septin 9 (SEPT9), which was purified after expression in bacteria [9]. In vitro TIRF microscopy assays with lysates from cells expressing the constitutively active green fluorescent protein (GTP)-tagged motor domains of kinesins KIF5C, KIF1A, and KIF17 revealed that the motility of these motors was differentially regulated by SEPT9 [9]. This unexpected finding led to the discovery of a septin-mediated mechanism, which reinforces the polarity of neuronal membrane traffic on the microtubules of the proximal dendrites of hippocampal neurons [9]. Here, we describe a procedure for reconstituting the traffic of the neuronal motor KIF1A on microtubules associated with the heteromeric SEPT2/6/7 complex (Fig. 1). This method can be adapted for various septin complexes and kinesin motors, complementing functional studies that aim to uncover the function of microtubule-associated septins in neuronal compartments such as axons, dendrites, dendritic spines, and neurite growth cones.

2 Materials

All reagents should be prepared fresh (unless otherwise indicated) and stored at the specified temperature. Molecular biology grade water or ultrapure deionized water should be used unless otherwise indicated.

2.1 Kinesin Expression and Lysate Collection Materials

1. Kinesin expression plasmid (KIF1A(1-393)-LZ-3xmCitrine) transfected into COS-7 cells on a 35 mm cell culture dish (*see* **Note 1**) [11].
2. COS-7 lysis buffer in BRB12 (0.1% Triton-X 100, 1 mM PMSF, and 1X mammalian protease inhibitor cocktail). Prepare in ice-cold buffer immediately preceding lysis and store on ice for one-time use (*see* **Notes 2** and **3**).
3. Adenosine-5′-triphosphate (ATP): 100 mM in Tris–HCl, pH 7.5. Store at −80 °C or − 20 °C for up to 6 months.
4. Plastic cell scraper.
5. Platform rocker.
6. Microcentrifuge tube rotator.
7. 1.5 mL microcentrifuge tubes.
8. Refrigerated table-top centrifuge (max speed: 16,000*g*).
9. Liquid nitrogen.

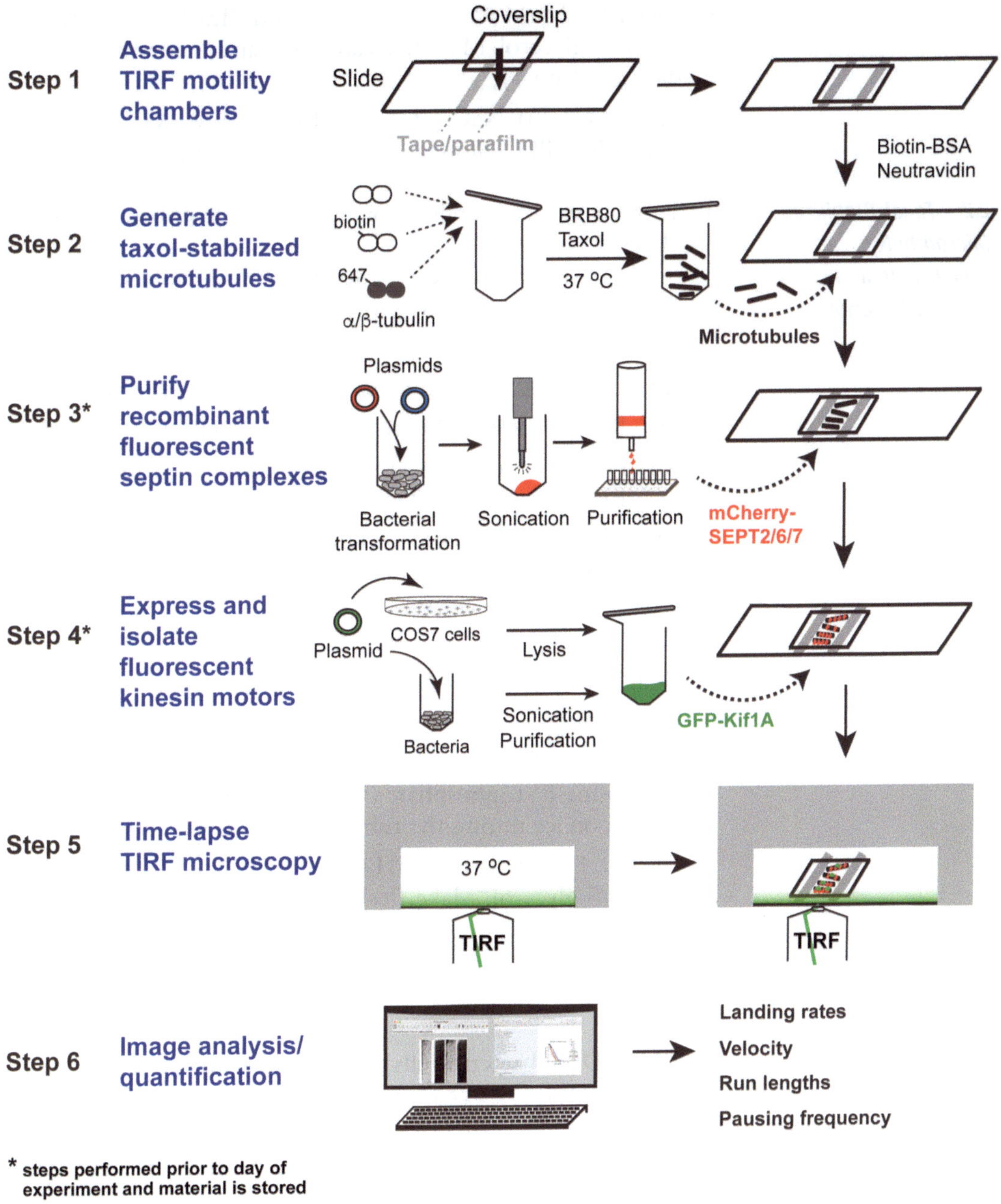

Fig. 1 A step-by-step workflow diagram outlines the procedure of in vitro reconstitution of kinesin KIF1A motility on septin-coated microtubules. The sequence of each step is illustrated in the order by which material is used on the day of the experiment. The recombinant septins and kinesin motors, which are introduced into the motility chamber in **steps 3** and **4**, respectively, are prepared prior to the day of the experiment and stored. Analysis of the time-lapse series in **step 6** can be performed at any point after the collection of data

10. BRB12: 12 mM PIPES, pH 6.9, 2 mM $MgCl_2$, 1 mM EGTA, and 10% glycerol. Make fresh and store an aliquot on ice for the day (*see* **Note 4**).
11. PBS: 137 mM NaCl, 2.7 mM KCl, 8 mM Na_2HPO_4, and 2 mM KH_2PO_4. Store at 4 °C or room temperature.

2.2 Taxol-Stabilized Microtubule Preparation and Storage Materials

1. Tubulin stocks (purchase or prepare as described previously [12]):

   Long-term storage at −80 °C for 2–3 months. Store an aliquot on ice during the protocol.

   Avoid freeze–thaw cycles.

   (a) Unlabeled tubulin: 5 μg/μL in BRB80 + 1 mM GTP.

   (b) Biotin-labeled tubulin: 2 μg/μL in BRB80 + 1 mM GTP.

   (c) Far-red dye–labeled tubulin: 2 μg/μL in BRB80 + 1 mM GTP.
2. Paclitaxel (taxol): 3 mM in DMSO. Long-term storage at −80 °C. The aliquot is stable at room temperature for several weeks.
3. 37 °C water bath.
4. Wide-bore pipette tips (200 μL) for resuspending microtubules (*see* **Note 5**).
5. Aluminum foil.
6. Guanosine-5′-triphosphate (GTP): 1 mM in BRB80. Store an aliquot on ice during the tubulin preparation protocol.
7. BRB80: 80 mM PIPES, pH 6.9, 2 mM $MgCl_2$, 1 mM EGTA, and 10% glycerol. Make fresh and store at room temperature for the day.
8. BRB80 supplemented with 1 mM GTP. Store on ice until use (*see* **Note 6**).

2.3 Septin Expression and Purification Materials

1. BL21 (DE3) bacterial cells transformed with two septin expression plasmids (pET28a-mCherry-SEPT2 and pnCS SEPT6/7-Strep(+1-57 bp SEPT7 N-term)) (*see* **Notes** 7 and **8**) [13].
2. LB broth, Miller (Luria-Bertani) (*see* **Note 9**). Store at room temperature or 4 °C.
3. Kanamycin sulfate: 50 mg/mL in water. Store at −20 °C.
4. Spectinomycin sulfate: 50 mg/mL in water. Store at −20 °C.
5. Isopropyl ß-D-1-thiogalactopyranoside (IPTG). Store at −20 °C.
6. Dithiothreitol (DTT). Store at −20 °C.
7. Bovine serum albumin: 2 mg/mL in water. Store at 4 °C.
8. 2 L Erlenmeyer flasks (3).

9. 250 mL Erlenmeyer flask (1).
10. Sterile 0.45 μm filter.
11. Sterile 15 mL syringe.
12. Ni-NTA resin. Store at 4 °C.
13. Strep-Tactin resin. Store at 4 °C.
14. Gravity flow chromatography columns (10 mL reservoir, 2 mL bed volume) (*see* **Note 10**).
15. Wide-bore pipette tips (200 μL) for dispensing affinity chromatography resins.
16. Dialysis tubing (10,000 MWCO, 22 mm × 35 feet dry diameter).
17. Dialysis tubing clamps (2).
18. Aluminum foil.
19. Temperature-controlled orbital bacterial shaker.
20. Probe sonicator (suitable for volumes >10 mL).
21. Ultracentrifuge (capable of speeds 5000*g*–20,000*g*).
22. 500 mL Ultracentrifuge bottles (2).
23. 50 mL Ultracentrifuge tubes (2).
24. Liquid nitrogen.
25. Tube rotator.
26. SDS-PAGE gel equipment.
27. Lysis buffer: 50 mM Tris–HCl, pH 8.0, 300 mM KCl, 5 mM $MgCl_2$, 10 mM Imidazole, 0.25 mg/mL lysozyme, 10 μg/mL DNAse, 1 mM PMSF, and 1X Bacterial Protease Arrest cocktail. Make fresh and store on ice (*see* **Note 11**).
28. Equilibration buffer: 50 mM Tris–HCl, pH 8.0, 300 mM KCl, 5 mM $MgCl_2$, and 10 mM Imidazole. Make fresh and store on ice.
29. Ni-NTA elution buffer: 50 mM Tris–HCl, pH 8.0, 300 mM KCl, 5 mM $MgCl_2$, and 250 mM imidazole. Make fresh and store on ice.
30. Strep-Tactin elution buffer: 50 mM Tris–HCl, pH 8.0, 300 mM KCl, 5 mM MgCl2, and 2.5 mM d-Desthiobiotin. Store on ice. Make fresh and store on ice.
31. Dialysis buffer: 50 mM Tris–HCl, pH 8.0, 300 mM KCl, and 5 mM $MgCl_2$. Make fresh and store at 4 °C.

2.4 Kinesin Motility Assay and Septin Coating Materials

1. Kinesin lysate. Store aliquots on ice (*see* **Note 12**).
2. Purified septin complex. Store aliquots on ice.
3. Guanosine-5′-triphosphate (GTP): 1 mM in BRB80. Store aliquots on the bench at room temperature for 1–2 days.

4. Polymerized taxol-stabilized microtubules (10X stock). Protect from light by covering with aluminum foil (or by storing in a closed drawer). Remains stable at room temperature for 1–2 days.
5. Acid washed coverslips (22 × 22 mm, 0.15 mm thick). Store in 70% ethanol at room temperature for 1 month.
6. Glass slides (25 mm × 75 mm × 1 mm).
7. Double-sided tape (0.5 in × 900 in).
8. Scalpel or razor blade.
9. Kimwipes (single ply, 4.4 × 8.4 in) or Whatman filter paper (cut to 1 × 2.5 in).
10. Bovine biotin-labeled albumin (Bio-BSA): 5 mg/mL in water. Store at 4 °C for several weeks.
11. Neutravidin: 5 mg/mL in PBS. Store at 4 °C for several weeks.
12. Bovine serum albumin (BSA): 10 mg/mL in water. Make fresh and store on ice.
13. Pluronic F-127: 2% in water (*see* **Note 13**). Make fresh and store on ice.
14. Blocking buffer: 1% pluronic F-127, 1 mg/mL BSA, and 10 μM paclitaxel (taxol). Make fresh and store at room temperature.
15. Adenosine-5′-triphosphate (ATP): 100 mM in Tris–HCl, pH 7.5. Store an aliquot on ice for the day. Remains stable even after several freeze–thaw cycles.
16. Taxol: 3 mM in DMSO. The aliquot is stable at room temperature for several weeks.
17. Oxygen scavenging system (OSS) 10X in BRB12: 0.35 mg/mL catalase, 45 mg/mL D-glucose, 2 mg/mL glucose oxidase, and 715 mM β-mercaptoethanol (BME). Make fresh immediately preceding motility assay. Remains stable on ice for 1–3 h (*see* **Note 14**).
18. Sterile 0.22 μm filter.
19. Sterile 1 mL syringe.
20. BRB12: 12 mM PIPES, pH 6.9, 2 mM $MgCl_2$, 1 mM EGTA, and 10% glycerol. Make fresh immediately preceding assay and store at room temperature for the day (*see* **Note 15**).
21. Super resolution TIRF microscope with a 60x objective.

3 Methods

3.1 Kinesin Expression and Lysate Collection

1. Precool the table-top centrifuge at 4 °C for approximately 30 min.
2. After 24–48 h expression of kinesin construct, remove the cells from the incubator and place on ice. Wash the plate twice with 500 μL of cold PBS, gently swirling during each wash. Use a vacuum to thoroughly remove the remaining buffer.
3. Add 50 μL of COS-7 lysis buffer to the plate dropwise in different regions, making sure to cover the entire plate. Incubate the plate on a platform rocker for 5–10 min at 4 °C.
4. Put back the plate on ice and scrape the cells off using a plastic cell scraper. Transfer the lysate to a new 1.5 mL microcentrifuge tube on ice and repeat until all lysate is removed (*see* **Note 16**).
5. Pipette up and down to mix the lysate, which should yield a cloudy homogenous solution.
6. Incubate the tube at 4 °C on a rotator for 10 min. Pipette up and down to mix and rotate for another 10 min.
7. Remove the tube from the rotator and centrifuge at 16,000*g* for 10 min at 4 °C.
8. Chill fresh the microcentrifuge tube on ice.
9. Carefully avoid the pellet and transfer the supernatant, which contains kinesin motors, to a new microcentrifuge tube on ice. Add ATP (final concentration 1 mM) and pipette to mix (*see* **Note 17**).
10. Transfer the aliquot of the supernatant (10–12 μL) to new tubes on ice. Flash-freeze the aliquots in liquid nitrogen and maintain them at −80 °C for long-term storage. Aliquots remain stable even after several freeze–thaw cycles.

3.2 Microtubule Polymerization

1. Prepare fresh BRB80 and supplement with 1 mM GTP and place on ice.
2. In a new microcentrifuge tube on ice, combine 0.5 μL of biotin-labeled tubulin, 0.5 μL of far-red dye–labeled tubulin, and 1.6 μL of unlabeled tubulin (10% biotin-labeled tubulin, 10% far-red dye–labeled tubulin, and 80% unlabeled tubulin = 10 μg total tubulin). Gently swirl the pipette tip after expelling unlabeled tubulin to mix (*see* **Notes 18** and **19**).
3. Incubate both the tubulin mixture and the 1 mM GTP in BRB80 for 30 min in a 37 °C water bath.
4. Remove both tubes from the water bath. Add 0.2 μL of taxol (3 mM stock) to the tubulin mix tube (10 μM final concentration) and 3.4 μL of warm 1 mM GTP in BRB80 for a total of

6 μL. Return the tubulin mixture to the 37 °C water bath for 15 min and store 1 mM GTP in BRB80 on the bench at room temperature.

5. Remove the tubulin mixture from the water bath and store at room temperature. This is now a 100X stock (final tubulin concentration is approximately 20 μM). Protect the mixture from light by covering with aluminum foil or by storing in a closed drawer (*see* **Notes 20** and **21**).

3.3 Septin Expression and Purification

1. Sterilize LB media by autoclave and allow it to cool to room temperature.
2. Add kanamycin and spectinomycin (final concentration of 50 μg/mL) to a 250 mL Erlenmeyer flask containing 20 mL of LB media and inoculate with a single colony from the double septin transformation plate. Incubate the starter culture for 16–18 h in a bacterial shaker at 37 °C (220 rpm) (*see* **Note 22**).
3. Remove the culture from the shaker and store it at 4 °C until proceeding with **step 4** later in the day.
4. Add kanamycin and spectinomycin (final concentration of 50 μg/mL) to three 2 L Erlenmeyer flasks, each containing 400 mL of LB media, and inoculate at 1:100 with 4 mL of overnight starter culture. Incubate on shaker at 37 °C (220 rpm) for 2–3 h or until OD600 is between 0.6 and 0.8.
5. Induce the bacterial culture with 0.5 mM IPTG by shaking for 16–18 h at 18 °C (200 rpm).
6. The following day, prechill the ultracentrifuge rotor compatible with 500 mL bottles at 4 °C for approximately 30 min.
7. Prepare lysis buffer, equilibration buffer, Ni-NTA elution buffer, and Strep-Tactin elution buffer and store them on ice (*see* **Note 23**).
8. Prepare Ni-NTA and Strep-Tactin gravity flow columns (0.8 mL- 1 mL column volume) using a wide-bore pipette tip and then wash with ultrapure water (5 mL) followed by equilibration buffer (10 mL). Cap the bottom to prevent the column bed from drying out.
9. Remove the cultures from the shaker. Pour the contents of a single flask into a 500 mL ultracentrifuge bottle and spin at 5000*g* for 10 min at 4 °C. Repeat this for the remaining flasks, pooling each induced culture into a single bottle (*see* **Note 24**).
10. Prechill the ultracentrifuge rotor compatible with 50 mL bottles at 4 °C for approximately 15 min.
11. While keeping the bottle at a 45° angle on ice, resuspend the pellet with 10 mL of lysis buffer.

12. Transfer the resuspended pellet mixture to a 50 mL ultracentrifuge bottle. Lyse the cells by sonication (8 rounds of pulsing for 30 s followed by a 30 s pause between each round).
13. Centrifuge the sonicated mixture at 4 °C for 30 min at 20,000*g* and then pass through a 0.45 μm pore syringe filter into a fresh 15 mL conical tube.
14. Load the filtered supernatant onto the Ni-NTA column, gently invert several times to resuspend the resin, incubate on a rotator for 1 h at 4 °C, and then allow the supernatant to flow through (*see* **Note 25**).
15. Wash the Ni-NTA column with 10 mL of equilibration buffer and then collect a single fraction (2 mL) with Ni-NTA elution buffer.
16. Load this fraction onto a Strep-Tactin column (0.8–1 mL), gently invert several times to resuspend the resin, incubate on a rotator for 1 h at 4 °C, and then allow the supernatant to flow through.
17. Wash the Strep-Tactin column with 10 mL of equilibration buffer and use Strep-Tactin elution buffer to collect six 250 μL fractions into separate 1.5 mL microcentrifuge tubes. Place each fraction on ice after collection (*see* **Note 26**).
18. Pool the two fractions that contain the most mCherry-SEPT2/6/7 into a single 1.5 mL microcentrifuge tube. Transfer the pooled protein to dialysis tubing with clamps and incubate overnight in 1 L of dialysis buffer at 4 °C with gentle stirring to encourage buffer exchange.
19. The following day, remove the protein from dialysis tubing, add DTT (final concentration of 1–3 mM), and flash-freeze small aliquots (20–40 μL) for long-term storage at −80 °C (*see* **Note 27**).
20. Validate complex purification and determine the concentration by running samples of protein (5 μL, 10 μL, 15 μL) alongside BSA standards (total mass of 0.5 mg, 1 mg, 2 mg, and 3 mg) on an SDS-PAGE gel (*see* **Note 28**).

3.4 Motility Assay Conditions and Septin Coating

1. Acid-washed glass coverslips (0.15 mm thick) are mounted on a glass slide using double-sided tape in order to create 8–10 μL motility chambers (*see* **Note 29**) [14].
2. Fill the chamber with biotin-labeled BSA and incubate for 5 min (*see* **Note 30**).
3. Wash the chamber with 80 μL of BRB80 (or at least 8–10x chamber volume) by flowing in buffer at one opening and using Kimwipe or Whatman filter paper to pull liquid through the opposite end (*see* **Note 31**).

4. Add 20 μL of neutravidin and incubate for 5 min.
5. While waiting, add 100 μL of BRB80 + GTP (1 mM) to a fresh 1.5 mL microcentrifuge tube followed by taxol (final concentration 10 μM) and pipette to mix. Then, add 1 μL of concentrated microtubule stock into this tube. Use a wide-bore tip to slowly pipette up and down (10–15x) and mix the microtubule dilution. Set aside and protect the mixture from light by covering with aluminum foil (*see* **Note 32**).
6. After neutravidin incubation, wash the chamber with 80 μL of BRB80 (or at least 8–10x chamber volume).
7. Add 20 μL of diluted microtubules to the chamber and incubate for 10 min.
   (a) (*see* **Note 33**).
8. Add 80 μL of blocking solution and incubate for 10 min (*see* **Note 34**).
9. Wash the chamber with 80 μL of BRB12 (or at least 8–10x chamber volume) (*see* **Note 35**).
10. Dilute mCherry-SEPT2/6/7 into 50 μL of BRB12 (supplemented with 0.1 mg/mL BSA, 0.1% pluronic F-127, and 10 μM taxol) and incubate for 10 min (Fig. 2A; *see* **Note 36**).
11. Wash the chamber with 80 μL of BRB12 (or at least 8–10x chamber volume).
12. Prepare 100 μL OSS (10X) in BRB12 immediately before making the final motility mix. Add glucose oxidase last as it is sensitive to oxygen exposure.
13. Dilute kinesin lysate (0.5 μL–5 μL) into the final motility mix containing 50 μL of BRB12 supplemented with (2 mM ATP, 10 μM taxol, 0.1 mg/mL BSA, 0.1% pluronic, and 5 μL OSS), add motility mix to the chamber and seal with vacuum grease (*see* **Note 37**).
14. Observe kinesin motility using TIRF microscopy (Fig. 2B; *see* **Notes 38–40**).

4 Notes

1. HEK293T or any other easily transfected mammalian cells commonly used to collect lysates are acceptable.
2. Other common motility assay buffers can be used as a lysis base buffer (e.g., BRB80- or HEPES-based buffers) as long as the kinesin lysates, septin coating, and final motility assay are performed in the same buffer. Note that both septin–microtubule and kinesin–microtubule binding are enhanced in lower-ionic strength buffers. Due to enhanced binding, kinesins will

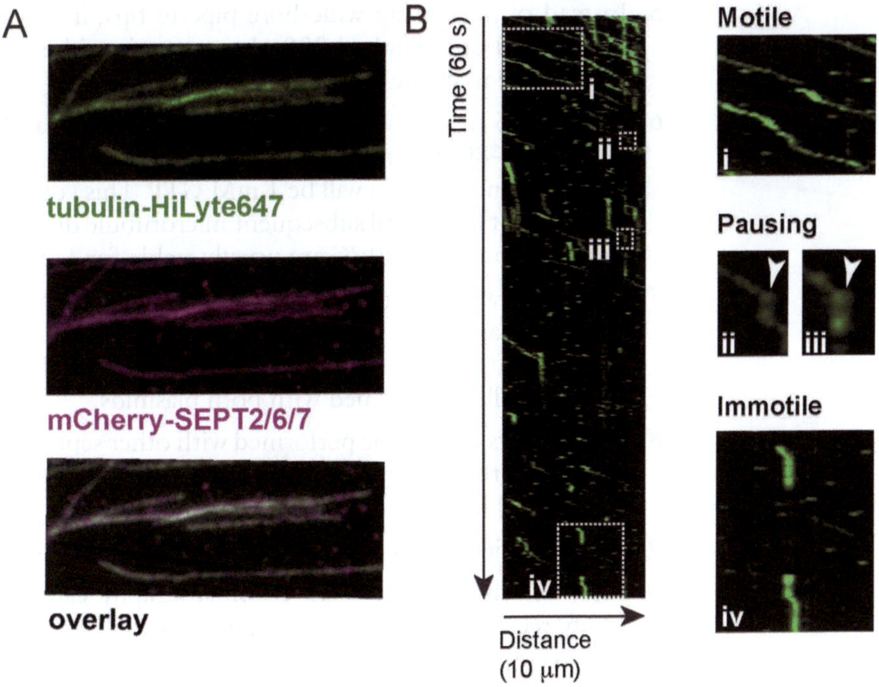

Fig. 2 (**a**) TIRF microscopy images show microtubules (green) that contain biotin- and HiLyte647-conjugated tubulin subunits (magenta) after immobilization on neutravidin-treated glass chambers and incubation with recombinant mCherry-SEPT2/6/7 complexes. (**b**) Kymograph of time-lapse TIRF microscopy of KIF1A(1-393)-LZ-3xmCitrine (green). The localization of KIF1A(1-393)-LZ-3xmCitrine along a 10 μm-long microtubule lattice (x-axis) is projected for every frame of the 60 s time-lapse series (y-axis). Outlined regions (i–iv) are shown in higher magnification as examples of motile (i), pausing (ii and iii), and immotile (iv) KIF1A(1-393)-LZ-3xmCitrine motors. The diagonal and vertical lines correspond to motile and immotile motors, respectively. Pausing events (arrowheads) are identified by the transition of motile diagonal trajectories into stationary particles that appear as vertical lines

hydrolyze ATP faster, meaning that the total amount of time a single chamber is imaged will be reduced. In lower-ionic strength buffers such as those that are HEPES-based, microtubule–kinesin binding is visibly reduced after 10–15 min.

3. PMSF is commonly prepared in isopropanol; however, it is also soluble in ethanol and goes into solution more easily in this medium. Vortex to make sure the PMSF crystals are completely dissolved before adding to the lysis buffer.

4. It is useful to prepare concentrated stocks of BRB80 and BRB12 buffer components and diluted glycerol: 500 mM PIPES, pH 6.9, 500 mM EGTA, pH 6.9, 1 M $MgCl_2$, and 50% glycerol. PIPES dipotassium salt can also be used in place of PIPES and may slightly improve kinesin–microtubule binding and motility.

5. Instead of purchasing wide-bore pipette tips, it is possible to cut the end of a standard 200 μL pipette tip with scissors or a razor blade to create a wider opening.
6. Typically, 5 μL aliquots of 100 mM GTP stock can be made (stored at −80 °C) and with the addition of 500 μL of BRB80, the final concentration will be 1 mM GTP. This volume should be sufficient for several subsequent microtubule dilutions. GTP aliquots stored at −80 °C are usually stable for 1–3 months.
7. Bacterial transformations of two expression plasmids are less efficient than single transformations, so typically 200 ng per plasmid is used. Be sure to plate onto double antibiotic plates to select cells transformed with both plasmids.
8. These assays can also be performed with other septin complexes including the octameric SEPT2/6/7/9 complex purified previously [15].
9. Other nutrient broths, such as Terrific Broth, can also be used.
10. Alternatively, size exclusion chromatography can be used to purify recombinant septin complexes [16, 17].
11. Add PMSF and protease inhibitor immediately before use.
12. Motility assays with septin coating can also be performed with the purified kinesin or dynein/dynactin/bicd2 complex [3, 12].
13. It is helpful to make a 2% stock solution of pluronic F-127 in water by incubating at 50 °C until dissolved and then passing through a filter (0.22 μm). Pluronic F-127 does not go into solution easily when compacted at the bottom of a tube. To avoid this issue, weigh out the required solid along the side of a microcentrifuge tube. Keep the microcentrifuge tube at a 30° angle, add water slowly and place the tube in the heat block. Pluronic F-127 acts as a crowding reagent.
14. Oxygen scavenging system (OSS) is an antibleaching solution composed of D-glucose, glucose oxidase, and catalase and improves fluorescent dye stability.. Typically, aliquots of 100X stock solutions of the individual reagents are made (12 μL) in BRB80 (3.5 mg/mL catalase, and 450 mg/mL D-glucose) and water (20 mg/mL glucose oxidase) and stored at −80 °C for several months. Immediately preceding the addition of the final motility mix, stock OSS reagents should be diluted 1:10 into the chosen base motility base buffer supplemented with 715 mM BME for a final volume of 100 μL. Add glucose oxidase last, as it is sensitive to oxygen exposure.
15. Do not filter sterilize $MgCl_2$ as this negatively impacts the septin–microtubule binding.

16. When scraping, it is helpful to prop up the dish at 45° angle and rotate. This should also encourage lysate to settle at the bottom edge of the dish, aiding in lysate removal.
17. Often the final volume of lysate is either slightly below or slightly above the 50 μL volume of COS-7 lysis buffer. Use a 200 μL pipette to estimate the volume and calculate the amount of ATP needed for a final concentration of 1 mM.
18. If tubulin aliquots are older than 4–6 months microtubules tend to appear dim under the TIRF microscope; however, 20% fluorescently labeled tubulin can be used to compensate for this without sacrificing microtubule integrity or polymerization.
19. When adding each tubulin element, be sure to quickly return to ice in order to avoid polymerization before adding all of the components.
20. This protocol has been tested for both bovine and porcine tubulin.
21. Any exposure to cold buffer after this step will depolymerize microtubules. Keep this stock and any microtubule dilutions at room temperature and protect them from light.
22. It is highly recommended to use a fresh double transformation plate and fresh overnight starter culture for subsequent large-scale induction. Using starter cultures that have been stored at 4 °C for several days can increase the likelihood of spontaneous loss of one or both plasmids and negatively impact protein induction.
23. Add PMSF and protease inhibitor immediately before use. Adding DNAse to the lysis buffer allows the cell lysate to pass through the syringe filter smoothly.
24. Before centrifuging the large-scale-induced cultures, check whether the culture appears pink in color as this indicates successful induction of the mCherry-tagged construct. If no obvious color is observed, it can be helpful to spin a small test volume (1 mL) in a 1.5 mL microcentrifuge tube to determine if a successful induction occurred. If a light to dark pink pellet is not observed, then protein induction has failed.
25. All of the filtered supernatant may not fit into the column at first. If this is the case, add the remaining volume during the column flow-through step.
26. Typically, fractions 3 and 4 contain most of the purified mCherry-SEPT2/6/7 and therefore appear redder based on visual inspection. However, if preparing for the first time, check each fraction by running an SDS-PAGE gel to confirm which fraction contains the most protein. Store septin aliquots at −80 °C for 2–3 months. The protein that is older than this can

exhibit markedly reduced microtubule binding and aggregation.

27. With the addition of 1–3 mM DTT, septin stability is not severely affected by 2–3 freeze–thaw cycles in the context of this particular assay.

28. If excessive protein degradation is observed via SDS-PAGE gel, an 8 h starter culture can be used instead of an overnight culture (**step 3.3.2**) and 1 mM PMSF can be added to each purification buffer immediately preceding use (equilibration/wash, Ni-NTA elution, and Strep-Tactin elution) (**step 3.3.7**).

29. Only a minimal amount of uniform pressure is needed to seal the coverslip to the double-sided tape. Too much pressure can crack the coverslip. Alternatively, chambers can be made by arranging two strips of parafilm across a glass slide, placing the acid-washed coverslip on top and incubating on a heat block for a few seconds to create a seal [18, 19].

30. Two to three chambers can be prepared simultaneously and it is safe to pause at either **step 2** or **step 8** as long as the chambers do not dry out. If pausing at **step 8**, microtubules will be stable for 1–2 h.

31. Removing the buffer faster than it enters the chamber can cause formation of bubbles. Should this occur, try flowing in an additional buffer in order to dislodge the bubble. As long as the chamber is wide enough such that the bubble does not cover the entire chamber, it is still safe to proceed.

32. More than 1 μL can be added depending on the efficiency of microtubule polymerization. This microtubule dilution is stable for up to 3 h. Be sure to occasionally pipette up and down gently to mix.

33. It is recommended to check for stable microtubule binding with the TIRF microscope before adding septins. If microtubules are not visible under the microscope, it is possible that there is an issue with the initial polymerization steps, an issue with adherence to the chamber or an issue in mixing the microtubule dilution. Consider making new 100X microtubules using fresh reagents (tubulin, GTP, BRB80, etc.), preparing fresh bio-BSA, fresh neutravidin, and mixing the microtubule dilution more thoroughly as settling can occur over time. If most microtubules appear very short (less than 10 μm), **step 3** of the microtubule polymerization protocol can be increased by 15–30 min and **step 4** can be increased by 10–15 min. In general, longer water bath incubations result in longer microtubules.

34. Microtubules are stable in the chamber for 1–1.5 h as long as the chamber is moist.

35. Other buffers can be used. However, from this point on, it is recommended that the chosen motility buffer be used for all subsequent steps (e.g., septin coating, OSS preparation, and the final motility mix). If for some reason the septin coating buffer is different from the final motility base buffer, wash the chamber with the motility base buffer prior to the addition of the final motility mix. Under all circumstances it is crucial that the OSS stock (10X) be made in the final motility base buffer.
36. It can be helpful to check for septin coating after washing the chamber and before adding the final kinesin motility mix.
37. While flowing the final motility mix into the chamber, move as quickly as possible as the glucose oxidase in the OSS is sensitive to oxygen exposure.
38. Typical time-lapse acquisition settings are as follows:

    (KIF1A) 488 laser, 31.1% power, acquired every 200 ms.

    (SEPT2/6/7) 594 laser, 31.1% power, acquired every 150 ms.

    (Tubulin) 647 laser, 10% power, acquired every 50–100 ms.

    Acquire (1–3) 2 min time-lapses per chamber for a total imaging time of 15 min.
39. Very few visible motors could be the result of poor kinesin plasmid transfection or inadequate COS-7 cell lysis. To address this, consider preparing fresh plasmid DNA, fresh kinesin lysates, or increasing triton percentage up to 0.25. Alternatively, manual cell lysis can be used by repeatedly passing kinesin lysate through a 25–30 gauge needle [20].
40. If kinesin motors do not bind to microtubules, first always consider preparing fresh motility buffer components (500 mM PIPES and 500 mM EGTA). Using more than 1-month-old buffers can significantly impact motility. $MgCl_2$ seems to be stable for a longer time. Additionally, increasing the concentration of pluronic F-127 in the final motility mix up to 0.5% and adding 0.2 mg/mL casein can improve kinesin-microtubule binding and motility. Lastly, preparing fresh tubulin could address the issue. If tubulin is purchased from a vendor, consider buying a new stock or switching vendors.

Acknowledgments

Research in the Spiliotis lab is supported by NIH grant 5R35GM136337-04. TIRF imaging was performed at Drexel University's Cell Imaging Center (CIC; RRID:SCR_022689). We thank Dr. Harini Sreenivasappa (CIC) for help with imaging and former Spiliotis lab graduate student Dr. Eva Karasmanis for technical advice and training.

References

1. Vale RD, Reese TS, Sheetz MP (1985) Identification of a novel force-generating protein, kinesin, involved in microtubule-based motility. Cell 42(1):39–50
2. Vale RD et al (1985) Organelle, bead, and microtubule translocations promoted by soluble factors from the squid giant axon. Cell 40(3):559–569
3. McKenney RJ et al (2014) Activation of cytoplasmic dynein motility by dynactin-cargo adapter complexes. Science 345(6194): 337–341
4. Reck-Peterson SL et al (2018) The cytoplasmic dynein transport machinery and its many cargoes. Nat Rev Mol Cell Biol 19(6):382–398
5. Brady ST, Lasek RJ, Allen RD (1985) Video microscopy of fast axonal transport in extruded axoplasm: a new model for study of molecular mechanisms. Cell Motil 5(2):81–101
6. Vale RD et al (1996) Direct observation of single kinesin molecules moving along microtubules. Nature 380(6573):451–453
7. Belyy V, Yildiz A (2014) Processive cytoskeletal motors studied with single-molecule fluorescence techniques. FEBS Lett 588(19): 3520–3525
8. Axelrod D (2008) Total internal reflection fluorescence microscopy. In: Correia JJ, Detrich HW (eds) Methods in cell biology. Biophysical tools for biologists, volume two: in vivo techniques. Academic/Elsevier, Cambridge, MA, pp 169–221
9. Karasmanis EP et al (2018) Polarity of neuronal membrane traffic requires sorting of kinesin motor cargo during entry into dendrites by a microtubule-associated Septin. Dev Cell 46(2): 204–218 e7
10. Spiliotis ET, Kesisova IA (2021) Spatial regulation of microtubule-dependent transport by septin GTPases. Trends Cell Biol 31(12): 979–993
11. Soppina V et al (2014) Dimerization of mammalian kinesin-3 motors results in superprocessive motion. Proc Natl Acad Sci USA 111(15): 5562–5567
12. Uppalapati M et al (2009) Microtubule motors in microfluidics. In: Zahn JD (ed) Methods in bioengineering: biomicrofabrication and biomicrofluidics. Artech House, Boston, MA, pp 311–337
13. Nakos K, Radler MR, Spiliotis ET (2019) Septin 2/6/7 complexes tune microtubule plus-end growth and EB1 binding in a concentration- and filament-dependent manner. Mol Biol Cell 30(23):2913–2928
14. Tanenbaum ME, Vale RD, McKenney RJ (2013) Cytoplasmic dynein crosslinks and slides anti-parallel microtubules using its two motor domains. elife 2:e00943
15. Iv F et al (2021) Insights into animal septins using recombinant human septin octamers with distinct SEPT9 isoforms. J Cell Sci 134(15)
16. Mavrakis M, Pourquie O, Lecuit T (2010) Lighting up developmental mechanisms: how fluorescence imaging heralded a new era. Development 137(3):373–387
17. Mavrakis M, Tsai FC, Koenderink GH (2016) Purification of recombinant human and drosophila septin hexamers for TIRF assays of actin-septin filament assembly. Methods Cell Biol 136:199–220
18. Reid TA et al (2016) Suppression of microtubule assembly kinetics by the mitotic protein TPX2. J Cell Sci 129(7):1319–1328
19. Nakos K, Rosenberg M, Spiliotis ET (2019) Regulation of microtubule plus end dynamics by septin 9. Cytoskeleton (Hoboken) 76(1): 83–91
20. Rogers SL et al (1997) Regulated bidirectional motility of melanophore pigment granules along microtubules in vitro. Proc Natl Acad Sci USA 94(8):3720–3725

Chapter 9

BioID Analysis of Actin-Binding Proteins

E. Emily Joo and Michael F. Olson

Abstract

Proteins often exist and function as part of higher-order complexes or networks. A challenge is to identify the universe of proximal and interacting partners for a given protein. We describe how the high-activity promiscuous biotin ligase called TurboID is fused to the actin-binding peptide LifeAct to label by biotinylation proteins that bind, or are in close proximity, to actin. The rapid enzyme kinetics of TurboID allows the profiles of actin-binding proteins to be compared under different conditions, such as acute disruption of filamentous actin structures with cytochalasin D.

Key words Proximity ligation, Biotin ligase, Actin, LifeAct, Cytoskeleton, Affinity purification

1 Introduction

The actin cytoskeleton is one of the three major cytoskeletal systems and its proper organization and regulation are critically important for the development and health of living organisms. Actin proteins constantly cycle between filamentous (F-actin) and globular (G-actin) states and the dynamic transformations between these two states are crucial for the normal functioning of numerous cellular processes including morphological transitions, cell migration, and cytokinesis [1]. In addition to its structural roles, the actin cytoskeleton also acts to modulate the activity of cell signaling pathways. Dysregulation of actin cytoskeleton organization or changes in the ability to undergo F-actin/G-actin dynamics, has been linked to a variety of pathological conditions including cancer, neurodegenerative disorders, and cardiovascular diseases [2]. Many actin-binding proteins have been identified as contributing to the regulation of actin dynamics at varying subcellular locations. In particular, microscopy-based assays revealed the details of precise spatial and temporal regulation of actin dynamics by the Rho GTPase family [3–6]. However, these microscopy-based studies require a priori knowledge of candidate proteins, which limits

Koh-ichi Nagata (ed.), *Cerebral Cortex Development: Methods and Protocols*, Methods in Molecular Biology, vol. 2794,
https://doi.org/10.1007/978-1-0716-3810-1_9,

their ability to assay the multitude of proteins known to bind and regulate the actin cytoskeleton.

The lack of comprehensive spatial and temporal information about protein–protein interactions may be attributed to the limitations of conventional assays, which are predominantly designed for studying steady-state interactions. For example, although several protein–protein interaction assays, including immunoprecipitation, yeast-2-hybrid, and GST-protein pulldown, have been used to identify many actin-binding proteins, these methods are primarily focused on characterizing interactions at equilibrium [7–10]. Biochemical cross-linkers have been used to identify protein–protein interactions [11–13]; however, the utility of these chemical tools is somewhat constrained by issues such as the need for crossing plasma membranes and over or under cross-linking.

To address these limitations, we present a robust method that combines the actin-binding peptide LifeAct with the protein-proximity ligation approach called BioID to tag actin-binding proteins in a temporally dynamic manner. Specifically, we used the 17 amino acid peptide derived from the N-terminus of *Saccharomyces cerevisiae* actin-binding protein 140 (ABP140) that was named LifeAct due to its ability to associate with F-actin without affecting actin cytoskeleton dynamics in live cells [14]. Structural studies revealed that LifeAct binds to F-actin at hydrophobic pockets formed by two adjacent actin subunits (Fig. 1a) [15]. The original BioID made use

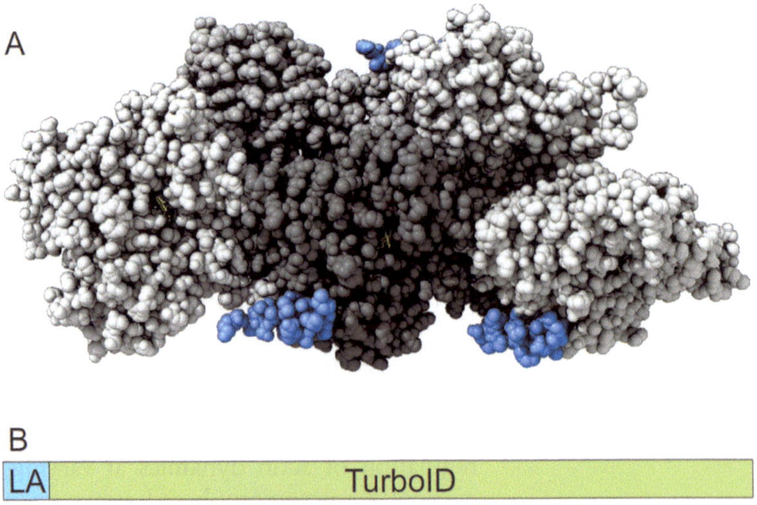

Fig. 1 Structure of LifeAct associated with F-actin and schematic diagram of the LifeAct–TurboID fusion protein. (**a**) The X-ray crystal structure of an actin filament composed of five subunits associated with three associated LifeAct protein molecules indicated in blue (PDF reference 7 AD9; [PMID: 33216759]). (**b**) Schematic diagram drawn to the scale of the amino-terminal LifeAct (LA) fused to the TurboID version of the BirA biotin ligase

of a promiscuous version of the *E. coli* biotin ligase BirA, designated BirA* to indicate the mutant form that releases it from its strict substrate specificity [16], to label proteins associated with or in close proximity (estimated to be within ~10 nm) to fused bait proteins, making it possible to identify the components of higher-order protein complexes. To overcome the relatively slow reaction kinetics of BirA*, a higher-activity version named TurboID was evolved in yeast cells to have increased enzyme kinetics to enable more rapid biotinylation in live cells [17]. In this study, LifeAct was fused to the higher-activity TurboID biotin ligase (Fig. 1b; LA = LifeAct). As shown in Fig. 2a, the LifeAct-TurboID method enabled efficient labeling of proteins in transfected HEK293T human embryonic kidney cells above the background levels of

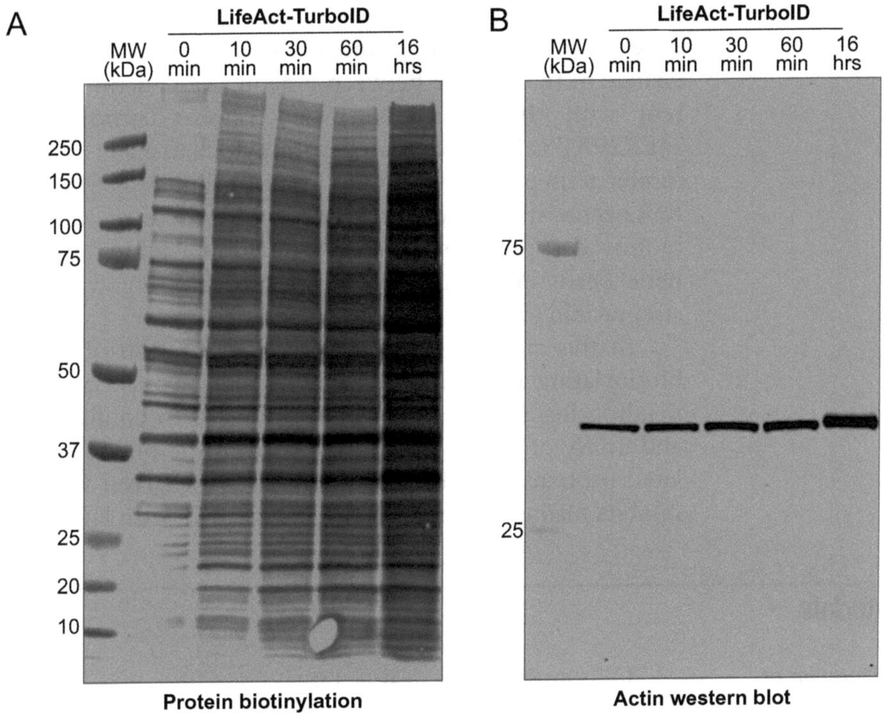

Fig. 2 Protein biotinylation by LifeAct–TurboID. (**a**) HEK293T cells were transiently transfected with the LifeAct–TurboID plasmid and ~40 h later the cells were treated with 50 μM biotin for the indicated number of times. Biotinylated proteins were enriched from cell lysates using streptavidin-conjugated magnetic beads and then separated on polyacrylamide gels. Following the transfer onto nitrocellulose membranes, biotinylated proteins were visualized with fluorescently labeled streptavidin and imaged on a Licor Odyssey DLx infrared scanner. Increasing the time of incubation with supplemented biotin resulted in increased protein biotinylation, with evidence of increased biotinylated proteins from the earliest 10 min time point. The molecular weights of the protein marker bands are indicated on the left. (**b**) To visualize biotinylated actin that had been enriched from cell lysates by streptavidin-conjugated magnetic beads, the membrane was reprobed with the anti-actin antibody, visualized with the fluorescent secondary antibody and imaged on a Licor Odyssey DLx infrared scanner. Actin biotinylation was evident from 10 min and increased over time. The molecular weights of the protein marker bands are indicated on the left

biotinylation within 10 min of biotin addition to tissue culture media, as detected by the enrichment of biotinylated proteins with streptavidin-conjugated magnetic beads followed by western blotting with fluorescently labeled streptavidin. Biotinylation continued to increase over time, with extensive labeling after 16 h of incubation in media supplemented with biotin. Accompanying the generalized increase in protein biotinylation, the levels of biotinylated actin protein enriched by isolation with streptavidin-conjugated magnetic beads also increased over time, with evident increased labeling in 10 min and most extensive labeling after 16 h (Fig. 2b). It was determined that LifeAct has ~30-fold higher affinity for G-actin than for F-actin [14], meaning that the proteins biotinylated by LifeAct–TurboID would most likely be a mixture of G-actin- and F-actin-binding proteins at equilibrium. As a result, the greatest utility of this tool would be in examining the differences in protein labeling under conditions that would promote a change in the balance between F-actin and G-actin forms. Consistent with this possibility, when LifeAct-TurboID expressing HEK293T cells were treated with cytochalasin D for 75 min to inhibit actin polymerization to induce a shift away from the F-actin to G-actin state, there were clear differences in the patterns of biotinylated proteins enriched with streptavidin-conjugated magnetic beads and visualized by western blotting with fluorescent streptavidin (Fig. 3).

In this chapter, we describe the generation of a plasmid for biotinylating actin-binding proteins and the pulse labeling of actin neighboring proteins with biotin. Finally, to confirm the validity and utility of this approach, we assessed the pulldown of biotinylated proteins after the treatment of cytochalasin D, a drug that inhibits filament formation and increases G-actin levels.

2 Materials

2.1 Generation of LifeAct–BioID

1. Oligonucleotides (Table 1):
2. pcDNA3.2-V5-DEST.
3. C1(1–29)-TuboID-V5_pLX304.
4. pDONR223.
5. Gateway BP and LR cloning kits.

2.2 Cell Growth and Treatment

1. 10 cm tissue culture dishes.
2. Cytochalasin D (5 mM stock, used at 10 μM).
3. HEK293T cells.
4. DMEM high glucose media.

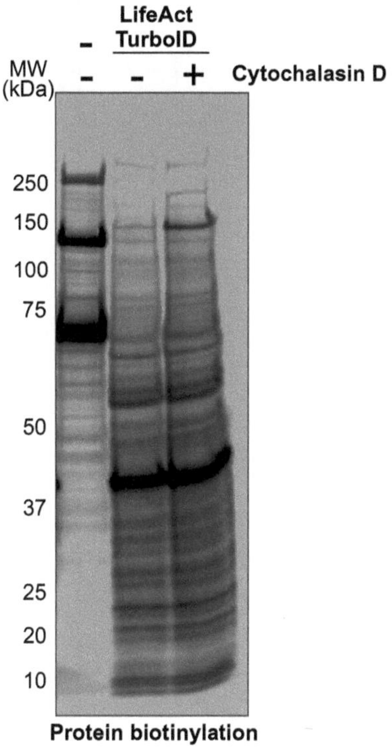

Fig. 3 F-actin disruption with cytochalasin D changed the pattern of LifeAct–TurboID-mediated protein biotinylation. Untransfected HEK293T cells, or LifeAct–TurboID transfected cells without or with cytochalasin D treatment for 75 min as indicated, were incubated with 50 μM biotin for 30 min. Cells were lysed and biotinylated proteins were enriched with streptavidin-conjugated magnetic beads, followed by polyacrylamide gel electrophoresis and transfer onto nitrocellulose membranes. Western blotting with fluorescently labeled streptavidin revealed that there were differences in the banding patterns between the untreated and cytochalasin D-treated cells, indicating that LifeAct–TurboID differentially labels proteins depending on the state of the actin cytoskeleton

5. Complete DMEM: DMEM supplemented with 10% fetal bovine serum.
6. Rubber cell scraper.
7. PBS without magnesium or calcium.
8. Falcon 15 mL tubes.
9. Tabletop centrifuges.
10. Polyjet in vitro DNA transfection reagent.
11. DMSO.
12. Biotin (stock solution made at 1 mM concentration).

Table 1
Oligonucleotides

Name of the primer	Sequence (5′ to 3′)
Flag-lifeact-forward	ctagcggatcctacaaggacgatgacgacaaggccatgggcgtggccgacttgatcaagaagttcgagtccatctccaaggaggagg
Flag-lifeact-reverse	ctagcctcctccttggagatggactcgaacttcttgatcaagtcggccacgcccatggccttgtcgtcatcgtccttgtaggatccg

2.3 Pulldown Using Streptavidin Magnetic Beads

1. Dynabeads MyOne Streptavidin T1.
2. Dynabeads MyOne Streptavidin C1.
3. DynaMag-2 Magnet (magnetic stand).
4. 25× cOmplete protease inhibitor cocktail.
5. DTT (stock solution made at 1 M concentration).
6. Modified RIPA buffer: 50 mM Tris–HCl, pH 7.5, 150 mM NaCl, 1 mM EDTA, 1 mM EGTA, 0.1% SDS, 1% Triton ×-100, 0.5% sodium deoxycholate, 1 mM DTT, and 1× cOmplete protease inhibitor cocktail.
7. 4× Bolt LDS Sample Buffer.

2.4 Western Blotting with Fluorescent Streptavidin and Actin

1. Odyssey DLx Imager.
2. BSA.
3. PBST: PBS supplemented with 0.05% Tween-20.
4. Alexa Fluor 790 conjugated streptavidin.
5. Mouse monoclonal Anti-β-Actin antibody.
6. IRDye 680RD Donkey anti-Mouse IgG.
7. Bolt Bis-Tris Plus Mini Protein gel, 4–12%.
8. MOPS buffer.

3 Methods

3.1 LifeAct–TurboID Construct

Oligonucleotides were designed and synthesized as indicated in Table 1 (flag-lifeact-forward and flag-lifeact-reverse) and first cloned into C1(1–29)-TuboID-V5_pLX304 and subsequently gateway cloned into pcDNA3.2-V5-DEST.

3.2 Transient Transfection into HEK293T Cells and Biotin Labeling After Disruption of Actin Polymerization

1. Dilute 5 μg of pcDNA3.2-V5-DEST-LifeAct into 250 μL of serum-free DMEM. Dilute 15 μL of PolyJet into 250 μL of serum-free DMEM. Mix the two diluted solutions and wait for 10–20 min. Add the mixture dropwise to 70% confluent HEK293T cells seeded in 10 mm tissue culture dishes the day before (*see* **Note 1**).
2. Incubate in a 37 °C/5% CO_2 incubator for 4 h, then replace with fresh complete DMEM media, and incubate for another 40 h (*see* **Note 2**).
3. Treat with cytochalasin D for 45 min (*see* **Note 3**).
4. At the end of 45 min, add 50 μM biotin (final concentration) and incubate for another 30 min to reach effective labeling (Fig. 2).
5. At the end of the incubation, wash with PBS by adding 10 mL of fresh PBS and then aspirating, repeating at least 3 times.

Proceed immediately to **step 1** described in Subheading 3.3. Alternatively, scrape the cells in media and transfer them to 15 mL tubes. Spin at 800 × *g* for 5 min. Wash 3× with room temperature PBS and spin at 800 × *g* for 5 min to pellet cells (*see* **Note 4**).

6. Snap-freeze or proceed to isolate biotinylated proteins (*see* **Note 5**).

3.3 Isolation of Biotinylated Proteins Using Magnetic Streptavidin Beads

1. Lyse the cells following **step 6** described in Subheading 3.2 with 1 mL of modified RIPA buffer and scrape/transfer into 1.5 mL Eppendorf tubes. Supplement the modified RIPA buffer with cytochalasin D or vehicle control (DMSO). Pass through a 27.5G needle of a 1 mL syringe 3 times to ensure complete lysis. Rotate on an end-over-end rotator for 30 min at 4 °C (*see* **Note 6**).
2. Spin at 10,000 × *g* for 10 min at 4 °C.
3. Carefully add the supernatant to the 40 μL prewashed 50:50 Dynabeads MyOne Streptavidin T1/C1 mixture. The bead–lysate slurry should be rotated end-to-end at 4 °C overnight (~18 h) (*see* **Note 7**).
4. Wash the beads 3–5 times with ice-cold 1 mL of modified RIPA.
5. Elute the bound proteins with 30 μL of 2× Bolt LDS sample buffer supplemented with fresh DTT (0.2 M final concentration) and heat for 5 min at 95 °C. The supernatants can be separated by electrophoresis on 4–12% precast polyacrylamide protein gels (Figs. 2 and 3) (*see* **Note 8**).

3.4 Western Blotting with Fluorescent Streptavidin and Actin Antibody

1. On a precast Bolt gel, load 2 μL of protein standards, 10% total cleared lysate or 50% eluted pulldown protein mixture (*see* **Note 9**).
2. Run gel until the dye front is at the bottom of the gel.
3. Transfer the gel onto nitrocellulose membranes.
4. Block with 5% BSA/PBS for 1 h at room temperature. Incubate with 1:1000 dilution of mouse monoclonal Anti-β-Actin antibody. Then, incubate with 1:15,000 dilution of IRDye 680RD donkey anti-mouse IgG and diluted Streptavidin Alexa Fluor 790 Conjugate in 5% BSA/PBST for 1 h (*see* **Note 10**).
5. Wash extensively with PBST and then rinse once in water.
6. Scan using a Licor Odyssey DLx Imager.

4 Notes

1. If cells are seeded for more than a day, replace the media with fresh complete DMEM media.
2. Alternatively, HEK293T cells could be incubated overnight (~12–18 h) before changing the media; the total time of transfection should be approximately 48 h.
3. If another actin-disrupting drug is used, the incubation time should be changed as appropriate for the specific drug to disrupt F-actin structures.
4. Scraping and washing cells with PBS will be more beneficial for cells that are loosely attached to the plate (e.g., HEK293 cells). For cells that are attached to the tissue culture plates, washing them with PBS will save time.
5. Cell pellet can be snap-frozen for long-term storage.
6. Sonication is another method to ensure complete lysis. However, the condition for sonication must be tested to ensure that the power setting and heat produced do not degrade any proteins.
7. Single pulldowns show that Dynabeads MyOne T1 and C1 have different selectivity toward biotinylated proteins. Thus, mixing the two differently modified beads ensured that more biotinylated proteins were pulled down.
8. Any precast or homemade protein gel will work. Similarly, the beads could be sent for mass spectrometry analysis. Alternatively, streptavidin-conjugated Sepharose beads could be used depending on the mass spectrometry facility's requirements for processing the samples.
9. Any protein gel (precast or homemade) and protein standards will serve the purpose.
10. HRP-conjugated streptavidin along with chemiluminescent reagents could be used for detection. Membranes could be incubated with the mouse monoclonal anti-β-actin antibody and then with the HRP-conjugated anti-mouse antibody separately, followed by detection with chemiluminescent reagents for detection.

Acknowledgments

This work was supported by funding to M.F.O. from the Canadian Institutes of Health Research (PJT-169125), Natural Sciences and Engineering Research Council of Canada (RGPIN-2020-05388), and Canada Research Chairs Program (950-231665).

References

1. Pollard TD, Cooper JA (2009) Actin, a central player in cell shape and movement. Science 326:1208–1212
2. Lappalainen P, Kotila T, Jégou A, Romet-Lemonne G (2022) Biochemical and mechanical regulation of actin dynamics. Nat Rev Mol Cell Biol 23:836–852
3. El-Sibai M, Pertz O, Pang H et al (2008) RhoA/ROCK-mediated switching between Cdc42- and Rac1-dependent protrusion in MTLn3 carcinoma cells. Exp Cell Res 314: 1540–1552
4. Weitzman M, Hahn KM (2014) Optogenetic approaches to cell migration and beyond. Curr Opin Cell Biol 30:112–120
5. Marston DJ, Anderson KL, Swift MF et al (2019) High Rac1 activity is functionally translated into cytosolic structures with unique nanoscale cytoskeletal architecture. Proc Natl Acad Sci USA 116:1267–1272
6. Zawistowski JS, Sabouri-Ghomi M, Danuser G et al (2013) A RhoC biosensor reveals differences in the activation kinetics of RhoA and RhoC in migrating cells. PLoS One 8:e79877
7. Berggård T, Linse S, James P (2007) Methods for the detection and analysis of protein–protein interactions. Proteomics 7:2833–2842
8. Phizicky EM, Fields S (1995) Protein-protein interactions: methods for detection and analysis. Microbiol Rev 59:94–123
9. Lin J-S, Lai E-M (2017) Protein-protein interactions: co-immunoprecipitation. Methods Mol Biol 1615:211–219
10. Shoemaker BA, Panchenko AR (2007) Deciphering protein-protein interactions. Part I. Experimental techniques and databases. PLoS Comput Biol 3:e42
11. Matzinger M, Mechtler K (2021) Cleavable cross-linkers and mass spectrometry for the ultimate task of profiling protein-protein interaction networks in vivo. J Proteome Res 20: 78–93
12. Sinz A (2017) Divide and conquer: cleavable cross-linkers to study protein conformation and protein-protein interactions. Anal Bioanal Chem 409:33–44
13. Agou F, Véron M (2015) In vivo protein cross-linking. Methods Mol Biol 1278:391–405
14. Riedl J, Crevenna AH, Kessenbrock K et al (2008) Lifeact: a versatile marker to visualize F-actin. Nat Methods 5:605–607
15. Belyy A, Merino F, Sitsel O, Raunser S (2020) Structure of the Lifeact-F-actin complex. PLoS Biol 18:e3000925
16. Roux KJ, Kim DI, Raida M, Burke B (2012) A promiscuous biotin ligase fusion protein identifies proximal and interacting proteins in mammalian cells. J Cell Biol 196:801–810
17. Feldman JL, Sanchez AD, Svinkina T et al (2018) Efficient proximity labeling in living cells and organisms with TurboID. Nat Biotechnol 36:880–887

Chapter 10

Proteomic Analyses of the Mouse Brain

Jing Li

Abstract

Proteomics is a scientific field that aims to identify and characterize all proteins within a biological system, including their posttranslational modifications (PTMs), quantitative changes, and protein–protein interactions. Over the last two decades, proteomic approaches have been widely used in neuroscience research, providing multidimensional insights into the biology and pathology of the brain.

Here, we present a basic protocol for profiling protein expression in the mouse brain, which involves total protein extraction, fractionation, digestion, and identification through liquid chromatography coupled with tandem mass spectrometry (LC-MS/MS). This method is compatible with many prevalent techniques used for protein quantitation, PTM analysis, and protein–protein interaction mapping.

Key words Proteomics, Mouse brain, Mass spectrometry, Protein identification, In-gel digestion

1 Introduction

Proteins are the building blocks of life. Proteomics is a scientific discipline that aims to characterize and study all proteins within a biological system, such as an organelle, cell, tissue, or organism, providing multidimensional insights into the organization and dynamics of life.

Since the first proteomic analysis of a rodent brain in 1999 [1], the past two decades have seen an increasingly widespread application of proteomic techniques in neuroscience, from the discovery of biomarkers for neurological and psychiatric disorders to the investigation of synaptic signaling networks, which has dramatically enhanced our comprehension of the structure and function of the brain [2–4]. From a technical standpoint, proteomic analyses of the brain can be classified into four categories based on their respective research goals. The first category, expression profiling, seeks to identify the protein composition of the brain or a specific brain region/cell type. The second category, quantitative proteomics, analyzes protein expression changes under various physiological and pathological conditions. The third category,

Koh-ichi Nagata (ed.), *Cerebral Cortex Development: Methods and Protocols*, Methods in Molecular Biology, vol. 2794, https://doi.org/10.1007/978-1-0716-3810-1_10,

modification-specific proteomics, focuses on the study of particular posttranslational modifications, such as phosphorylation or glycosylation, and their respective changes. The last category, protein interactome research, studies protein–protein interactions and the components and organization of multiprotein complexes. Each category of proteomic analysis has its typical methods and techniques, and a comprehensive study often involves techniques from more than one of the subdivisions of proteomics. For example, a study investigating the effects of long-term potentiation on synaptic signaling integrated analyses of the postsynaptic interactome, label-free protein quantitation, and phosphoproteomics [5].

A typical proteomic experiment involves several steps, including protein extraction, sample fractionation, protein digestion (with or without enrichment for modified peptides), peptide detection using liquid chromatography coupled with tandem mass spectrometry (LC-MS/MS), protein identification (and quantitation) via spectra processing and database search, and data mining using various bioinformatic tools (Fig. 1). Researchers have the option to choose from multiple methods and techniques at each step based on the study purpose and other considerations such as the availability of specific instruments.

We present here a detailed procedure for mouse brain protein identification, which is the fundamental process for mouse brain proteomic research. This gel-based strategy works for most proteins, is readily adaptable in an ordinary biochemical laboratory, and is compatible with many prevalent approaches for further investigation of protein quantitation, posttranslational modification analysis, and protein–protein interaction mapping. Interested readers looking for a more comprehensive description of the techniques and applications of proteomics in brain research are encouraged to refer to the extensive reviews available in neuroproteomics [2, 3, 6, 7], given the space limitations.

2 Materials

The protocol described here applies to the preisolated mouse brain (or a particular part of the mouse brain, e.g., the prefrontal cortex). Solutions used before the "protein digestion" step can be prepared using double-distilled water (ddH_2O) or ultrapure water and analytical-grade reagents. Starting from the "gel excision and protein digestion" step, prepare all solutions using mass spectrometry (MS)-grade reagents and solvents. Prepare and store the solutions and solvents at room temperature unless otherwise specified.

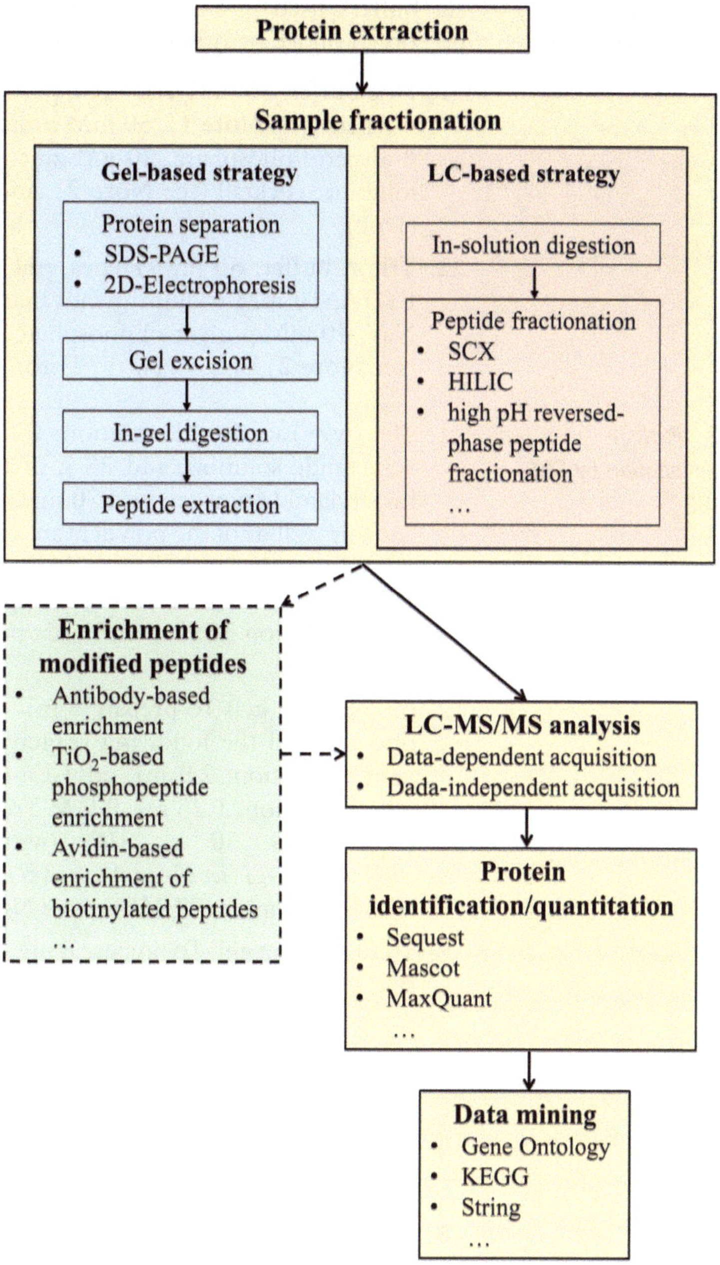

Fig. 1 Typical workflow for proteomic analysis

Before starting the experiment, read the Safety Data Sheets of each reagent carefully and take appropriate protective measures during the operation. Diligently follow all local laws and regulations on animal use and waste disposal to ensure the safety of all personnel involved in the experiment and the environment.

2.1 Protein Extraction from the Mouse Brain

All buffers used in this section should be freshly prepared and precooled on ice before the experiment:

1. Doc buffer: 50 mM Tris-HCl, pH 9.0, 10 mM sodium orthovanadate (*see* **Note 1**), 50 mM sodium fluoride (NaF), 40 mM β-glycerol phosphate, 20 μM zinc chloride ($ZnCl_2$), protease inhibitor cocktail (see **Note 2**), and 1% sodium deoxycholate (Doc).
2. Triton buffer: 50 mM Hepes, pH 7.4, 2 mM EGTA, 2 mM EDTA, 10 mM sodium orthovanadate (see **Note 1**), 30 mM NaF, 40 mM β-glycerol phosphate, protease inhibitor cocktail (see **Note 2**), and 1% (w/v) Triton.

2.2 Protein Fractionation by SDS-PAGE

1. 40% (w/v) acrylamide solution: To prepare 100 mL 40% (w/v) acrylamide solution, add 38 g of acrylamide powder, 2 g of bisacrylamide powder, and 70 mL of ddH_2O to a clean beaker and stir well until the powders are completely dissolved. Dilute the solution to 100 mL with ddH2O. Filter the solution using a 0.45 μm filter to remove any impurities and store the acrylamide solution in a brown bottle protected from light at 4 °C (*see* **Note 3**).
2. 6% Stacking gel: To prepare 5 mL of stacking gel solution for one gel, add the following reagents into a clean beaker with gentle agitation: 2.9 mL ddH_2O, 0.75 mL 40% (w/v) acrylamide solution, 1.25 mL 0.5 M Tris-HCl, pH 6.8, 50 μL 10% (w/v) SDS, 50 μL 10% (w/v) ammonium persulfate ($(NH_4)_2S_2O_8$) (*see* **Note 4**), and 5 μL N,N,N′,N′- tetramethylethylenediamine (TEMED) (*see* **Note 5**).
3. 10% resolving gel: To prepare 8 mL of resolving gel solution for one gel, add the following reagents into a clean beaker with gentle agitation: 3.8 mL ddH_2O, 2 mL 40% (w/v) acrylamide solution, 2 mL 1.5 M Tris-HCl, pH 8.8, 80 μL 10% (w/v) SDS, 80 μL 10% (w/v) $(NH_4)_2S_2O_8$ (*see* **Note 4**), and 8 μL TEMED (*see* **Note 5**).
4. SDS-PAGE running buffer: 0.025 M Tris-HCl, 0.25 M glycine, and 0.1% (w/v) SDS.
5. 5× SDS-PAGE loading buffer: 250 mM Tris–HCl, pH 6.8, 10% (w/v) SDS, 0.5% (w/v) bromophenol blue, and 50% (w/v) glycerol.
6. Reducing solution: 400 mM dithiothreitol (DTT). This stock solution can be aliquoted and stored at −20 °C for up to 1 month.
7. Alkylation solution: 400 mM Iodoacetamide (IAM). This stock solution can be aliquoted and stored at −20 °C for up to 1 month. Care should be taken to protect IAM from light to avoid its decomposition.

8. Gel staining solution: To prepare 1 L Coomassie brilliant blue staining solution, dissolve 1 g of Coomassie brilliant blue R-250 reagent in a clean beaker containing 250 mL isopropanol, 100 mL acetic acid, and 650 mL ddH_2O. Use a filter paper to remove particulates, and store the solution at room temperature. Organic solvents should be prepared and used in a fume hood.
9. Destaining solution: 10% acetic acid (v/v), 5% ethanol (v/v), and 85% ddH_2O. Organic solvents should be prepared and used in a fume hood.

2.3 Gel Excision and Protein Digestion

1. ABC buffer: 25 mM ammonium bicarbonate (NH_4HCO_3) in water, LC-MS grade, pH 8.0 (*see* **Note 6**).
2. Acetonitrile (ACN), LC-MS grade.
3. Trypsin stock solution: Reconstitute a vial of MS-grade trypsin powder in 50 mM acetic acid to make a 1 μg/μL stock solution. This stock solution can then be aliquoted and stored at −20 °C for up to 1 month (*see* **Note 7**).

2.4 Peptide Extraction

1. 0.5% formic acid (FA) solution: 0.5% (v/v) FA in water, LC-MS grade.
2. Reconstitution solution: 3% (v/v) ACN and 0.1% (V/V) FA in water, LC-MS grade.

2.5 LC-MS/MS Analysis

1. Solvent A: 0.1% (v/v) FA in water, LC-MS grade.
2. Solvent B: 80% (v/v) ACN/0.1% (v/v) FA in water, LC-MS grade.

3 Methods

3.1 Protein Extraction from the Mouse Brain

All procedures in this section should be carried out on ice unless otherwise specified:

1. For proteomic analysis, the mouse brain samples should be either freshly obtained brain tissues or frozen tissues that were snap-frozen in liquid nitrogen immediately after dissection and stored at −80 °C until the experiment.
2. Homogenize the brain samples using a glass tissue grinder with 20 up-and-down passes. Use the Triton buffer for embryonic mouse brains; otherwise, use the Doc buffer as the lysis buffer (*see* **Note 8**). The volume of lysis buffer to be used should be determined based on the sample amount. For example, using 2–3 mL lysis buffer for half of an adult cortex or 300–400 μL buffer for two embryonic day 14 mouse brains can yield a protein concentration of 3–4 mg/mL.

3. Incubate the homogenates at 4 °C for 30 min with end-to-end rotation.
4. Sonicate the samples using the following parameters: A frequency of 20 kHz, with each sonication cycle including 2 s of vibration followed by a 3 s interval. Depending on the sample volume and density, 20–30 sonication cycles may be necessary (*see* **Note 9**).
5. After sonication, incubate the samples on ice for an additional 5 min.
6. Centrifuge the lysates at 18,000 × *g* and 4 °C for 30 min.
7. Collect the supernatants and determine the protein concentration using the BCA assay.
8. Aliquot the samples for subsequent experiments. The aliquoted lysates can be used directly or stored at −80 °C for protein identification, quantitation, and posttranslational modification analyses (*see* **Note 10**).

3.2 Protein Fractionation by SDS-PAGE (See Note 11)

The following procedures should be carried out at room temperature unless otherwise specified:

1. Cast an SDS-PAGE gel or use a commercial precast gel. For gel casting, set up a casting chamber, pour the resolving gel solution into the chamber (for a 1.5 mm thick gel, approximately 6 mL of resolving gel solution is used), and add butanol or ddH_2O immediately to remove any unwanted air bubbles. After the polymerization of the resolving gel, pour out the butanol or ddH_2O, add stacking gel solution into the chamber (approximately 2 mL of stacking gel solution is used for a 1.5 mm thick gel), and insert a comb with an appropriate thickness in between the two glass plates. Once the stacking gel is polymerized, remove the gel cassette from the casting stand, secure it in the electrode assembly, and place the electrode assembly in the tank. Pull out the comb, wash the resulting loading wells using ddH_2O, and pour the SDS running buffer into the electrode assembly and the tank up to an appropriate level (*see* **Note 12**).
2. Prepare the samples for electrophoresis (*see* **Note 13**). Take one tube of the aliquoted protein lysates, add 5× loading buffer in a ratio of 1:4 (loading buffer: lysate, v/v), mix well using a vortex mixer, and boil the sample at 95 °C for 5 min (*see* **Note 14**). Spin down the liquid using a microcentrifuge.
3. Reduce the disulfide bonds of the proteins by adding the Reducing solution in proportion to the sample tubes to achieve a DTT working concentration of 10 mM after the boiled samples have been cooled down to below 60 °C. Incubate the samples at 56 °C for 1 h in a thermomixer set to 300 rpm. Spin down the liquid using a microcentrifuge.

4. Alkylate the exposed sulfhydryl groups of the proteins by adding the Alkylation solution in proportion to the sample tubes to achieve an IAM working concentration of 40 mM after the reduced samples have been cooled down to room temperature. Incubate the samples in the dark at room temperature for 45 min in a thermomixer set to 300 rpm.
5. Centrifuge the samples at 12,000 × *g* for 5 min and collect the supernatants for SDS-PAGE analysis.
6. Load equal amounts of proteins (20–30 μg) into each loading well (*see* **Note 15**).
7. Place the lid on the electrophoresis tank and connect it to a power supply. Make sure the anode and cathode plugs are correctly connected. Perform electrophoresis in the constant voltage mode at 80 mV for the stacking gel and 120 mV for the resolving gel.
8. Stop electrophoresis when the bromophenol blue tracking line reaches the bottom of the gel.
9. Carefully remove the gel cassette from the tank and open it. Submerge the gel in a clean glass dish containing the Coomassie brilliant blue staining solution and stain it by shaking it gently at room temperature for approximately 1 h in a fume hood (*see* **Note 16**). Starting from this step, wear disposable hats, masks, and gloves when handling the gel to prevent contamination of human proteins from hairs, saliva, and skin.
10. Transfer the gel to a clean glass dish containing the Destaining solution. Destain the gel by shaking it gently at room temperature for 2–4 h (*see* **Note 17**).

3.3 Gel Excision and Protein Digestion

To prevent contamination with foreign proteins, make sure to wipe the bench and labware with 75% ethanol and wear disposable hats, masks, and gloves during the operation:

1. Carefully place the gel on a clean lightbox for gel excision. Using a sterile scalpel, cut the protein bands into small pieces of length and width no more than 1.5 cm. Place up to six gel pieces in each of a series of 1.5 mL Eppendorf tubes (the batch A tubes) labeled as A1, A2, A3, etc. This way, an entire lane can be divided into 8–12 fractions (*see* **Note 18**).
2. Add 100 μL of ABC buffer to each tube, ensuring that the buffer covers all gel pieces, and incubate these tubes at 37 °C for 15 min in a thermomixer set to 450 rpm.
3. Add 50 μL of ACN to each tube and incubate the tubes at room temperature for 15 min in a thermomixer set to 450 rpm.
4. Carefully remove the liquid from each tube, taking care not to disturb the gel pieces (*see* **Note 19**).

5. Repeat **steps 2–4** until the gel pieces are completely decolorized.
6. Add 75 μL of ABC buffer to each tube and incubate the tubes at 37 °C for 15 min in a thermomixer set to 450 rpm.
7. Add 75 μL of ACN to each tube and incubate the tubes at room temperature for 15 min in a thermomixer set to 450 rpm.
8. Carefully remove the liquid from each tube, taking care not to disturb the gel pieces.
9. Add 100 μL of ACN to each tube and incubate the tubes at room temperature for 30 min in a thermomixer set to 450 rpm.
10. Remove the liquid from the tubes and use a centrifuge concentrator to dry the gel pieces until they shrink into small white pieces. Prepare a batch of low-binding tubes labeled as B1, B2, B3, etc., and transfer the dried gel pieces into the corresponding tubes (the batch B tubes).
11. Thaw a tube of aliquoted Trypsin stock solution on ice and dilute it with the ABC buffer to make a working solution with a concentration of 5 ng/μL.
12. Add 20 μL of the diluted trypsin solution into each tube containing the gel pieces. Place the tubes on ice for 30 min to allow complete absorption of the trypsin solution by the gels. The gel pieces become transparent again at this step.
13. Add 150 μL of ABC buffer to each tube and check the pH of the solutions in each tube with pH test strips. If necessary, add more ABC buffer to ensure that all gel pieces are submerged in the solution and the pH of the solution is above 8.0. Incubate the samples at 37 °C overnight with interrupted oscillation, for example, by shaking at 500 rpm for 1 min every 15 min.

3.4 Peptide Extraction

1. Next day in the morning, spin down the liquid to the bottom of the tubes and add 50 μL of 0.5% FA solution to each tube to terminate the digestion. Incubate the tubes at 37 °C in a thermomixer set to 450 rpm for 30 min.
2. Add 100 μL of ACN to each tube and incubate the tubes at room temperature in a thermomixer set to 450 rpm for 30 min.
3. Prepare a new batch of low-binding tubes labeled as C1, C2, C3, etc., and transfer the peptide solutions in the batch B tubes to the corresponding batch C tubes.
4. Add 50 μL of 0.5% FA solution to each of the batch B tubes and incubate them at 37 °C in a thermomixer set to 450 rpm for 30 min.
5. Add 100 μL of ACN to each of the batch B tubes and incubate them at room temperature in a thermomixer set to 450 rpm for 30 min.

6. Transfer the peptide solutions in the batch B tubes to the corresponding batch C tubes to combine with the previously collected peptide solutions.
7. Dry out the solutions in the batch C tubes with a centrifuge concentrator.
8. Reconstitute the samples in each of the batch C tubes with 10 μL Reconstitution solution at room temperature in a thermomixer set to 600 rpm for 15 min (*see* **Notes 20** and **21**).
9. Centrifuge the samples at 18,000 × g for 15 min and then transfer the supernatants to the sample vials for LC-MS/MS analysis or to the new low-binding tubes labeled as batch D tubes for storage at −80 °C for up to 2 weeks.

3.5 LC-MS/MS Analysis and Data Processing

1. Set the instrument parameters according to the operation manuals of the LC-MS/MS system. Choose a reverse phase (RP) capillary high-performance liquid chromatography (HPLC) column that is compatible with the system (*see* **Note 22**), and use the system's data acquisition and interpretation software for instrument control and data collection.
2. Use a gradient of Solvent B for the sequential elution of peptides. An example of the LC gradient setup is shown in Table 1.
3. Set the parameters for MS analysis as follows: Use a data-dependent acquisition mode with dynamic exclusion set to 30 s. One full MS scan, with a mass range of 300–1500 m/z, is followed by ten MS/MS scans. The isolation window for MS/MS scan can be set to 2.0 m/z, and the normalized collision energy can be set to 28.
4. Load each extracted peptide sample onto the LC-MS/MS system and perform the analysis to collect data.
5. Choose the appropriate software for protein identification (and quantitation) (*see* **Notes 23** and **24**). Search the MS data from a mouse protein database, such as the reference proteome for

Table 1
Example settings of LC gradient for proteomic analysis

Time (min)	Duration (min)	Flow (nL/min)	%B
0	0	300	3
3	3	300	3
83	80	300	28
93	10	300	40
95	2	300	100
100	5	300	100

Mus musculus by Uniprot. Set the parameters for database search as follows: No more than two missed cleavages are allowed for trypsin digestion; and cysteine carbamidomethyl should be set as static modification and methionine oxidation and protein N-terminal acetylation may be set as dynamic modifications. The setting of mass tolerance depends on the MS instrument's performance. For high-resolution mass spectrometers, mass tolerance can be set to 10 ppm for full scans and 0.02 Da for MS/MS scans. Usually, valid identification of a protein requires at least two confidentially identified unique peptides.

6. Proteomic data mining: Several online resources are available for further investigation and interpretation of the identification results, such as the Gene Ontology knowledgebases [8, 9] (http://geneontology.org) for protein function and subcellular location annotation, the Kyoto Encyclopedia of Genes and Genomes [10, 11] (https://www.genome.jp/kegg) for pathway mapping, and the String database [12, 13] (https://cn.string-db.org) for functional protein association network analysis. There are also resources that are specifically helpful for brain proteomic studies, such as the Allen Brain Atlas [14–16] (https://atlas.brain-map.org), which features a distribution illustration of brain proteins, and the Synaptic Gene Ontologies [17] (https://syngoportal.org), which focuses on synaptic proteins.

4 Notes

1. To achieve maximal inhibition of protein tyrosine phosphatases, sodium orthovanadate should be activated. Briefly, depolymerize the vanadate by boiling the solution and adjusting its pH to 10.0 once it has been cooled to room temperature. Repeat these steps several times until the solution becomes colorless and the pH stabilizes at 10.0 [18].
2. Protease inhibitor cocktails from various manufacturers can differ in composition and concentration. Some cocktails are designed for specific sample types or applications, such as phosphoproteomic analysis. Carefully read the manufacturer's instructions before use.
3. The acrylamide solution can be stored at 4 °C for up to 1 month. Acrylamide and bisacrylamide are potent neurotoxins and possible carcinogens. Therefore, handle them with care and avoid direct contact with skin and inhalation of dust or fumes.

4. The $(NH_4)_2S_2O_8$ solution can be stored at 4 °C for up to 2 weeks. Note that $(NH_4)_2S_2O_8$ is a potent oxidizing agent and can be hazardous. Handle it with care and follow appropriate safety precautions.
5. Polymerization of the gel occurs quickly after the addition of TEMED, so add this reagent only when the casting chamber is set and the gel is ready to be poured. TEMED is also a potent neurotoxin and should be handled with caution.
6. Prepare the ABC buffer at least 12 h in advance to ensure that the salts are fully dissolved. A 1 M stock solution can be prepared and stored protected from light at 4 °C for up to 3 months. During prolonged storage, the buffer's pH may slightly decrease due to the presence of CO_2 in the environment. Check the pH of the ABC buffer before use. If the pH is below 8.0, increase the NH_4HCO_3 concentration to no more than 50 mM.
7. For long-term storage, aliquot and freeze the Trypsin stock solution at −80 °C.
8. The Doc buffer containing the detergent sodium deoxycholate can solubilize most membrane proteins and is the preferred lysis buffer for the membrane-associated subcellular structure of postsynaptic density in neurons. However, homogenizing embryonic mouse brains in the Doc buffer results in the gelation of the tissue lysate, so the Triton buffer is recommended for use in this situation.
9. Sonication may disrupt weak protein–protein interactions. Consider skipping the sonication step if subsequent immunoprecipitation experiments are planned.
10. It is always best to start the follow-up procedures immediately after protein extraction, especially for posttranslational modification research. Even with good protease inhibitor cocktails, the shedding of modification groups and the degradation of some proteins can still occur during prolonged storage. In our experience, storage at −80 °C for up to 2 weeks has a little effect on the identification and quantitation of most proteins. Lysates intended for protein–protein interaction studies should not be frozen at any time and must be used immediately.
11. Methods for sample fractionation in proteomic analysis can be generally classified into two categories: those working at the protein level (e.g., SDS-PAGE, two-dimensional electrophoresis [1], and RP LC [19]) and those working at the peptide level (e.g., strong cation exchange (SCX) cartridges, hydrophilic interaction liquid chromatography (HILIC) [20], and high-pH reversed-phase peptide fractionation [21]). Each method has its advantages and disadvantages, and the choice of fractionation method (or combination of methods) depends on

the samples, research objectives, and laboratory resources. Here, we describe the SDS-PAGE method, which works for most proteins, requires minimal equipment, and can be easily implemented in most biological and biochemical laboratories.

12. Although the polymerized acrylamide gel is nontoxic, it should still be handled with caution as unpolymerized residual acrylamide may be present in the solution, which poses a safety hazard to humans.
13. The sample preparation step takes approximately 2.5 h and can be performed simultaneously with the gel-casting procedure to save time.
14. To prevent air bursts, loosen the caps of the sample tubes during heating. Do not open the tube caps immediately after heating to avoid splashing of the liquid.
15. To facilitate the subsequent gel-cutting process and to prevent any potential interference, a blank lane loaded with an equal volume of the 1× loading buffer can be placed between each sample.
16. In recent years, several fast-staining solutions have been developed that can stain a gel within minutes without the need for destaining.
17. The duration of the destaining step can be adjusted based on the experimental purpose. To analyze differential bands in different samples, set a shorter destaining time to reveal as many protein bands as possible; if the entire lane needs to be analyzed, extend the destaining time to fade the bands and facilitate the destaining step in the subsequent gel digestion procedures.
18. Specially designed grid cutters are available to assist in the excision of the gels. Rinse the blades thoroughly between samples to avoid contamination.
19. Take care not to puncture the gels as it can be challenging to remove the broken gel particles, which may clog the capillary chromatography columns in subsequent LC-MS analysis.
20. The volume of the Reconstitution solution used here is limited by the LC-MS/MS instrument's autosampler sample loop size. If the sample loop allows, a larger volume can be used to facilitate the complete reconstitution of the sample.
21. Enrichment of peptides of certain modifications (e.g., phosphopeptides) can be applied to the extracted peptides for post-translational modification analysis. In such cases, reconstitute the samples according to the requirement of the specific methods used.

22. There are commercially available capillary analytical columns for each LC system. In-house-packed C18 capillary columns are an alternative option with high performance and cost-effectiveness. In this case, a quartz capillary with an inner diameter of 75 μm and a length of 15–30 cm and filled with C18 resins featuring a particle diameter of 2–3 μm and a pore size of 5 Å is typically used. Mounting parts such as nuts and unions for the homemade capillary columns are available from chromatograph manufacturers or other companies producing capillary fittings.
23. There are numerous commercial or free database search engines available for peptide identification, including Sequest [22], Mascot [23], X!Tandem [24, 25], Andromeda [26], and MSFragger [27]. Software suites such as MaxQuant [28, 29], Trans-Proteomic Pipeline (TPP) [30], and Proteome Discovery [31] offer an integrated running environment, extended toolkits, and a user-friendly interface to simplify protein identification, result display, and data processing.
24. Quantitative proteomic approaches are generally classified into two categories: methods based on stable isotope labeling (e.g., using the isobaric Tag for Relative and Absolute Quantitation (iTRAQ) [32] or the Tandem Mass Tag (TMT) [33, 34] reagents) and label-free methods. Relative quantitation can be achieved without extra isotopic labeling steps by comparing the peak intensity (area) or spectra count [35]. Most proteomic software suites, including MaxQuant, TPP, and Proteome Discoverer, offer a label-free quantitation module that can be easily incorporated into the workflow. For label-free quantitative analysis using the protocol described in this chapter, it is essential to ensure that the lanes are correctly aligned and cut down between the samples during the gel excision step.

Acknowledgments

This work was supported by the Natural Science Foundation of China (grant No. 32101020) and the Natural Science Foundation of Shandong Province (CN) (grant No. ZR2020MC071).

References

1. Fountoulakis M, Schuller E, Hardmeier R, Berndt P, Lubec G (1999) Rat brain proteins: two-dimensional protein database and variations in the expression level. Electrophoresis 20(18):3572–3579. https://doi.org/10.1002/(SICI)1522-2683(19991201)20:18<3572::AID-ELPS3572>3.0.CO;2-T
2. Bayes A, Grant SG (2009) Neuroproteomics: understanding the molecular organization and complexity of the brain. Nat Rev Neurosci 10(9):635–646. https://doi.org/10.1038/nrn2701
3. Cookson MR (2019) Proteomics: techniques and applications in neuroscience. J Neurochem 151(4):394–396. https://doi.org/10.1111/jnc.14867
4. Xu Y, Song X, Wang D, Wang Y, Li P, Li J (2021) Proteomic insights into synaptic signaling in the brain: the past, present and future. Mol Brain 14(1):37. https://doi.org/10.1186/s13041-021-00750-5
5. Li J, Wilkinson B, Clementel VA, Hou J, O'Dell TJ, Coba MP (2016) Long-term potentiation modulates synaptic phosphorylation networks and reshapes the structure of the postsynaptic interactome. Sci Signal 9(440): rs8. https://doi.org/10.1126/scisignal.aaf6716
6. Liao L, McClatchy DB, Yates JR (2009) Shotgun proteomics in neuroscience. Neuron 63(1):12–26. https://doi.org/10.1016/j.neuron.2009.06.011
7. Fernandez-Irigoyen J, Corrales F, Santamaria E (2019) The human brain proteome project: biological and technological challenges. Methods Mol Biol 2044:3–23. https://doi.org/10.1007/978-1-4939-9706-0_1
8. Ashburner M, Ball CA, Blake JA, Botstein D, Butler H, Cherry JM, Davis AP, Dolinski K, Dwight SS, Eppig JT, Harris MA, Hill DP, Issel-Tarver L, Kasarskis A, Lewis S, Matese JC, Richardson JE, Ringwald M, Rubin GM, Sherlock G (2000) Gene ontology: tool for the unification of biology. Gene Ontol Consort Nat Genet 25(1):25–29. https://doi.org/10.1038/75556
9. Gene Ontology C (2021) The Gene Ontology resource: enriching a GOld mine. Nucleic Acids Res 49(D1):D325–D334. https://doi.org/10.1093/nar/gkaa1113
10. Ogata H, Goto S, Sato K, Fujibuchi W, Bono H, Kanehisa M (1999) KEGG: Kyoto encyclopedia of genes and genomes. Nucleic Acids Res 27(1):29–34. https://doi.org/10.1093/nar/27.1.29
11. Kanehisa M, Goto S (2000) KEGG: Kyoto encyclopedia of genes and genomes. Nucleic Acids Res 28(1):27–30. https://doi.org/10.1093/nar/28.1.27
12. Snel B, Lehmann G, Bork P, Huynen MA (2000) STRING: a web-server to retrieve and display the repeatedly occurring neighbourhood of a gene. Nucleic Acids Res 28(18): 3442–3444. https://doi.org/10.1093/nar/28.18.3442
13. Szklarczyk D, Gable AL, Nastou KC, Lyon D, Kirsch R, Pyysalo S, Doncheva NT, Legeay M, Fang T, Bork P, Jensen LJ, von Mering C (2021) The STRING database in 2021: customizable protein-protein networks, and functional characterization of user-uploaded gene/measurement sets. Nucleic Acids Res 49(D1): D605–D612. https://doi.org/10.1093/nar/gkaa1074
14. Lein ES, Hawrylycz MJ, Ao N, Ayres M, Bensinger A, Bernard A, Boe AF, Boguski MS, Brockway KS, Byrnes EJ, Chen L, Chen L, Chen TM, Chin MC, Chong J, Crook BE, Czaplinska A, Dang CN, Datta S, Dee NR, Desaki AL, Desta T, Diep E, Dolbeare TA, Donelan MJ, Dong HW, Dougherty JG, Duncan BJ, Ebbert AJ, Eichele G, Estin LK, Faber C, Facer BA, Fields R, Fischer SR, Fliss TP, Frensley C, Gates SN, Glattfelder KJ, Halverson KR, Hart MR, Hohmann JG, Howell MP, Jeung DP, Johnson RA, Karr PT, Kawal R, Kidney JM, Knapik RH, Kuan CL, Lake JH, Laramee AR, Larsen KD, Lau C, Lemon TA, Liang AJ, Liu Y, Luong LT, Michaels J, Morgan JJ, Morgan RJ, Mortrud MT, Mosqueda NF, Ng LL, Ng R, Orta GJ, Overly CC, Pak TH, Parry SE, Pathak SD, Pearson OC, Puchalski RB, Riley ZL, Rockett HR, Rowland SA, Royall JJ, Ruiz MJ, Sarno NR, Schaffnit K, Shapovalova NV, Sivisay T, Slaughterbeck CR, Smith SC, Smith KA, Smith BI, Sodt AJ, Stewart NN, Stumpf KR, Sunkin SM, Sutram M, Tam A, Teemer CD, Thaller C, Thompson CL, Varnam LR, Visel A, Whitlock RM, Wohnoutka PE, Wolkey CK, Wong VY, Wood M, Yaylaoglu MB, Young RC, Youngstrom BL, Yuan XF, Zhang B, Zwingman TA, Jones AR (2007) Genome-wide atlas of gene expression in the adult mouse brain. Nature 445(7124):168–176. https://doi.org/10.1038/nature05453
15. Oh SW, Harris JA, Ng L, Winslow B, Cain N, Mihalas S, Wang Q, Lau C, Kuan L, Henry AM, Mortrud MT, Ouellette B, Nguyen TN, Sorensen SA, Slaughterbeck CR, Wakeman W, Li Y, Feng D, Ho A, Nicholas E, Hirokawa KE,

Bohn P, Joines KM, Peng H, Hawrylycz MJ, Phillips JW, Hohmann JG, Wohnoutka P, Gerfen CR, Koch C, Bernard A, Dang C, Jones AR, Zeng H (2014) A mesoscale connectome of the mouse brain. Nature 508(7495): 207–214. https://doi.org/10.1038/nature13186

16. Daigle TL, Madisen L, Hage TA, Valley MT, Knoblich U, Larsen RS, Takeno MM, Huang L, Gu H, Larsen R, Mills M, Bosma-Moody A, Siverts LA, Walker M, Graybuck LT, Yao Z, Fong O, Nguyen TN, Garren E, Lenz GH, Chavarha M, Pendergraft J, Harrington J, Hirokawa KE, Harris JA, Nicovich PR, McGraw MJ, Ollerenshaw DR, Smith KA, Baker CA, Ting JT, Sunkin SM, Lecoq J, Lin MZ, Boyden ES, Murphy GJ, da Costa NM, Waters J, Li L, Tasic B, Zeng H (2018) A suite of transgenic driver and reporter mouse lines with enhanced brain-cell-type targeting and functionality. Cell 174(2):465–480 e422. https://doi.org/10.1016/j.cell.2018.06.035

17. Koopmans F, van Nierop P, Andres-Alonso M, Byrnes A, Cijsouw T, Coba MP, Cornelisse LN, Farrell RJ, Goldschmidt HL, Howrigan DP, Hussain NK, Imig C, de Jong APH, Jung H, Kohansalnodehi M, Kramarz B, Lipstein N, Lovering RC, MacGillavry H, Mariano V, Mi H, Ninov M, Osumi-Sutherland D, Pielot R, Smalla KH, Tang H, Tashman K, Toonen RFG, Verpelli C, Reig-Viader R, Watanabe K, van Weering J, Achsel T, Ashrafi G, Asi N, Brown TC, De Camilli P, Feuermann M, Foulger RE, Gaudet P, Joglekar A, Kanellopoulos A, Malenka R, Nicoll RA, Pulido C, de Juan-Sanz J, Sheng M, Sudhof TC, Tilgner HU, Bagni C, Bayes A, Biederer T, Brose N, Chua JJE, Dieterich DC, Gundelfinger ED, Hoogenraad C, Huganir RL, Jahn R, Kaeser PS, Kim E, Kreutz MR, McPherson PS, Neale BM, O'Connor V, Posthuma D, Ryan TA, Sala C, Feng G, Hyman SE, Thomas PD, Smit AB, Verhage M (2019) SynGO: an evidence-based, expert-curated knowledge base for the synapse. Neuron 103(2):217–234 e214. https://doi.org/10.1016/j.neuron.2019.05.002

18. Gordon JA (1991) Use of vanadate as protein-phosphotyrosine phosphatase inhibitor. Methods Enzymol 201:477–482. https://doi.org/10.1016/0076-6879(91)01043-2

19. Li J, Xie Z, Shi L, Zhao Z, Hou J, Chen X, Cui Z, Xue P, Cai T, Wu P, Guo S, Yang F (2012) Purification, identification and profiling of serum amyloid A proteins from sera of advanced-stage cancer patients. J Chromatogr B Analyt Technol Biomed Life Sci 889–890:3–9. https://doi.org/10.1016/j.jchromb.2012.01.002

20. Shu Q, Li M, Shu L, An Z, Wang J, Lv H, Yang M, Cai T, Hu T, Fu Y, Yang F (2020) Large-scale identification of N-linked intact glycopeptides in human serum using HILIC enrichment and spectral library search. Mol Cell Proteomics 19(4):672–689. https://doi.org/10.1074/mcp.RA119.001791

21. Batth TS, Francavilla C, Olsen JV (2014) Off-line high-pH reversed-phase fractionation for in-depth phosphoproteomics. J Proteome Res 13(12):6176–6186. https://doi.org/10.1021/pr500893m

22. Eng JK, McCormack AL, Yates JR (1994) An approach to correlate tandem mass spectral data of peptides with amino acid sequences in a protein database. J Am Soc Mass Spectrom 5(11):976–989. https://doi.org/10.1016/1044-0305(94)80016-2

23. Perkins DN, Pappin DJ, Creasy DM, Cottrell JS (1999) Probability-based protein identification by searching sequence databases using mass spectrometry data. Electrophoresis 20(18):3551–3567. https://doi.org/10.1002/(SICI)1522-2683(19991201)20:18<3551::AID-ELPS3551>3.0.CO;2-2

24. Craig R, Beavis RC (2003) A method for reducing the time required to match protein sequences with tandem mass spectra. Rapid Commun Mass Spectrom 17(20):2310–2316. https://doi.org/10.1002/rcm.1198

25. Craig R, Beavis RC (2004) TANDEM: matching proteins with tandem mass spectra. Bioinformatics 20(9):1466–1467. https://doi.org/10.1093/bioinformatics/bth092

26. Cox J, Neuhauser N, Michalski A, Scheltema RA, Olsen JV, Mann M (2011) Andromeda: a peptide search engine integrated into the MaxQuant environment. J Proteome Res 10(4):1794–1805. https://doi.org/10.1021/pr101065j

27. Kong AT, Leprevost FV, Avtonomov DM, Mellacheruvu D, Nesvizhskii AI (2017) MSFragger: ultrafast and comprehensive peptide identification in mass spectrometry-based proteomics. Nat Methods 14(5):513–520. https://doi.org/10.1038/nmeth.4256

28. Cox J, Mann M (2008) MaxQuant enables high peptide identification rates, individualized p.p.b.-range mass accuracies and proteome-wide protein quantification. Nat Biotechnol 26(12):1367–1372. https://doi.org/10.1038/nbt.1511

29. Cox J, Hein MY, Luber CA, Paron I, Nagaraj N, Mann M (2014) Accurate proteome-wide label-free quantification by delayed normalization and maximal peptide ratio extraction, termed MaxLFQ. Mol Cell

Proteomics 13(9):2513–2526. https://doi.org/10.1074/mcp.M113.031591

30. Deutsch EW, Mendoza L, Shteynberg D, Slagel J, Sun Z, Moritz RL (2015) Trans-proteomic pipeline, a standardized data processing pipeline for large-scale reproducible proteomics informatics. Proteomics Clin Appl 9(7–8):745–754. https://doi.org/10.1002/prca.201400164
31. Orsburn BC (2021) Proteome discoverer-a community enhanced data processing suite for protein informatics. Proteomes 9(1). https://doi.org/10.3390/proteomes9010015
32. Wiese S, Reidegeld KA, Meyer HE, Warscheid B (2007) Protein labeling by iTRAQ: a new tool for quantitative mass spectrometry in proteome research. Proteomics 7(3):340–350. https://doi.org/10.1002/pmic.200600422
33. Thompson A, Schafer J, Kuhn K, Kienle S, Schwarz J, Schmidt G, Neumann T, Johnstone R, Mohammed AK, Hamon C (2003) Tandem mass tags: a novel quantification strategy for comparative analysis of complex protein mixtures by MS/MS. Anal Chem 75(8):1895–1904. https://doi.org/10.1021/ac0262560
34. McAlister GC, Huttlin EL, Haas W, Ting L, Jedrychowski MP, Rogers JC, Kuhn K, Pike I, Grothe RA, Blethrow JD, Gygi SP (2012) Increasing the multiplexing capacity of TMTs using reporter ion isotopologues with isobaric masses. Anal Chem 84(17):7469–7478. https://doi.org/10.1021/ac301572t
35. Zhu W, Smith JW, Huang CM (2010) Mass spectrometry-based label-free quantitative proteomics. J Biomed Biotechnol 2010:840518. https://doi.org/10.1155/2010/840518

Chapter 11

Generation of iPSCs Using Sendai Virus Vectors

Hitomi Aoki

Abstract

Induced pluripotent stem cells (iPSCs) are in vitro–derived cells capable of giving rise to several different cell types. The generation of iPSCs holds great promise for regenerative medicine and drug discovery research because it allows mature cells to be reprogrammed into a state of pluripotency. These highly versatile cells can then be induced to produce a variety of cell lineages and tissues by activating specific regulatory genes that drive their differentiation along distinct lineages. The great potential of these cells was recognized by Shinya Yamanaka who was awarded the 2012 Nobel Prize for the discovery of iPSCs. Following their discovery, various methods have now been developed for generating iPSCs. Here, we describe a method for deriving iPSCs from human dental pulp using Sendai virus vectors.

Key words Induced pluripotent stem cells, Dental pulp, Sendai virus, Oct3/4, Sox2, Klf4, c-Myc

1 Introduction

In 2006, Kazutoshi Takahashi and Shinya Yamanaka tested the genes specifically expressed in embryonic stem (ES) cells for their ability to reprogram terminally differentiated cells into an ES-like pluripotent state. From 24 candidate genes, Oct3/4, Sox2, c-Myc, and Klf4 were found to be sufficient for the conversion of embryonic fibroblast cells or adult skin fibroblasts into a pluripotent state. These cells were named induced pluripotent stem cells (iPSCs) [1]. iPSCs are morphologically similar to ES cells and possess the same limitless self-renewal potential and the pluripotency to differentiate into numerous cell types. iPSCs were first developed from murine cells but now are also being produced from human cells [2, 3].

Initially, iPSCs were generated using gene transfer methods that relied on retroviruses [4], before more convenient lentivirus systems were introduced [5]. However, both these approaches have the potential to induce mutations and increase the risk of tumor formation because they integrate exogenous reprogramming genes into host genomes. For this reason, more recent methodologies

Koh-ichi Nagata (ed.), *Cerebral Cortex Development: Methods and Protocols*, Methods in Molecular Biology, vol. 2794,
https://doi.org/10.1007/978-1-0716-3810-1_11,

using episomal plasmid vectors and Sendai virus–based vectors have been developed to establish iPSCs without requiring genome insertions [6–8].

The Sendai virus (SeV) vector allows the four exogenous genes required for efficient nuclear reprogramming (Oct3/4, Sox2, Klf4, and c-Myc) to be carried within a single vector [8], providing an efficient means to generate iPSCs from human somatic cells. The advantage of the SeV vector is that it provides cytoplasmic transcription of exogenous genes without nuclear transition, thereby preventing the integration of vector DNA into the genome of a target cell [9]. This minimizes genotoxicity and damage to iPSC chromosomes. SeV vectors can be used to transfect the dividing and nondividing cells of various animal species, such as birds, amphibians, and mammals, including human cells [10].

Dental pulp (DP) cells derived from wisdom teeth are highly efficient starting tissue with which iPSCs can be induced [11]. The cell bank within Gifu University now possesses DP cell lines established from approximately 400 Japanese people, with one line (Cell No. RCB5063) available through the RIKEN BioResource Research Center (BRC) [12]. In addition, the RIKEN BRC also has an iPSC line (Cell No. HPS0077) derived from these DP cells, which was generated using episomal plasmid vectors containing six reprogramming factors (Oct3/4, Sox2, Klf4, L-Myc, Lin28, and p53 shRNA) [7]. The efficiency of iPSC induction from dental pulp cells can be further enhanced using the DLX4 gene [13] or through the use of hypoxic culture conditions [14]. Furthermore, chemically defined, serum-free culture protocols have also been established for the generation of iPSCs from DP cells [15]. Here, we describe a protocol for efficient induction of iPSCs from DP cells using SeV vectors.

2 Materials

2.1 Solutions and Reagents for Inducing iPSCs from Human DP Cells Using SeV Vectors

1. SeV vectors for iPSC induction: SeVdp (KOSM) or SeVdp (KOSM302L) [16] (Provided by the AIST Biotherapeutics Research Team) (*see* **Notes 1** and **2**) (Fig. 1). Vectors should be suspended in DMEM +10% FCS and frozen at −80 °C (*see* **Note 3**).
2. SNL feeder cell medium: DMEM (high glucose) supplemented with 7% fetal bovine serum (FBS), 2 mM L-glutamine, 50 units/mL penicillin, and 50 μg/mL streptomycin. Store at 4 °C for up to 2 weeks (*see* **Note 4**).
3. hiPS medium: Primate the ES cell culture medium supplemented with 5 ng/mL basic fibroblast growth factor (bFGF) (*see* **Note 5**). To prepare 10 μg/mL bFGF solution, dissolve 50 μg of bFGF in 5 mL of PBS containing 0.1% BSA. Make aliquots

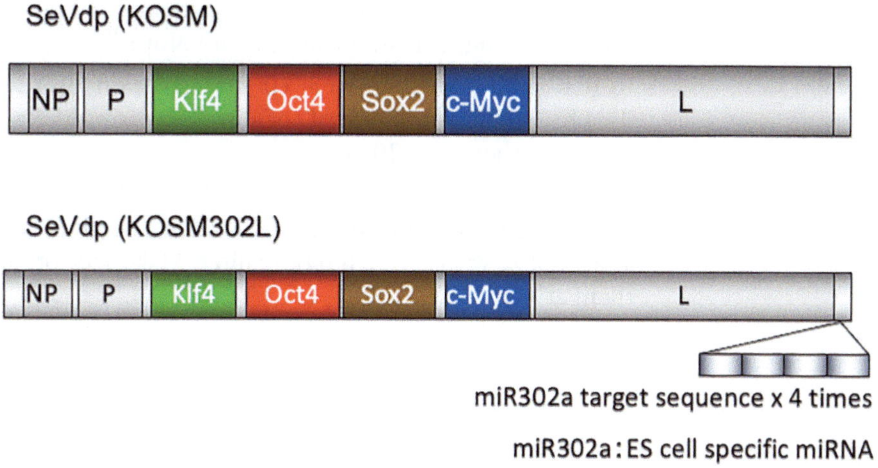

Fig. 1 The structures of SeV vectors used for the induction of iPS cells. The structures of the SeVdp (KOSM) and SeVdp (KOSM302L) vectors provided by the AIST Biotherapeutics Research Team are shown. NP, P, and L indicate the NP, P, and L genes of SeV, which encode a major NP and two subunits of RNA-dependent RNA polymerase (P and L), respectively

and store them at −20 °C. Before use, add 0.25 mL of 10 μg/mL bFGF for every 500 mL of the Primate ES cell culture medium. Store at 4 °C for up to 2 weeks.

4. Phosphate-buffered saline (20 × PBS): Dissolve 160 g of NaCl, 4 g of KCl, 23 g of Na_2HPO_4, and 4 g of KH_2PO_4 in 1 L of H_2O.
5. Human embryonic stem (hES) dissociation solution (CKT solution): Dissolve 0.25 g of trypsin, 10 mL of 1 mg/mL collagenase IV, 20 mL of KnockOut Serum Replacement (KSR), and 1 mL of 0.1 M $CaCl_2$ and make up to 100 mL using 1 × PBS (*see* **Note 6**). Sterilize using a 0.22 μm pore filter. Make aliquots and store at −20 °C. Avoid repeated freezing and thawing. To prepare 1 mg/mL collagenase IV, dissolve 10 mg of collagenase IV in 10 mL of distilled water and sterilize using a 0.22 μm pore filter. Make aliquots and store them at −20 °C. To prepare 0.1 M $CaCl_2$, dissolve 0.11 g of $CaCl_2$ in 10 mL of distilled water and sterilize with a 0.22 μm pore filter. Store at 4 °C.
6. hiPS preservation solution (DAP213 solution): Add 1.43 mL of DMSO, 1 mL of 10 M acetamide, and 2.2 mL of propylene glycol to 5.37 mL of hES medium. Make aliquots and store them at −80 °C. Avoid repeated freezing and thawing. To prepare 10 M acetamide, dissolve 5.9 g of acetamide in 10 mL of distilled water. Store at room temperature. To prepare the hES medium, supplement the DMEM/F12 medium with 20% KSR, 2 mM L-glutamine, 1×10^{-4} M nonessential

amino acids, 1×10^{-4} M 2-mercaptoethanol, 50 units penicillin, and 50 mg/mL streptomycin (*see* **Note 7**).

7. ROCK inhibitor: Prepare a 10 mM solution of Y27632 by dissolving 5 mg in 1.48 mL of distilled water. Make aliquots and store them at −20 °C.
8. 0.05% trypsin–EDTA solution (for SNL cells): Dissolve 0.5 g of porcine trypsin and 0.2 g of $EDTA.Na_4$ in 1 L of 1 × PBS and sterilize with a 0.22 μm pore filter. Make aliquots and store them at −20 °C.
9. 0.25% trypsin–EDTA solution (for dental pulp cells): Dissolve 2.5 g of porcine trypsin and 0.4 g of $EDTA.Na_4$ in 1 L of 1 × PBS and sterilize with a 0.22 μm pore filter. Make aliquots and store them at −20 °C.
10. DP medium: Mesenchymal Stem Cell Growth Medium (MSCGM).
11. DP preservation medium: CELLBANKER cell-freezing media.
12. SNL preservation medium: 10% DMSO in the MEF medium. Mix 1 mL of DMSO, 1 mL of FBS, and 8 mL of DMEM and sterilize with a 0.22 μm pore filter.
13. Mitomycin C solution: 0.2 mg/mL mitomycin C in PBS.
14. 0.1% Gelatin solution: Dissolve 0.5 g of gelatin powder in 500 mL of distilled water and then sterilize in an autoclave. To prepare a gelatin-coated culture dish, add a sufficient volume of 0.1% gelatin solution to cover the entire bottom of the dish. For example, 1, 2, or 5 mL of gelatin solution is used for a 35-, 60-, or 100 mm dish, respectively. Incubate the dish overnight at 37 °C for at least 30 min. Before use, aspirate excess gelatin solution.
15. SNL cells (*see* **Note 8**) [17].
16. Dental pulp cells (*see* **Note 9**) (Fig. 2a) [11, 12].

2.2 SeV Vector Removal and Detection Reagents

1. L527 siRNA #1: 5′-GGUUCAGCAUCAAAAUGAAG-3′.
2. L527 siRNA #2: 5′-UCAUAUUUGAUGCUGAACCAU-3′.
3. RNAi MAX.
4. Opti-MEM.
5. Anti-SeV NP mouse monoclonal antibody #2E4 [8, 16].
6. Fluorescently labeled anti-mouse IgG.
7. 4% Paraformaldehyde (PFA).
8. 0.1% Tween-20 in PBS.
9. 0.5% Bovine serum albumin (BSA) in PBS.
10. DAPI or Hoechst.

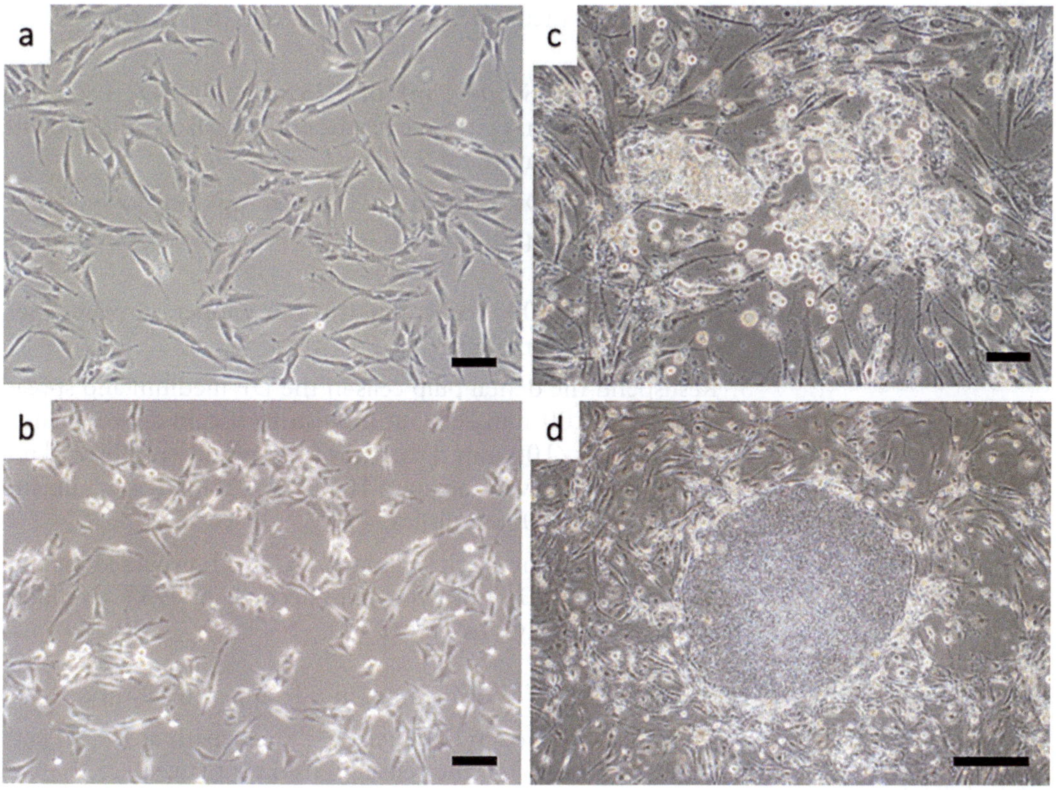

Fig. 2 Generation of iPS cells from dental pulp cells. (**a**) Image of dental pulp cells. (**b**) Dental pulp cells 5 days after infection in which morphological changes and an increase in the cell number can be observed. (**c**) Dental pulp cells 9 days after infection. Morphological changes are now easily recognized and the number of cells has further increased. (**d**) Dental pulp cells 16 days after infection. The cells have flattened and formed a hES-like colony. Scale bars = 100 μm (**a**, **b**, and **c**) and 400 μm (**d**)

3 Methods

3.1 Preparation of Dental Pulp Cells

3.1.1 Thawing Cryopreserved Dental Pulp Cells

1. Prepare 9 mL of DP medium in a 15 mL conical tube.
2. Thaw a frozen vial of dental pulp cells in a water bath at 37 °C until most (but not all) cells are thawed.
3. Transfer the cell suspension to the conical tube prepared in **step 1**. Pipette up and down several times to mix.
4. Centrifuge at 200*g* for 4 min at room temperature and then discard the supernatant.
5. Resuspend the dental pulp cells in 10 mL of DP medium and spread on a 100 mm cell culture dish to obtain approximately 1×10^6 cells/100 mm dish.
6. Incubate the dental pulp cells at 37 °C in a humidified incubator with 5% CO_2 until the cells become 80–90% confluent (typically 3–4 days) (*see* **Note 10**).

3.1.2 Passaging Dental Pulp Cells

1. Discard the medium and wash the cells twice with 10 mL of PBS.
2. Aspirate the PBS, add 1 mL of 0.25% trypsin–EDTA solution per 100 mm dish, swirl to cover the entire surface, and incubate at room temperature for 3 min.
3. Add 9 mL of DP medium and pipette up and down several times to break up the clumps of cells to obtain a suspension of individual cells.
4. Centrifuge at 200*g* for 4 min at room temperature and then discard the supernatant.
5. Resuspend the dental pulp cells in the DP medium and spread them on a 100 mm cell culture dish to obtain approximately 1×10^6 cells/100 mm dish. Cells are typically split by making 1/3–1/4 of the volume up to 10 mL with the DP medium and plating onto 100 mm dishes.
6. Incubate the dental pulp cells at 37 °C in a humidified incubator with 5% CO_2 until the cells become 80–90% confluent (typically 3–4 days) (*see* **Note 11**).

3.2 Preparation of SNL Feeder Cells

3.2.1 Thawing Cryopreserved SNL Feeder Cells

1. Prepare 9 mL of SNL medium in a 15 mL conical tube.
2. Thaw a vial of frozen SNL cells in a water bath at 37 °C until most (but not all) of the cell suspension is thawed.
3. Transfer the cell suspension to the conical tube prepared in **step 1** and mix by pipetting up and down several times.
4. Centrifuge at 200*g* for 4 min at room temperature and then discard the supernatant.
5. Resuspend the SNL cells in 10 mL of SNL medium and spread on a 100 mm cell culture dish to obtain approximately 1×10^6 cells/100 mm dish.
6. Incubate the SNL cells at 37 °C in a humidified incubator with 5% CO_2 until the cells become 80–90% confluent (typically 3–4 days) (*see* **Note 12**).

3.2.2 Passaging SNL Cells

1. Discard the medium and wash the cells once with 10 mL of PBS.
2. Aspirate the PBS, add 1 mL per 100 mm dish of 0.05% trypsin–EDTA solution, swirl to cover the entire surface, and incubate at room temperature for approximately 2 min.
3. Add 9 mL of SNL medium and pipette up and down several times to break up the clumps of cells to obtain a suspension of individual cells.
4. Adjust the volume up to 160 mL with the SNL medium and transfer 1/16 cell suspension volumes in 10 mL of SNL medium to 16 gelatin-coated 100 mm cell culture dishes.

5. Incubate the SNL cells at 37 °C in a humidified incubator with 5% CO_2 until the cells become 80–90% confluent (typically 3–4 days). Cells up to passage number 20 are used for mitomycin C treatment.

3.2.3 Mitomycin C Treatment of SNL Feeder Cells

1. Add 0.5 mL of 0.2 mg/mL mitomycin C solution directly drop by drop to the culture medium of 80–90% confluent SNL cells in a 100 mm dish. Swirl 2–3 times to mix. The final concentration of mitomycin C will be approximately 10 μg/mL.
2. Incubate at 37 °C in a humidified incubator with 5% CO_2 for 2.5 h.
3. After incubation, aspirate the medium containing mitomycin C and wash the cells twice with 10 mL of PBS.
4. Aspirate the PBS, add 1 mL of 0.05% trypsin–EDTA solution per 100 mm dish, swirl to cover the entire surface, and incubate at room temperature for approximately 2 min.
5. Add 5 mL of SNL medium and pipette up and down several times to break up the clumps of cells to obtain a suspension of individual cells.
6. Transfer the cell suspension to a 50 mL conical tube and calculate the cell number.
7. Centrifuge at 200*g* for 4 min at room temperature and then discard the supernatant.
8. To make frozen stocks of mitomycin C–treated SNL cells, resuspend the cells in the SNL preservation medium at 6×10^6 cells/mL and aliquot 1 mL of cells into cryovials (*see* **Note 13**).
9. Place the cryovials into a freezing container and store overnight at −80 °C.
10. Transfer the frozen cryovials into the vapor phase of a liquid nitrogen tank.

3.2.4 Using Cryopreserved Mitomycin C–Treated SNL Feeder Cells

1. Add 0.1% gelatin solution to a culture dish and incubate at room temperature or at 37 °C for at least 30 min. The volume of 0.1% gelatin solution required for a 60 mm dish is 2 mL; for a 100 mm dish, 5 mL; for a 6-well plate, 1 mL/well; for a 12-well plate, 0.5 mL/well; and for a 24-well plate, 0.25 mL/well.
2. Prepare 9 mL of SNL medium in a 15 mL conical tube.
3. Rapidly thaw a frozen cryovial of mitomycin C–treated SNL cells in a water bath at 37 °C until most (but not all) of the cell suspension is thawed.
4. Transfer the cell suspension to the conical tube prepared in **step 2** and pipette up and down several times to mix.

5. Centrifuge at 200*g* for 4 min at room temperature and then discard the supernatant.
6. Resuspend the cells in the SNL medium. Adjust the concentration to 2×10^5 cells/mL when iPSCs are induced or 1×10^5 cells/mL when iPSCs are subcultured. Seed the cells in a gelatin-coated cell culture dish or plate. The volume required for seeding a 60 mm dish is 4 mL; for a 100 mm dish, 10 mL; for a 6-well plate, 2 mL/well; for a 12-well plate, 1 mL/well; and for a 24-well plate, 0.5 mL/well (*see* **Notes 14** and **15**).

3.3 Induction of iPSCs from Dental Pulp Cells

3.3.1 Preparation of Dental Pulp Cells

Day 0

1. Culture dental pulp cells to ~90% confluency in 100 mm dishes (~2×10^6 cells per dish).
2. Aspirate the culture medium and wash twice with 10 mL of PBS.
3. Discard the PBS, add 1 mL of 0.25% trypsin–EDTA solution per dish, swirl to cover the entire surface, and incubate at room temperature for approximately 3 min.
4. Add 9 mL of DP medium and pipette up and down several times to break up the cell clumps to obtain a suspension of individual cells.
5. Transfer the cell suspension to a 50 mL conical tube and calculate the cell number.
6. Centrifuge at 200*g* for 4 min at room temperature and then discard the supernatant.
7. Adjust the concentration to 2.0×10^5 cells/mL using the DP medium and transfer 1 mL to a 12-well plate (*see* **Note 16**).

3.3.2 Sendai Virus Infection

Day 1

1. Rapidly thaw a frozen vial of Sendai virus vector for iPSC induction in a 37 °C water bath and place it on ice once thawed.
2. Dilute the SeV vector with the MSCGM medium to a final volume of 300 μL and adjust the MOI to 5. MOI is adjusted to 3–10 depending on the cell type (*see* **Note 17**).
3. Thoroughly aspirate the culture medium from the cultured dental pulp cells and add the SeV vector–containing medium prepared in **step 2**.
4. Incubate at room temperature for 2 h (preferably in a biosafety cabinet) (*see* **Note 18**).
5. Option: Add 150 μL of MSCGM medium and incubate at 32 °C in a humidified incubator with 5% CO_2 for 2 h (*see* **Note 19**).
6. Add the MSCGM medium to a final volume of 500 μL and incubate at 37 °C overnight in a humidified incubator with 5% CO_2.
7. Prepare mitomycin C–treated SNL feeder cells on a gelatin-coated plate.

Day 2

8. Remove the SeV vector–containing medium from the dental pulp cells and wash with 500 μL PBS (*see* Note 20).
9. Discard PBS, add 200 μL of 0.25% trypsin–EDTA solution per well, and incubate at 37 °C in an incubator with 5% CO_2 for approximately 2 min.
10. Once the cells have detached, add 1 mL of culture medium, and pipette up and down several times to break up the cell clumps to obtain a suspension of individual cells.
11. Transfer the cell suspension to a 15 mL conical tube and calculate the cell number.
12. Centrifuge at 200*g* for 4 min at room temperature and discard the supernatant.
13. Suspend the cells in the hiPS medium and adjust the concentration to 1.0×10^5 cells per mL with the hiPS medium. Seed 0.5–2.0 mL of the cell suspension on mitomycin C–treated SNL feeder cells prepared the day before. Adjust the number of cells seeded per 60 mm cell culture dish such that each contains between 0.5×10^4 and 2.0×10^5 (*see* **Note 21**). Add the hiPS medium to obtain a volume of 4 mL per 60 mm dish.
14. The remaining cells are cryopreserved in the DP preservation medium. These stored cells can be thawed and reused.
15. To check the rate of infection, spread the surplus cells across a 24-well plate and culture them in the DP medium.

After Day 3

16. Replace the medium with a fresh hiPS medium every other day until the colonies are big enough to be picked from the culture dish (approximately 0.5 mm in diameter). Incubate at 37 °C in a humidified incubator with 3% CO_2.
17. Minimal morphological changes are expected 5 days after infection, but the number of cells should be increasing (Fig. 2b) (*see* **Note 22**).
18. Nine days postinfection, the number of cells should increase further, and morphological changes should be seen (Fig. 2c) (*see* **Note 23**).
19. A flat colony is formed approximately 2 weeks after infection (Fig. 2d).
20. Passaging can be performed approximately 2–3 weeks after infection when human ES cell-like colonies have grown to a sufficient size to be picked up (approximately 0.5–1 mm in diameter).

3.4 Picking and Passaging Primary iPS Cell-Like Colonies

1. Seed mitomycin C–treated SNL feeder cells at a density of 0.5×10^5 cells/well on a gelatin-coated 24-well plate on the day before passage.

2. Aspirate the culture medium from a SNL plate and add 0.5 mL of hiPS medium prior to picking the colonies.
3. Aliquot 50 μL hiPS medium per well of 96-well plate or in PCR tubes.
4. Remove the medium from the iPSC-inducing 60 mm dish and wash the colonies with 4 mL of PBS.
5. Aspirate the PBS and add 2 mL of fresh PBS.
6. Observe the colonies to be subcultured under a microscope and use a pen or needle to mark the circumference of the colonies (*see* **Note 24**).
7. Attach a tip to a P20 pipette-man set at 2 μL and use it to pick a colony by poking the edge of the colony with the tip and peeling it from the surface of the dish.
8. Collect the detached colony using the P20 pipette-man and transfer it to a well of the 96-well plate or a PCR tube prepared in **step 3**.
9. Break the colony into small clumps by vigorously pipetting it up and down.
10. Spread the cell suspension into the wells of the 24-well plate containing mitomycin C–treated SNL feeder cells prepared in **step 2** (*see* **Note 25**).
11. Incubate the plates at 37 °C in an incubator with 3% CO_2 until the colonies have grown to a sufficient size to be passaged (approximately 1 mm in diameter).
12. When using SeVdp (KOSM), SeVdp needs to be removed using an siRNA suspension (*see* **Note 26**). If SeVdp (KOSM302L) is used, the abovementioned step is not required as SeVdp is automatically removed [16].
13. Replace media with the hiPS medium after 2 days if siRNA-treated, or every other day if untreated (*see* **Note 27**).
14. Approximately 3–4 weeks after infection, flat human ES cell-like colonies appear, from which the SeV vector can be removed.

3.5 Confirmation of SeV Vector Infection and Elimination (See Note 28)

1. Aspirate the culture medium and wash the cells with PBS.
2. Fix the cells with 4% PFA at room temperature for 10 min.
3. Remove the PFA and wash twice with PBS.
4. Add 0.1% Tween-20 to PBS and leave at room temperature for 5 min.
5. Remove the 0.1% Tween-20 in PBS and wash the cells twice with PBS.
6. Add 0.5% BSA to PBS and leave at room temperature for 10 min.

7. Remove the 0.5% BSA in PBS and add the anti-SeV NP antibody to 0.5% BSA in PBS.
8. Leave at room temperature for 60 min.
9. Remove the anti-SeV NP antibody solution and wash twice with PBS.
10. Add fluorescently labeled anti-mouse IgG solution diluted 500-fold with 0.5% BSA in PBS.
11. Leave at room temperature for 30 min.
12. Remove the fluorescently labeled anti-mouse IgG solution and wash twice with PBS.
13. Stain the nuclei with DAPI or Hoechst and examine under a microscope (*see* **Note 29**) (Fig. 3).

3.6 Passage of iPSCs

1. The day before passage, coat a 6-well plate with gelatin and seed the SNL feeder cells treated with mitomycin C.
2. Aspirate the SNL medium from a SNL plate or dish and replace with 1.5 mL of hiPS medium before passage.
3. Remove the medium from the iPSCs plate and wash with 0.5 mL of PBS.
4. Remove the PBS and add 0.1 mL of hES dissociation solution.
5. Incubate for 1–2 min at 37 °C (*see* **Note 30**).
6. Aspirate the hES dissociation solution and wash with 0.5 mL of PBS.
7. Remove PBS, add 0.5 mL of hiPS medium, and suspend the cells by pipetting up and down with a P-1000 pipette until they resemble small clams (*see* **Note 31**).
8. Transfer the cell suspension to a well of the 6-well plate prepared in **step 1**.
9. Incubate at 37 °C in a humidified incubator with 3% CO_2 until the colonies have grown to a sufficient size to be passaged (approximately 1 mm or greater in diameter). This typically takes 4–7 days after passaging.
10. Replace the medium with a fresh hiPS medium every other day (*see* **Note 32**).
11. If differentiated colonies are observed after passaging, it is necessary to remove them during the culture or prior to passaging and only replate the undifferentiated colonies onto new mitomycin C–treated SNL feeder cell plates.

3.7 Expansion of iPSCs

1. One day prior to passaging, coat a dish or plate with gelatin and seed the SNL feeder cells treated with mitomycin C.
2. Aspirate the SNL medium from an SNL plate or dish and replace with the hiPS medium before passage. The final amount

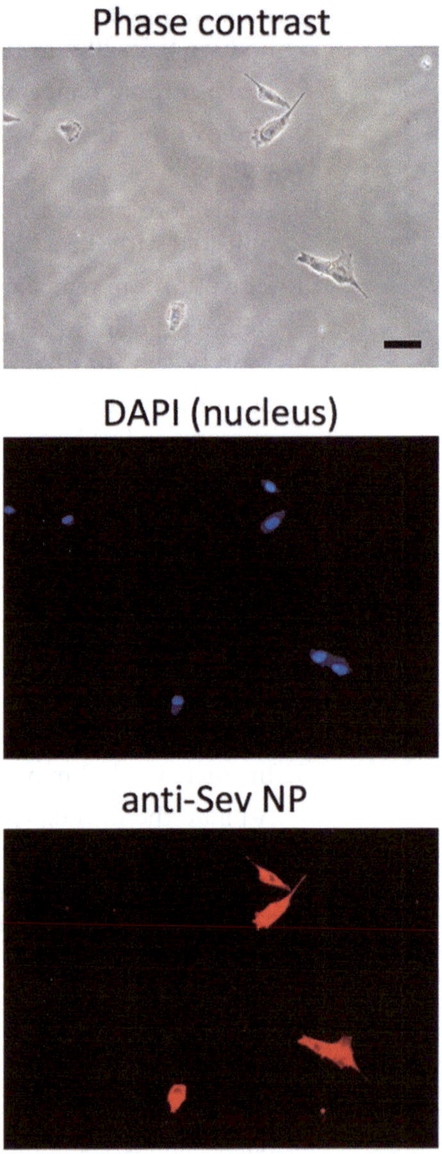

Fig. 3 Immunocytochemistry staining of dental pulp cells with an anti-SeV NP antibody. Dental pulp cells infected by Sendai virus are detected with an anti-SeV NP antibody. The top panel displays a phase-contrast image, the middle panel shows DAPI staining of the cell nuclei, and the bottom panel displays anti-Sav NP antibody staining. Scale bar = 50 μm

of medium required for a 60 mm dish is 4 mL; a 100 mm dish, 10 mL; a 6-well plate, 2 mL; a 12-well plate, 1 mL; and a 24-well plate, 0.5 mL.

3. Remove the medium from the iPSCs and wash with 2 mL of PBS.

4. Remove PBS and replace with the hES dissociation solution. The amount of hES dissociation solution for a 60 mm dish is 0.5–1 mL; a 100 mm dish, 1–2 mL; a 6-well plate, 0.3–0.5 mL; a 12-well plate, 0.1–0.2 mL; and a 24-well plate, 0.1 mL.
5. Incubate at 37 °C for 2–3 min (*see* **Note 30**).
6. Aspirate the hES dissociation solution and wash with 2 mL of PBS.
7. Remove PBS and add another 2 mL of PBS. Most of the SNL feeder cells will be detached during these washes and will be removed by discarding the PBS (*see* **Note 33**).
8. Remove the PBS, add 1 mL of hiPS medium per well of the 6-well plate, and detach the iPS colonies using a cell scraper.
9. Break up the colonies into small clamps by pipetting up and down (*see* **Note 31**).
10. Seed the cell suspension onto the SNL feeder cells prepared in **step 1** at a dilution ratio of about 1/5 to 1/2.
11. Replace the medium with a fresh hiPS medium every other day (*see* **Note 32**).
12. When the colonies grow to a sufficient size (approximately 1 mm or greater in diameter), subculture them once or twice per week.
13. If differentiated colonies are observed after passaging, remove them from the cultures or select only the undifferentiated colonies for passaging and replate them onto new mitomycin C–treated SNL feeder cell plates.

3.8 Preservation of iPSCs

1. Culture iPSCs to confluency in 60 mm dishes.
2. To treat with the ROCK inhibitor, add 10 μL of 10 mM Y27632 to the culture medium (final concentration is 10 μM) and incubate at 37 °C in a humidified incubator with 5% CO_2 for 2 h before harvesting iPSCs (*see* **Note 34**).
3. Prepare a liquid nitrogen bath by safely decanting some into a stainless steel Dewar.
4. Aspirate the medium from the iPSCs and wash with 4 mL of PBS.
5. Aspirate the PBS and add 0.5 mL of hES dissociation solution.
6. Incubate at 37 °C for 2–3 min (*see* **Note 30**).
7. Aspirate the hES dissociation solution and wash with 4 mL of PBS.
8. Remove the PBS and add another 4 mL of PBS. Most of the SNL feeder cells will be removed during these wash steps (*see* **Note 33**).

9. Remove the PBS, add 4 mL of hiPS medium, and use a cell scraper to detach the iPS colonies from the dish.
10. Transfer the cell suspension to a 15 mL conical tube using a P-1000 pipette (*see* **Note 35**).
11. Centrifuge the cells at 200*g* for 4 min at room temperature and discard as much of the supernatant as possible.
12. Suspend the pellet in 0.2 mL of hiPS preservation solution by gently pipetting 2–3 times with a P-1000 pipette (*see* **Note 36**).
13. Transfer the entire cell suspension volume to a cryotube.
14. Immediately immerse the cryotube in liquid nitrogen in the stainless steel Dewar and leave for 1 min to freeze the storage solution.
15. Transfer to a cryopreservation liquid nitrogen tank while still soaked in liquid nitrogen (*see* **Note 37**).

3.9 Thawing Frozen iPSCs

1. One day prior to thawing iPSCs, coat a 60 mm dish with gelatin and seed with mitomycin C–treated SNL feeder cells.
2. Transfer 10 mL of 37 °C prewarmed hiPS medium to a 15 mL conical tube.
3. Remove a cryotube of iPSCs from the liquid nitrogen tank, immediately immerse it in liquid nitrogen in a stainless steel Dewar, and move it to a clean bench (*see* **Note 38**).
4. Remove the cryotube from liquid nitrogen, add 0.8 mL of the prewarmed medium prepared in **step 2** to the cryotube, and quickly thaw by pipetting up and down 5–7 times using a P-1000 pipette.
5. Return the thawed cell suspension to the 15 mL conical tube prepared in **step 2**.
6. Centrifuge at 200*g* for 4 min at room temperature and remove the supernatant.
7. Aspirate the SNL medium from a 60 mm SNL dish prepared the previous day.
8. Add 4 mL of fresh hiPS medium to the cell pellet.
9. Transfer 4 mL of cell suspension to the 60 mm dish seeded with mitomycin C–treated SNL feeder cells.
10. Add 4 μL of 10 mM Y27632 as a ROCK inhibitor to the culture medium (final concentration is 10 μM), and incubate at 37 °C in a humidified incubator with 3% CO_2 (*see* **Note 39**).
11. Replace the medium with a hiPS medium every other day from day 2 until iPSCs grow to a sufficient size to be passaged (approximately 1 mm or greater in diameter) (typically 4–10 days after passaging).

12. Do not move the dish for 48 h after thawing.
13. If only a few undifferentiated colonies are observed after thawing, it is necessary to select only those colonies for passaging and replate them in the same size well on a new mitomycin C–treated SNL feeder cell plate.

4 Notes

1. SeV vectors are available from the Research Center for Stem Cell Engineering, the National Institute of Advanced Industrial Science and Technology (Central 4, Tsukuba, Ibaraki, Japan). Alternatively, the Sendai virus vector kit CytoTune-iPS can be purchased from DNAVEC Corporation.
2. Since SeV vectors can infect mucous membranes, a safety cabinet should be used at all times, together with infection prevention measures, such as wearing gloves, using a virus protection mask, and a protective face shield. Used disposable equipment and liquid waste should always be autoclaved prior to disposal. Any used equipment that cannot be autoclaved should be treated with UV irradiation for 30 min to inactivate the vector.
3. It is advisable to avoid repeatedly freezing and thawing SeV vectors. The vector can be aliquoted after it is first thawed by quickly freezing the aliquotes in liquid nitrogen and storing them at −80 °C.
4. Mouse embryonic fibroblast (MEF) cells can be used instead of SNL. MEF culture medium: DMEM (high glucose) containing 10% FBS, 50 units of penicillin, and 50 mg/mL streptomycin. This media can be stored at 4 °C for up to 2 weeks.
5. bFGF solution should be adjusted to 0.5 mg/mL and stored at −20 °C in 0.5 mL aliquots to prevent the need for repeated cycles of freezing and thawing.
6. The concentration of collagenase IV varies from protocol to protocol. If the iPSCs are easy to peel off, use a lower concentration, but use a higher concentration if they are difficult to remove. Commercial products can be used in place of collagenase IV, such as the Dissociation Solution for human ES/iPSCs.
7. Commercial products are available that include freezing medium for human ES/iPSCs and VPL-A1.
8. SNL feeder cells are available from Dr. Allan Bradley of the Sanger Institute, Cambridge, England, or the European Collection of Cell Cultures, Wiltshire, England. SNL cells are derived as a subclone of the STO cell and stably express the neomycin resistance and LIF genes.

9. Dental pulp cells are available from the Riken Bioresource Center, Ibaraki, Japan. These cells were deposited by our research team at Gifu University Graduate School of Medicine.
10. Do not seed the cells at low density as replicative senescence may be induced.
11. Old dental pulp cells may not be suitable for generating iPSCs due to low induction efficiencies.
12. Do not let the cells get over-confluent or else their use as feeder cells might deteriorate.
13. If used infrequently or in small amounts, frozen stocks of the mitomycin C–treated SNL can be stored with a small number of cells to meet the frequency of use.
14. SNL cells should be evenly spread with small gaps between the cells. They should be ready for use by the next day.
15. When iPSCs are induced, the SNL cells gradually peel off, so more SNL cells might be present at seeding than when passaged. It is desirable to use the mitomycin C–treated SNL cells within 3 days after seeding, and they should be seeded at approximately 1 to 4 × 10^5 cells/mL since the properties of the cells differ depending on the specific frozen stock used. Within 3 days of when they are to be used, these stocks should be thawed and seeded onto a gelatin-coated dish or plate. Older SNL feeder cells (greater than 3 days after mitomycin C treatment) may peel away from the dish during the process of iPSC generation, which takes over 3 weeks.
16. When preparing dental pulp cells in a 12-well plate the day before infection, the standard number of cells to be seeded is approximately 2.0 × 10^5 cells/well, but this can range from 1.0 to 5.0 × 10^5 cells/well depending on the exact cells used.
17. Adjust the MOI depending on the cell. Infection is performed at MOI = 3 to 10. Be careful not to damage the cells when increasing the MOI.
18. Adjust the infection times according to the cells (1–3 h) while making sure not to damage the cells. In the case of human monocytes, cell damage is observed when the infection times exceed 2 h.
19. If the SeV infection efficiency or the iPSC induction efficiency is low, this step may improve the situation. In addition, if it is difficult to establish iPSCs from human dermal fibroblasts or human normal keratinocytes, changing the culture conditions to 38–39 °C in 5% CO_2 until the first colony passage may lead to improvements.

20. Some cells may curl up or peel off easily after infection, which may be due to the highly infectious nature of the SeV vector and subsequent high expression of the transgene. This may be avoided by adjusting the cell density or through the use of collagen-coated plates.
21. Adjust the number of seeded cells according to the cells. It is desirable to adjust the number of cells at several points. A 6-well plate can also be used instead of the 60 mm dish. The number of cells seeded in a 6-well plate is adjusted in the range of about 0.3×10^4–8.0×10^4 cells/well.
22. Although dependent on the specific cells used, colonies should start appearing on day 5 of infection.
23. If the cells proliferate rapidly, show morphological changes, or if colonies are observed, the entire well can be dissociated and passaged. The cells should be harvested in the same volume as SNL or dental pulp cell passages and suspended in primate ESM. Mitomycin C–treated feeder cells (SNL) should be prepared the day before and a half to a quarter of the cells should be seeded. ROCK inhibitors should be used when few colonies appear because they can improve cell engraftment.
24. It is better to choose large colonies or flattened colonies, and it is best to pick single colonies to establish clonal colonies. However, as the selected colonies may not be clones derived from a single cell, recloning may be necessary.
25. Do not break down large clumps into single cells because engraftment from colonies broken up too finely will be poor. The attachment and/or survival of the cells will be improved by passing a 23G or 26G needle through the colony several times to break it up. ROCK inhibitors can also improve cell survival, so they should be used when a few colonies appear or when the colonies are small.
26. L527 siRNA is dissolved in RNase-free H_2O, adjusted to 10 μM and stored at −80 °C.

    Preparation of siRNA suspension to remove SeVdp (for 1 well of the 24-well plate): add 2.5 μL of 10 μM L527 siRNA and 1.25 μL of RNAi MAX to 125 μL of Opti-MEM and incubate at room temperature for 15 min prior to adding the iPSC medium.
27. The siRNA treatment is repeated 3–4 times every 2 days. If the removal of the SeV vector is insufficient, it can readily be removed by breaking up the colony. ROCK inhibitors should be added when the colony is finely divided to improve the attachment and survival of the cells.

28. This section does not include volumes. The sizes of the plates or dishes used for immunostaining can be decided relative to the needs of the experiment.
29. When confirming that the SeV vector has been removed, the undifferentiated properties of the iPSCs should also be confirmed. This can be performed by immunostaining for the undifferentiated markers, such as SSEA-4, Nanog, Tra1–60, Tra1–81, and the differentiation markers SSEA-1 and alkaline phosphatase.
30. The incubation time may vary depending on the cell density and cell quality. Incubate until the edges of the colony are slightly detached. If the colonies of cells are exposed to the dissociation solution for an extended period of time, that is, beyond 1–2 mins, at 37 °C, subsequent PBS washes will cause both the SNL cells and iPS colonies to move away from the culture dishes. It is advisable to observe the colony with the naked eye every minute during incubation with a dissociation solution.
31. When breaking the colony, make sure that each clump contains several hundred cells. Take care not to break up the large clumps too much and make them too fine.
32. When bFGF in the media is effective, the SNL cells display a slightly elongated morphology. If such morphologies are not observed for SNL cells, immediately replace them with fresh media containing bFGF. If SNL cells have peeled off or appear to be deteriorated, more SNL cells can be added directly to the culture medium. The presence of too few or too many SNL feeder cells can be damaging for iPSCs, so care should be taken to maintain an appropriate density.
33. Washing with PBS will cause the SNL cells to peel away and be washed off while leaving the flattened hES-like iPS colonies attached to the dish. If the SNL cells do not readily peel off, tapping the dish or plate can make it easier to detach them.
34. The high osmotic pressure of the cryopreservation solution can significantly reduce the viability of iPSCs. To counteract this, iPSCs should be treated with ROCK inhibitors prior to storage to improve their survival.
35. Do not use a pipette for this step as there is no need to destroy the colonies.
36. Adding the cryopreservation solution to the iPSCs and immersing them in liquid nitrogen should be completed within 10 s as prolonging this time can greatly reduce cell viability.
37. Keep in mind that the cryopreservation solution will thaw even below 0 °C.

38. The process of defrosting the cells and transferring them to 15 mL tubes should be completed in less than 10 s to avoid significant reduction in cell viability.
39. Since the survival rate of cells typically ranges from 30 to 50%, frozen stocks of iPSCs can be made using all cells from a 60 mm dish in one cryotube and replating the stock onto a 60 mm dish.

Acknowledgments

SeVdp (KOSM) and SeVdp (KOSM302L) vectors were kindly provided by Drs. Mahito Nakanishi and Ken Nishimura. Dental pulp cells were provided by Drs. Ken-ichi Tezuka and Toshiyuki Shibata from the shizui project at Gifu University.

References

1. Takahashi K, Yamanaka S (2006) Induction of pluripotent stem cells from mouse embryonic and adult fibroblast cultures by defined factors. Cell 126(4):663–676. https://doi.org/10.1016/j.cell.2006.07.024
2. Takahashi K, Tanabe K, Ohnuki M et al (2007) Induction of pluripotent stem cells from adult human fibroblasts by defined factors. Cell 131(5):861–872. https://doi.org/10.1016/j.cell.2007.11.019
3. Yu J, Vodyanik MA, Smuga-Otto K et al (2007) Induced pluripotent stem cell lines derived from human somatic cells. Science 318(5858):1917–1920. https://doi.org/10.1126/science.1151526
4. Takahashi K, Okita K, Nakagawa M, Yamanaka S (2007) Induction of pluripotent stem cells from fibroblast cultures. Nat Protoc 2(12):3081–3089. https://doi.org/10.1038/nprot.2007.418
5. Sommer CA, Sommer AG, Longmire TA et al (2010) Excision of reprogramming transgenes improves the differentiation potential of iPSCs generated with a single excisable vector. Stem Cells 28(1):64–74. https://doi.org/10.1002/stem.255
6. Okita K, Hong H, Takahashi K, Yamanaka S (2010) Generation of mouse-induced pluripotent stem cells with plasmid vectors. Nat Protoc 5(3):418–428. https://doi.org/10.1038/nprot.2009.231
7. Okita K, Matsumura Y, Sato Y et al (2011) A more efficient method to generate integration-free human iPS cells. Nat Methods 8(5):409–412. https://doi.org/10.1038/nmeth.1591
8. Nishimura K, Sano M, Ohtaka M et al (2011) Development of defective and persistent Sendai virus vector: a unique gene delivery/expression system ideal for cell reprogramming. J Biol Chem 286(6):4760–4771. https://doi.org/10.1074/jbc.M110.183780
9. Lamb RA, Kolakofsky D (2001) In: Knipe DM, Howley PM (eds) Fundamental virology, 4th edn. Lippincott Williams & Wilkins, Philadelphia, pp 689–724
10. Griesenbach U, Inoue M, Hasegawa M, Alton EW (2005) Sendai virus for gene therapy and vaccination. Curr Opin Mol Ther 7(4):346–352
11. Tamaoki N, Takahashi K, Tanaka T et al (2010) Dental pulp cells for induced pluripotent stem cell banking. J Dent Res 89(8):773–778. https://doi.org/10.1177/0022034510366846
12. Takeda T, Tezuka Y, Horiuchi M et al (2008) Characterization of dental pulp stem cells of human tooth germs. J Dent Res 87(7):676–681. https://doi.org/10.1177/154405910808700716
13. Tamaoki N, Takahashi K, Aoki H et al (2014) The homeobox gene *DLX4* promotes generation of human induced pluripotent stem cells. Sci Rep 4:7283. https://doi.org/10.1038/srep07283
14. Iida K, Takeda-Kawaguchi T, Hada M et al (2013) Hypoxia-enhanced derivation of iPSCs from human dental pulp cells. J Dent Res

92(10):905–910. https://doi.org/10.1177/0022034513502204

15. Takeda-Kawaguchi T, Sugiyama K, Chikusa S et al (2014) Derivation of iPSCs after culture of human dental pulp cells under defined conditions. PLoS One 9(12):e115392. https://doi.org/10.1371/journal.pone.0115392
16. Nishimura K, Ohtaka M, Takada H et al (2017) Simple and effective generation of transgene-free induced pluripotent stem cells using an auto-erasable Sendai virus vector responding to microRNA-302. Stem Cell Res 23:13–19. https://doi.org/10.1016/j.scr.2017.06.011
17. McMahon AP, Bradley A (1990) The Wnt-1 (int-1) proto-oncogene is required for development of a large region of the mouse brain. Cell 62(6):1073–1085. https://doi.org/10.1016/0092-8674(90)90385-r

Chapter 12

Directed Differentiation of Neurons from Human iPSCs for Modeling Neurological Disorders

Cheng Wang, Jonas Cerneckis, and Yanhong Shi

Abstract

Human-induced pluripotent stem cell (hiPSC) technology has enabled comprehensive human cell-based disease modeling in vitro. Due to limited accessibility of primary human neurons as well as species-specific divergence between human and rodent brain tissues, hiPSC-derived neurons have become a popular tool for studying neuronal biology in a dish. Here, we provide methods for transcription factor-driven directed differentiation of neurons from hiPSCs via a neural progenitor cell (NPC) intermediate. Doxycycline-inducible expression of neuron fate-determining transcription factors neurogenin 2 (*NGN2*) and achaete-scute homolog 1 (*ASCL1*) enables rapid and controllable differentiation of human neurons for disease modeling applications. The provided method is also designed to improve the reproducibility of human neuron differentiation by reducing the batch-to-batch variation of NPC differentiation and lentiviral transduction.

Key words iPSCs, Stem cells, Neurons, Directed differentiation, Disease modeling, Alzheimer's disease, Neurological disorders

1 Introduction

Derivation of human-induced pluripotent stem cells (hiPSCs) that can be expanded almost indefinitely and differentiated into most terminal cell types of the human body has transformed in vitro research, enabling human cell-based disease modeling [1, 2]. hiPSCs can be derived from patients with disease-relevant genetic backgrounds as well as undergo multiple events of genetic manipulation [3–6]. Differentiation of hiPSCs into cell types implicated in a particular disease enables modeling of the effects of genetic mutations, risk variants, and environmental perturbations on disease progression. hiPSC-derived cellular products are especially useful for studying cell types that cannot be easily obtained from primary human tissues, such as neurons. hiPSC-derived neurons can be used to study genetic neurodevelopmental disorders, sporadic neurodegenerative diseases, and viral infections, among other types of

Koh-ichi Nagata (ed.), *Cerebral Cortex Development: Methods and Protocols*, Methods in Molecular Biology, vol. 2794,
https://doi.org/10.1007/978-1-0716-3810-1_12,

pathology [6–10]. Modeling human diseases with hiPSC-derived neurons has several advantages. First, human iPSC-derived neurons are more likely to faithfully recapitulate human-specific molecular mechanisms of disease as compared to their murine counterparts. Indeed, functional experiments and high-throughput profiling of rodent and human brain tissues have revealed species-specific divergence that hinders the translatability of rodent-based findings to human therapeutic applications [11, 12]. Human cell-based models may be particularly important for studying complex polygenic diseases, such as sporadic Alzheimer's disease, that are challenging to recapitulate in animal models [13–18]. Second, isogenic iPSC lines that differ in disease-relevant genetic variants only can be obtained by CRISPR/Cas9-mediated genetic editing. In this way, modeling of disease-relevant genetic variants can be achieved without the confounding effects associated with patient-to-patient variability. For example, multiple studies have investigated the effects of the apolipoprotein E variants *APOE3* and *APOE4* on the pathogenesis of diseases including Alzheimer's disease and COVID-19 using isogenic iPSC-derived neurons [9, 10, 19, 20]. Third, given the almost unlimited potential for iPSC expansion, large quantities of iPSC-derived neurons can be obtained for drug screening applications to accelerate the pipeline of drug development [21]. Finally, the ongoing development of iPSC biobanks and well-characterized reference iPSC lines enables large-scale population-based screening applications and highly reproducible collaborative studies, respectively [22–24].

In this chapter, we describe a protocol for the directed differentiation of human neurons from hiPSCs via a neural progenitor cell (NPC) intermediate. We begin by providing a detailed protocol for hiPSC maintenance because high quality of iPSCs is critical for the successful differentiation of iPSC-derived neurons [25]. Then, NPCs are differentiated from hiPSCs by dual SMAD inhibition (dSMADi), which promotes rapid specification of neuroectoderm and emergence of NPCs in as few as 7 days [26]. NPCs are subsequently transduced with a lentivirus carrying a doxycycline-inducible (Tet-On) cassette for expression of two neuron fate-determining transcription factors: neurogenin 2 (*NGN2*) and achaete-scute homolog 1 (*ASCL1*). NGN2 and ASCL1 are basic helix-loop-helix transcription factors that can access closed chromatin, indicating their pioneering activity, and initiate neuron-specific gene expression programs [27, 28]. Transcription factor-driven directed differentiation results in accelerated specification of neuronal fate and increased neuron purity [29]. Moreover, using a doxycycline-inducible system enables expansion of the transduced NPCs before induction of terminal neuron differentiation. Therefore, repeated transduction of the same line of NPCs is not required, whereas a single batch of transduced NPCs may be stocked and used for multiple experiments so that batch-to-batch variability is reduced [25]. We also provide a protocol for lentiviral production, concentration, and storage. Our transfection reagent made in-house is cost-effective, allowing the lentivirus production protocol to be scaled as needed. Lentiviral particles are concentrated by ultracentrifugation to remove the HEK293T medium before transduction so that serum used for HEK293T culture is not transferred to the fully-defined NPC culture medium. Finally, long-term storage of lentiviral particles enables multiple rounds of NPC transduction without the need for repeated production and titration of lentiviral particles.

2 Materials

1. Coating solutions:

   1% (v/v) Matrigel (Corning, Catalog #354230) in DMEM/F12 basal medium.

   20 μg/mL poly-L-ornithine (PLO) solution in DPBS (without Ca++/Mg++) (Thermo Fisher, Catalog #14190250).

   1 μg/mL laminin solution (Thermo Fisher, Catalog #23017-015) in DPBS.
2. mTeSR™1 complete medium (STEMCELL Technologies, Catalog #85850).
3. Neural induction medium-1 (NIM-1): DMEM/F12, 1× N2 supplement, 1× B27 supplement (serum-free), 1× NEAA, 1× GlutaMAX, 10 ng/mL epidermal growth factor (EGF), 10 ng/mL fibroblast growth factor (FGF), 0.1 μM retinoic acid (RA), 4 μM CHIR99021, 3 μM SB431542, and 2 μM dorsomorphin (*see* **Note 1**).
4. Neural induction medium-2 (NIM-2): DMEM/F12, 1× N2 supplement, 1× B27 supplement (serum-free), 1× NEAA, 1× GlutaMAX, 10 ng/mL EGF, 10 ng/mL FGF, 0.1 μM RA, 4 μM CHIR99021, and 3 μM SB431542 (*see* **Note 1**).
5. Neural progenitor maintenance medium (NPMM): DMEM/F12, 1× N2 supplement, 1× B27 supplement (serum-free), 1× NEAA, 1× GlutaMAX, 10 ng/mL EGF, 10 ng/mL FGF, 0.1 μM RA, 2 μM SB431542, and 3 μM CHIR99021 (*see* **Note 1**).
6. HEK293T culture medium (HCM): DMEM, 10% (v/v) fetal bovine serum (FBS), 2 mM L-glutamine (*see* **Note 2**).
7. Cell dissociation reagents: UltraPure™ 0.5M EDTA (Thermo Fisher, Catalog #15575020), Trypsin-EDTA (0.25%) (Thermo Fisher, Catalog #25200114), TrypLE™ Express (Thermo Fisher, Catalog #12604013), Accutase (Sigma-Aldrich, Catalog #A6964).

8. Plasmids:

   pLVX-UbC-rtTA-Ngn2:2A:Ascl1 (UNA) (Addgene, Catalog #127289).

   pMD2.G (Addgene, Catalog #12259).

   psPAX2 (Addgene, Catalog #12260).

9. Transfection solutions:

   2× HBS: 280 mM NaCl, 10 mM KCl, 1.5 mM Na_2HPO_4, 12 mM glucose, 50 mM HEPES.

   10× TE buffer: 1 M Tris-HCl, pH 7.5, 0.5 M EDTA (pH 8.0).

   1× TE buffer: 10 times dilution of the 10× TE buffer, pH 7.9.

   $CaCl_2$ solution: 2 M $CaCl_2$.

10. Neuron differentiation medium (NDM): DMEM/F12, 1× N2 supplement, 1× B27 supplement (serum-free), 1× NEAA, 1× GlutaMAX, 5 μM DAPT (γ-secretase inhibitor), and 2 μg/mL doxycycline (*see* **Note 1**).

11. Neuron maturation medium (NMM): BrainPhys™ neuronal medium (STEMCELL Technologies, Catalog #05790), 1× N2 supplement, 1× B27 Supplement (serum-free), 100 μg/mL dibutyryl cAMP, 10 ng/mL glial cell-derived neurotrophic factor (GDNF), and 10 ng/mL brain-derived neurotrophic factor (BDNF) (*see* **Note 1**).

12. Blocking buffer: 5% (v/v) donkey serum and 0.1% (v/v) Triton™ X-100 in DPBS.

3 Methods

3.1 iPSC Culture and Passaging

Here, we describe the protocol for feeder-free iPSC culture in six-well plates.

3.1.1 Preparation of Matrigel-Coated Six-Well Plates

1. Prepare 1% Matrigel coating solution: take one aliquot (0.5 mL) of Matrigel from a −80 °C freezer and place the aliquot on ice for 30 minutes (min) to 1 hour (h) to thaw completely. Transfer the thawed Matrigel into 50 mL of cold DMEM/F12 medium and invert the tube several times to mix thoroughly. Keep the 1% Matrigel coating solution at 4 °C for up to 1 month.
2. Prepare Matrigel-coated plates: add 1 mL of the Matrigel coating solution into each well of a six-well plate and incubate the plate at 37 °C for 1–2 h before seeding iPSCs (*see* **Note 3**).

3.1.2 Preparation of mTeSR™1 Complete Medium

1. Take out one bottle of the mTeSR™1 supplement (5×) from a −20 °C freezer and thaw the mTeSR™1 supplement (5×) at 2–8 °C overnight prior to use.

2. Add 100 mL of the mTeSR™1 supplement (5×) to 400 mL of cold mTeSR™1 basal medium to prepare the mTeSR™1 complete medium.
3. Aliquot the mTeSR™1 complete medium into 50 mL aliquots. The aliquots may be kept at 4 °C for up to 2 weeks or frozen and stored at −20 °C for up to 6 months (*see* **Note 4**).

3.1.3 Thawing hiPSCs

1. Take out one cryovial of hiPSCs from the liquid nitrogen tank and immediately place the cryovial into a 37 °C water bath. Swirl the cryovial gently in the water bath to ensure rapid thawing, which should take about 2 min (*see* **Note 5**).
2. Gently transfer hiPSCs from the cryovial into a 15 mL conical tube and slowly add 9 mL of cold DMEM into the tube.
3. Centrifuge the cells at 300 *g* for 3 min.
4. Aspirate the supernatant and resuspend the cells in the mTeSR™1 complete medium supplemented with 10 μM of ROCK inhibitor (Y-27632, RI).
5. Count the number of viable cells using a hemocytometer and seed the cells onto Matrigel-coated six-well plates at a density of 2×10^5–1×10^6 of viable cells per well. Add 2 mL of the mTeSR™1 complete medium supplemented with 10 μM of RI to each well of a six-well plate (*see* **Note 6**).
6. Incubate the cells at 37 °C in a 5% CO_2 incubator. Change the medium every day.

3.1.4 Passaging hiPSCs

Observe hiPSC colonies under the microscope each time when changing the medium. hiPSC colonies are ready for passaging when the cells reach approximately 80% confluency:

1. Aspirate the culture medium and wash the cells once with 1 mL of DPBS.
2. Add 1 mL of 0.5 mM EDTA solution into each well of a six-well plate to dissociate iPSC colonies.
3. Incubate the plate at room temperature for 4–5 min. Observe iPSC colonies under the microscope after incubation. If the iPSC colonies exhibit fractured morphology, proceed to the next step. If the iPSC colonies still exhibit compact morphology, incubate the cells with EDTA for a longer period of time and observe the iPSC colonies every 2 min until the colonies exhibit fractured morphology.
4. Carefully remove the EDTA solution after incubation (*see* **Note 7**).
5. Gently detach the iPSC colonies by flushing the well surface with 1 mL of the mTeSR™1 complete medium (*see* **Note 8**).

6. Add 2 mL of the mTeSR™1 complete medium into each well of a six-well plate prior to cell seeding.
7. Add 1–2 drops of the iPSC colony suspension into each well using a 1 mL pipette. Shake the plate to distribute the cells evenly.

3.2 Neural Progenitor Cell (NPC) Differentiation, Passaging, and Freezing

3.2.1 NPC Differentiation

1. Prepare Matrigel-coated plates as described in Subheading 3.1.1.
2. Day 0: Add 2 mL of the mTeSR™1 complete medium supplemented with 10 μM RI (mTeSR™1 + RI) into each well of iPSCs and incubate for 1–2 h at 37 °C prior to cell dissociation. After incubation, aspirate the medium and wash the cells once with 1 mL of DPBS for each well. Add 1 mL of the Accutase cell detachment solution into each well and incubate the plate at 37 °C for 3–5 min to dissociate the iPSC colonies.
3. Gently tap the plate to detach the iPSC colonies, add 1 mL DMEM/F12 medium to each well, and detach the remaining cells by gentle pipetting 1–2 times. Transfer the iPSC colony suspension into a 15 mL conical tube and pipette the suspension 2–3 times to dissociate the iPSC colonies into single cells. Centrifuge the cells at 200 *g* for 2 min. Aspirate the supernatant and resuspend the cells in 2 mL of mTeSR™1 + RI. Count viable cells and seed 1.0×10^5 iPSCs into each well of a six-well plate in mTeSR™1 + RI.
4. Day 1 and Day 2: aspirate the culture medium and add 2 mL NIM-1 into each well.
5. Day 3 to Day 7: Replace NIM-1 with NIM-2. Change the medium with fresh NIM-2 daily for 5 days. From Day 3 to Day 4, add 3 mL NIM-2 into each well; from Day 5 to Day 7, add 4–6 mL NIM-2 into each well of a six-well plate.
6. Day 8: NPCs should become confluent by Day 8 and need to be passaged. Aspirate the culture medium and wash the cells once with 1 mL DPBS. Add 1 mL of the TrypLE dissociation reagent into each well and incubate the cells at 37 °C for 5 min. After incubation, gently tap the plate to detach NPCs, add 1 mL DMEM/F12 medium to each well, and detach the remaining cells by gentle pipetting. Transfer the cell suspension into a 15 mL conical tube and pipette the suspension 2–3 times to dissociate NPCs into single cells. Centrifuge the cells at 200 *g* for 2 min. Aspirate the supernatant and resuspend the cells in 5 mL NPMM supplemented with 10 μM of RI (NPMM+RI). Count viable cells and transfer the NPCs into a T-75 suspension flask at a density of 2×10^5 cells/mL in 15 mL of NPMM +RI (*see* **Note 9**). Incubate the flask on a shaker at 37 °C in a 5% CO_2 incubator (*see* **Note 10**).

7. Day 9: Change the medium to remove RI. Check the NPCs under the microscope; neurospheres should have formed by now.
8. Beyond Day 9: To maintain the neurospheres, change half of the culture medium with fresh NPMM every other day (*see* **Note 11**).

3.2.2 NPC Passaging

Neurospheres should be passaged every 5–7 days although NPC growth rate may be variable across different iPSC lines.

1. Transfer the neurospheres into a 15 mL conical tube and centrifuge at 100 *g* for 1 min.
2. Aspirate the supernatant, add 3 mL of the Accutase cell detachment solution, and incubate the cells at 37 °C for 5 min. Gently tap the tube several times during incubation.
3. After incubation, gently pipette the neurospheres 3–5 times to dissociate the spheres into single cells.
4. Add 6 mL of DMEM/F12 into the single-cell suspension and centrifuge the cells at 200 *g* for 2 min.
5. Aspirate the supernatant and resuspend the cells in 5 mL NPMM+RI.
6. Count the cells and transfer the cell suspension into a new T-75 flask at a density of 2×10^5 cells/mL in NPMM+RI.
7. Incubate the flask on a shaker at 37 °C in a 5% CO_2 incubator.
8. Change the medium after 24 h to remove RI.

3.2.3 NPC Freezing

1. Dissociate neurospheres into single cells as described in **steps 1–4** of Subheading 3.2.2.
2. Resuspend NPCs in NPMM supplemented with 10% (v/v) DMSO (2×10^6–5×10^6 cells/mL) and aliquot 1 mL of the cell suspension into cryovials.
3. Place the cryovials into a Mr. Frosty™ Freezing Container and transfer the container to a −80 °C freezer. After at least 4 h, transfer the cryovials into a liquid nitrogen tank for long-term storage.

3.3 Preparation of UNA Lentivirus for Neuron Differentiation

3.3.1 HEK293T Cell Seeding for Transfection

1. Passage HEK293T cells when the cells reach 80% confluency (*see* **Note 12**).
2. Aspirate the culture medium and add 1 mL of 0.25% trypsin into each 10 cm dish. Incubate the cells at room temperature for 10–20 s. Carefully aspirate trypsin and incubate the cells at 37 °C for another 3 min.
3. Add 3 mL of HCM into each dish and transfer the cells into a 50 mL conical tube by gentle pipetting.

4. Transfer the HEK293T cell suspension into 10 cm dishes containing 9 mL of HCM. HEK293T cells should be split at a ratio of 1:2.5–1:3 (*see* **Note 13**).

3.3.2 HEK293T Cell Transfection

Transfection should be performed less than 24 h after seeding the cells as described in Subheading 3.3.1. The following transfection protocol is formulated for transfecting four 10 cm plates of HEK293T cells at 80% confluency.

1. Add 2.1 mL of 2× HBS into a 15 mL conical tube (tube A).
2. Dilute 31.5 μg of psPAX2, 10.5 μg of PMD2.G, and 42 μg of UNA plasmids in 1.764 mL of the 1× TE buffer in another 15 mL conical tube (tube B) and mix the solution by pipetting.
3. Slowly add 252 μL of $CaCl_2$ solution into the tube B, mixing the solution after each drop by gently shaking the tube (*see* **Note 14**).
4. Transfer the solution from tube B to tube A drop-by-drop. Mix the solution by pipetting after every several drops (*see* **Note 15**).
5. Incubate the prepared transfection solution at room temperature for 30 min.
6. Add 1 mL of transfection solution into each dish of HEK293T cells drop by drop and swirl the dish to mix the solution evenly (*see* **Note 16**).
7. Change the medium 5–6 h after transfection. Aspirate the culture medium and add 7 mL fresh HCM into each dish.

3.3.3 Lentivirus Collection and Purification

1. Transfer the lentivirus-containing medium into a 50 mL conical tube 48 h after HEK293T cell transfection (*see* **Note 17**).
2. Centrifuge the virus-containing medium at 600 *g* for 10 min.
3. After centrifugation, carefully filter the virus-containing medium using a syringe with a 0.22 μm filter.
4. Prepare ultracentrifuge tubes and a weight scale in the tissue culture hood. Label the ultracentrifuge tubes and transfer the tubes into ultracentrifuge buckets.
5. Transfer the virus-containing medium into the ultracentrifuge tubes. Weigh each bucket and balance the buckets by adding DPBS as needed (*see* **Note 18**).
6. Centrifuge the virus-containing medium at 40,000 *g* at 4 °C for 2 h.
7. Place a paper towel in the tissue culture hood and turn on the UV light for 10 min to sterilize the towel.

8. Take out the ultracentrifuge tubes, pour off the supernatant into a 10% bleach solution, and place the ultracentrifuge tubes upside down on the paper towel to dry the remaining medium.
9. Resuspend the lentiviral pellet using 60 μL PBS (*see* **Note 19**).
10. Aliquot the lentivirus into 1.5 mL tubes (10 μL/tube). Store the aliquots at −80 °C (*see* **Note 20**).

3.3.4 Virus Titration

1. Day 0: Seed 0.5×10^5 of HEK293T cells into each well of a 24-well plate (a total of four wells) in 500 μL of HCM.
2. Day 1: Prepare fivefold serial dilutions of the lentivirus in 1.5 mL tubes containing 500 μL of fresh HCM supplemented with 4 μg/mL polybrene (*see* **Note 21**). Aspirate the culture medium and add the lentivirus-containing medium into each well of HEK293T cells (*see* **Note 22**).
3. Day 2: Replace the lentivirus-containing medium with 500 μL of fresh HCM supplemented with 2 μg/mL of doxycycline.
4. Day 4: Wash the cells with DPBS, add 500 μL of 4% paraformaldehyde solution into each well, and incubate the plate at room temperature for 10–15 min to fix the cells. Wash the cells with 500 μL of DPBS three times. Add 500 μL of the Blocking Buffer into each well and incubate the plate at room temperature for 1 h. Prepare the primary antibody solution by adding the desired concentration of an anti-NGN2 primary antibody into the blocking buffer. Remove the blocking buffer and add 200 μL of the primary antibody solution into each well. Incubate the plate at 4 °C overnight.
5. Day 5: Prepare the secondary antibody solution by adding a secondary antibody at the desired concentration into DPBS containing a nuclear stain (for example, DAPI). Remove the primary antibody solution and wash the cells with 500 μL of DPBS three times. Add 200 μL of the secondary antibody solution into each well. Incubate the plate at room temperature for 1 h protected from light. Remove the secondary antibody solution and wash the cells with 500 μL of DPBS three times. Add the mounting solution into each well until it covers the entire surface area. Using a fluorescent microscope, calculate the percentage of $NGN2^+$ cells in each well. Determine the lentiviral titer defined here as transforming units per milliliter (TU/mL). Titer $= 1.0 \times 10^5/$ (virus aliquot volume) $\times$ ($NGN2^+$ cell percentage) $\times$ 1000.

3.4 Lentiviral Transduction of NPCs

1. Day 0: Thaw one vial of NPCs by incubating it in a 37 °C water bath (*see* **Note 23**). Transfer the cells into a 15 mL conical tube and add 9 mL DMEM. Centrifuge the cells at 200 *g* for 2 min. Aspirate the supernatant and resuspend the cells in 1 mL NPMM+RI. Count and transfer the cells into a T-25 flask at a

density of 2 × 10^5 cells/mL in NPMM+RI. Place the flask on a shaker at 37 °C in a 5% CO_2 incubator.

2. Day 1: Change medium to remove RI. Culture the neurospheres for 3–4 days to allow the cells to recover from freezing.
3. Day 5: Dissociate the neurospheres into single cells as described in **steps 1–4** of Subheading 3.2.2. Resuspend the cells in 1 mL of NPMM+RI. Count the cells and seed NPCs onto a Matrigel-coated six-well plate at a density of 2 × 10^4 cells/cm^2.
4. Day 6: Replace the culture medium with 2 mL of NPMM supplemented with 4 μg/mL polybrene and add the UNA virus directly into the NPC culture. Calculate the desired volume of lentivirus by referring to the virus titer of each batch of the virus.
5. Day 7: Replace the culture medium with 2 mL of fresh NPMM.
6. Day 9: Replace the culture medium with 2 mL of fresh NPMM supplemented with 2 μg/mL puromycin. Maintain puromycin selection for 7 days, changing the medium every 2 days (*see* **Note 24**).
7. Day 16: Remove the culture medium and add 1 mL of the Accutase cell detachment solution into each well of a 6-well plate. Incubate the cells at 37 °C in a 5% CO_2 incubator for 5 min. Add 2 mL of DMEM/F12 into each well and pipette gently to dissociate the NPCs. Transfer the cells into a 15 mL conical tube and add 6 mL of DMEM/F12. Centrifuge the cells at 200 *g* for 2 min. Aspirate the supernatant and resuspend the cells in 1 mL NPMM+RI. Count the cells and transfer the cell suspension into a T-25 flask at a density of 2 × 10^5 cells/mL in NPMM+RI.
8. Beyond Day 16: UNA-transduced NPCs (UNA-NPCs) can be expanded and stored in a liquid nitrogen tank before proceeding to neuron differentiation.

3.5 Neuronal Differentiation and Maturation

3.5.1 Neuronal Differentiation

1. Thaw one cryovial of UNA-NPCs and culture the cells for 3–4 days (*see* **Note 25**).
2. Day 0: Dissociate the neurospheres into single cells as described in Subheading 3.2.2. For neuronal differentiation, seed the UNA-NPCs into a Matrigel-coated 6-well plate at a density of 2 × 10^5 cells/cm^2.
3. Replace the culture medium with 2 mL of fresh NDM. Change medium every 2 days for 5–7 days.
4. Day 6: Cells with neuronal morphology should be apparent at this stage (Fig. 1a). Neurons may be split and seeded into poly-L-ornithine/laminin (PLO/laminin)-coated plates as

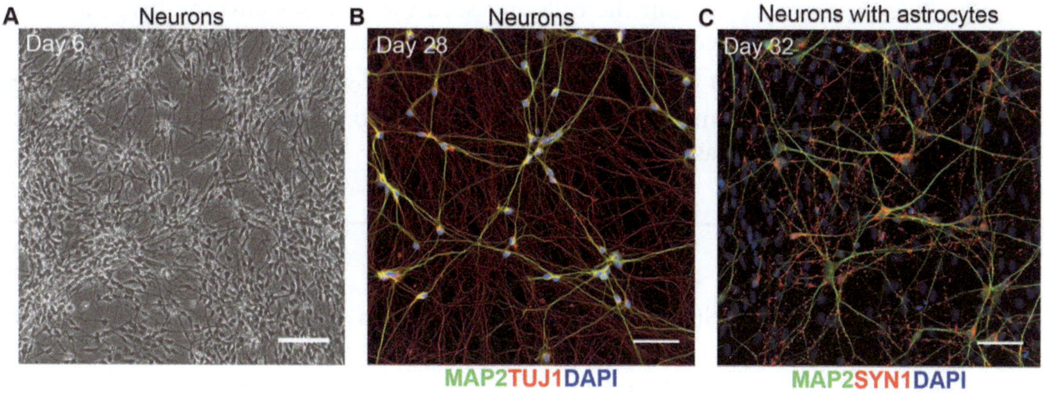

Fig. 1 (**a**) A representative bright field image of neurons derived from NPCs at day 6 of differentiation. Scale bar, 100 μm. (**b**) A representative image of neuron immunostaining for neuronal markers microtubule-associated protein 2 (MAP 2) and tubulin β3 (TUJ1) at day 28 of differentiation. Scale bar, 50 μm. (**c**) A representative image of neuron immunostaining for neuronal markers MAP 2 and synapsin 1 (SYN1) after co-culture with iPSC-derived astrocytes for 28 days (day 32 of differentiation). Scale bar, 50 μm

described in Subheading 3.5.2 or co-cultured with astrocytes for further maturation as described in our published study [9].

3.5.2 Neuronal Splitting and Maturation

1. To prepare PLO/laminin-coated plates, dilute PLO to a final concentration of 20 μg/mL in DPBS. For each well of a 96-well plate, add 50 μL of the PLO solution and incubate the plate at 37 °C overnight.
2. Aspirate the PLO solution, wash the wells with sterile water three times, and dry the wells completely under the laminar flow of a tissue culture hood.
3. Thaw laminin at 2–8 °C and dilute it to a final concentration of 10 μg/mL in DPBS.
4. Add 50 μL of the laminin solution into each PLO-coated well of the 96-well plate and incubate the plate at 37 °C for at least 1 h.
5. To dissociate neurons, aspirate the NDM and add 1 mL of TrypLE into each well of a 6-well plate of neuronal culture.
6. Incubate the plate at 37 °C for 5 min.
7. Add 1 mL of DMEM/F12 into each well and tap the plate to detach neurons. Transfer the cells into a 15 mL conical tube and dissociate the neurons into single cells by gentle pipetting 2–3 times.
8. Add another 5 mL of DMEM/F12 and centrifuge the cells at 300 *g* for 3 min.
9. Aspirate the supernatant and resuspend the cells in 1 mL of NMM supplemented with 10 μM of RI.

10. Count the cells and seed the neurons into the PLO/laminin-coated plate at the desired density (*see* **Note 26**).
11. Change half of the NMM every 3–4 days. Neurons can be maintained in NMM for up to 6 weeks and co-cultured with astrocytes (Fig. 1b, c).

4 Notes

1. Store the formulated medium at 4 °C protected from light (e.g., by wrapping the bottle with aluminum foil) for up to 2 weeks.
2. The HCM may be supplemented with antibiotics and antimycotics to prevent bacterial and fungal contamination, respectively.
3. Matrigel-coated plates may be stored in a 37 °C incubator for up to 1 week. To prevent evaporation of the Matrigel solution, add 1–2 mL of DMEM into each well of a 6-well plate.
4. The mTeSR™1 Complete Medium should be thawed only once.
5. Avoid submerging the cap of the cryovial into the water to prevent contamination. Wipe the cryovial with a 70% ethanol solution before opening.
6. To exclude dead cells, mix a small volume of cell suspension with trypan blue before cell counting.
7. Prolonged incubation with EDTA will result in cell detachment and cell loss when EDTA solution is removed.
8. Use a 1 mL pipette to gently blow off the iPSC colonies moving in a circle from the edge to the center of the wells.
9. For the initial 6 passages, NPCs should be treated with 10 μM RI for 24 h after dissociation to prevent excessive cell death.
10. Culturing NPCs on a shaker will promote neurosphere formation.
11. When changing the medium, allow the flask to stand still until the neurospheres sink to the bottom of the flask. Then, aspirate the culture medium and add 8 mL of fresh NPMM.
12. We use at least four 10 cm plates of HEK293T cells for packaging lentivirus.
13. HEK293T cells seeded at a ratio of 1:2.5 to 1:3 should reach 80% confluency in 18 to 20 h.
14. The concentrations of the plasmids are diluted to 1 μg/mL. The volume of the transfection solution is prepared at a ratio of 4.2× for transfecting four dishes of cells. $CaCl_2$ solution should be added slowly to prevent plasmid DNA precipitation.

15. After mixing the contents of tubes A and B, the solution should turn slightly turbid but no precipitation should occur.
16. The culture medium does not need to be changed at this step.
17. The lentivirus may also be collected for a second time, 72 h after HEK293T cell transfection. If the second collection is desired, add 6 mL of fresh HCM after the first collection and collect the virus-containing medium again after 24 h. Before discarding the plates, add 10 mL of 10% bleach solution into each plate to inactivate any remaining virus. The lentivirus-containing medium can be kept at 4 °C overnight before proceeding to ultracentrifugation to allow for two rounds of collection.
18. Handle the lentivirus-containing medium carefully and decontaminate plasticware and medium with a 10% bleach solution before discarding the lentiviral waste.
19. We usually use 60 μL of DPBS to resuspend a lentiviral pellet collected from four 10 cm plates. However, the pellet size will depend on the volume of the lentiviral-containing medium that is used in each ultracentrifuge tube. To resuspend the lentiviral pellet, wash the bottom of the ultracentrifuge tube 20 times with DPBS by pipetting. Avoid introducing bubbles. Although the lentiviral pellet will not be visible, the resuspended lentiviral suspension should appear cloudy.
20. The lentivirus aliquots should be thawed only once. Repeated thawing will result in a substantial decrease in the lentiviral titer.
21. To prepare fivefold serial dilution of lentivirus, add 7 μL lentivirus into 63 μL of cell culture medium; then, add 2 μL, 10 μL, or 50 μL of the diluted virus into each tube of HCM.
22. The HEK293T cell number should be 1.0×10^5 per well on Day 1. Use this estimate to calculate the viral titer as described in **step 5** of Subheading 3.3.4.
23. NPCs of passage 3 or earlier should be used for transduction with the UNA virus.
24. The cells may need to be passaged during puromycin selection if the culture becomes confluent.
25. NPCs of passage 6 or earlier should be used for neuronal differentiation.
26. Neurons can be seeded at a density of 8×10^4–3.2×10^5 cells/cm^2.

Acknowledgments

The authors would like to thank Louise and Herbert Horvitz, the Christopher Family, the Judy and Bernard Briskin Fund, and the Sidell Kagan Foundation for their generosity and forethought. This work was supported by the National Institute of Aging of the National Institutes of Health R01 AG056305, RF1 AG061794, R01 AG072291, and RF1 AG079307 to Y.S. J.C. is a predoctoral scholar in the Stem Cell Biology and Regenerative Medicine Research Training Program of the California Institute for Regenerative Medicine (CIRM).

References

1. Takahashi K et al (2007) Induction of pluripotent stem cells from adult human fibroblasts by defined factors. Cell 131(5):861–872
2. Yu J et al (2007) Induced pluripotent stem cell lines derived from human somatic cells. Science 318(5858):1917–1920
3. Li L et al (2018) Modeling neurological diseases using iPSC-derived neural cells: iPSC modeling of neurological diseases. Cell Tissue Res 371(1):143–151
4. Rowe RG, Daley GQ (2019) Induced pluripotent stem cells in disease modelling and drug discovery. Nat Rev Genet 20(7):377–388
5. Shi Y et al (2017) Induced pluripotent stem cell technology: a decade of progress. Nat Rev Drug Discov 16(2):115–130
6. Vadodaria KC et al (2020) Modeling brain disorders using induced pluripotent stem cells. Cold Spring Harb Perspect Biol 12(6): a035659
7. Cui Q et al (2022) Compound screen identifies the small molecule Q34 as an inhibitor of SARS-CoV-2 infection. iScience 25(1): 103684
8. Giacomelli E et al (2022) Human stem cell models of neurodegeneration: from basic science of amyotrophic lateral sclerosis to clinical translation. Cell Stem Cell 29(1):11–35
9. Wang C et al (2021) ApoE-isoform-dependent SARS-CoV-2 neurotropism and cellular response. Cell Stem Cell 28(2):331–342 e5
10. Wang C et al (2018) Gain of toxic apolipoprotein E4 effects in human iPSC-derived neurons is ameliorated by a small-molecule structure corrector. Nat Med 24(5):647–657
11. Fang R et al (2022) Conservation and divergence of cortical cell organization in human and mouse revealed by MERFISH. Science 377(6601):56–62
12. Hodge RD et al (2019) Conserved cell types with divergent features in human versus mouse cortex. Nature 573(7772):61–68
13. Knopman DS et al (2021) Alzheimer disease. Nat Rev Dis Prim 7(1):33
14. Long JM, Holtzman DM (2019) Alzheimer disease: an update on pathobiology and treatment strategies. Cell 179(2):312–339
15. Cerneckis J, et al (2023) Pushing the boundaries of brain organoids to study Alzheimer's disease. Trends Mol Med 29(8):659–672
16. Cerneckis J, Shi Y (2023) Modeling brain macrophage biology using human iPSC-derived neuroimmune organoids. Front Cell Neurosci 17:1198715
17. Cerneckis J, Shi Y (2023) Context matters: hPSC-derived microglia thrive in a humanized brain environment in vivo. Cell Stem Cell 30(7):909–910
18. Cerneckis J, Shi Y (2023) Myelin organoids for the study of Alzheimer's disease. Front Neurosci 17:1283742
19. Lin YT et al (2018) APOE4 causes widespread molecular and cellular alterations associated with Alzheimer's disease phenotypes in human iPSC-derived brain cell types. Neuron 98(6):1141–1154
20. Meyer K et al (2019) REST and neural gene network dysregulation in iPSC models of Alzheimer's disease. Cell Rep 26(5): 1112–1127 e9
21. Cheng C et al (2021) High-content image-based analysis and proteomic profiling identifies tau phosphorylation inhibitors in a human iPSC-derived glutamatergic neuronal model of tauopathy. Sci Rep 11(1):17029
22. Kondo T et al (2022) Dissection of the polygenic architecture of neuronal Aβ production using a large sample of individual iPSC lines

derived from Alzheimer's disease patients. Nature Aging 2(2):125–139

23. Pantazis CB et al (2022) A reference human induced pluripotent stem cell line for large-scale collaborative studies. Cell Stem Cell 29(12):1685–1702 e22
24. Workman MJ et al (2023) Large-scale differentiation of iPSC-derived motor neurons from ALS and control subjects. Neuron 111:1191
25. Engle SJ et al (2018) Best practices for translational disease modeling using human iPSC-derived neurons. Neuron 100(4):783–797
26. Chambers SM et al (2009) Highly efficient neural conversion of human ES and iPS cells by dual inhibition of SMAD signaling. Nat Biotechnol 27(3):275–280
27. Smith DK et al (2016) Small molecules modulate chromatin accessibility to promote NEUROG2-mediated fibroblast-to-neuron reprogramming. Stem Cell Rep 7(5):955–969
28. Wapinski OL et al (2013) Hierarchical mechanisms for direct reprogramming of fibroblasts to neurons. Cell 155(3):621–635
29. Zhang Y et al (2013) Rapid single-step induction of functional neurons from human pluripotent stem cells. Neuron 78(5):785–798

Chapter 13

Generating Homogeneous Brain Organoids from Human iPSCs

Xianwei Chen and Yanhong Shi

Abstract

There is a high demand for the development of in vitro models for human brain development and diseases due to the inaccessibility of human brain tissues. The human iPSC-derived brain organoids provide a promising in vitro model for studying human brain development and disorders. However, it is challenging to generate a large number of brain organoids with high consistency for modeling human neurological diseases. Here, we describe a method for generating high-yield brain organoids with high consistency by combining large-scale embryoid body (EB) generation and incorporating a quality control screening step during differentiation. The method described in this chapter provides a robust way to generate brain organoids for studying human brain development and modeling neurological diseases.

Key words Pluripotent stem cells (PSCs), Induced pluripotent stem cells (iPSCs), Brain organoids, Embryoid bodies (EBs), Differentiation, Disease modeling, Alzheimer's disease, Neurological diseases

1 Introduction

Human iPSCs provide an excellent platform for disease modeling, drug discovery, and cell therapy development [1–6]. Human iPSC-derived brain organoids are capable of self-assembling into an organized structure, which can recapitulate the features of brain development, such as ventricle formation, cortical layer organization, and neuronal migration [7–10]. Due to the inaccessibility of human brain tissues and the species difference between human and animal models, brain organoids have been widely used for studying human brain development and neurological diseases. The brain organoid technology was first reported by Lancaster et al. [7, 8, 10–12] for modeling early stages of human brain development up to 24 gestation weeks [10, 13]. Subsequently, other brain organoid differentiation protocols have been developed, including the ones for generating brain region-specific organoids [9, 14–17]. Brain organoids can also be engineered to model various neurological

Koh-ichi Nagata (ed.), *Cerebral Cortex Development: Methods and Protocols*, Methods in Molecular Biology, vol. 2794, https://doi.org/10.1007/978-1-0716-3810-1_13,

disorders, including neurodegenerative diseases such as Alzheimer's disease [18–22]. Therefore, the brain organoid platform provides a promising in vitro human cellular model for studying human neurological diseases and developing therapeutic strategies for these disorders.

Various challenges for studies using brain organoids include heterogeneity between individual organoids, the necrotic inner core of the organoid, and the lack of immune cells, such as microglia, and blood vessels in current organoids [13, 23, 24]. Sliced brain organoids have been used to delay cell death within the inner core [25, 26]. Microglia-containing brain organoids can be obtained by co-culturing microglia with neural organoids, although it is still challenging to maintain microglia in brain organoids for a prolonged period of time [27–29]. Transplantation of brain organoids with microglial progenitor cells promotes microglial survival and maturation [30, 31]. Efforts to introduce blood vessels into brain organoids include organoid transplantation into the mouse brain or directed endothelial cell differentiation in the developing organoids [32–34]. It is critical to keep optimizing brain organoid differentiation protocols to advance the study of human brain development and neurological diseases.

The protocol described in this chapter is aimed at reducing the heterogeneity between individual organoids and increasing the yield of brain organoids with high consistency. Although multiple protocols for making brain organoids have been reported, it is challenging to generate brain organoids with a consistent proportion of the cortical region with the SOX2+ layer, the TBR2+ layer, and the CTIP2+ layer in the inside-to-outside order in individual organoids. It is even harder to generate a large number of brain organoids with good consistency. In this study, we describe a method for generating brain organoids in large quantities with high consistency by combining large-scale EB generation and a screening strategy at days 20–40 of organoid differentiation. Brain organoids generated using this method will provide a robust platform for studying human brain development and modeling neurological diseases.

2 Materials

1. Essential 8™ Medium (E8).
2. 0.5 mM of EDTA solution: 0.5 mM of EDTA, pH 8.0, in Dulbecco's phosphate-buffered saline (DPBS).
3. Dulbecco's modified eagle medium/nutrient mixture F-12 with HEPES (DMEM/F12).
4. Coating solution: 1% Matrigel in DMEM/F12 basal medium.

5. iPSC freezing medium: E8 medium containing 10% dimethyl sulfoxide (DMSO) and 10 μM Y-27632.
6. Neural induction medium (NIM): DMEM/F12 basal medium supplied with 1× N2 supplement, 1× glutamax, 1× MEM non-essential amino acids (MEM-NEAA, and 2 μg/mL heparin.
7. Embedding solution: 25% Matrigel in NIM medium.
8. Brain organoid differentiation medium: DMEM-F12 basal medium supplied with 1× N2 supplement, 1× B27 supplement, 2.5 μg/mL human insulin, 1× glutamax, 0.5× MEM-NEAA, and 3.5 μL/L (V/V) 2-mercaptoethanol.

3 Methods

3.1 iPSC Culture and Passaging

We describe the protocol for the culture of human iPSCs in a 6-well plate under the feeder-free condition.

1. Prepare the coating solution: Take one aliquot (0.5 mL) of Matrigel from a − 80 °C freezer and thaw it on ice immediately. It will take about 30 min to 1 h to thaw Matrigel completely. Matrigel can be thawed in a much shorter time (within 1 min) (*see* **Note 1**). Mix the thawed Matrigel with ~50 mL of cold DMEM/F12 medium to make the coating solution and keep it at 4 °C for storage.
2. Prepare the coated plates: The tissue culture-treated six-well plates are used for iPSC culture. To coat the plate, add 1 mL of coating solution into each well and incubate at 37 °C for 1 to 2 h before using it for iPSC culture (*see* **Note 2**).
3. iPSCs are ready for passaging when the cells reach about 50%–60% confluency (*see* **Note 3**). Aspirate off the medium from iPSCs cultured in a six-well plate and add 1 mL of DPBS into each well to wash cells once.
4. Aspirate off DPBS and add 1 mL of 0.5 mM EDTA solution to dissociate iPSC colonies.
5. Incubate the plate at room temperature for 4–5 min.
6. Aspirate off the EDTA solution and detach iPSC colonies by gently blowing cells once using 1 mL of E8 medium (try to blow all the surface area by making cycles).
7. Plate several drops (usually 2–4 drops) of iPSC colonies per well into Matrigel-coated six-well plates containing 2 mL of fresh E8 medium per well using the 1 mL pipettor. Shake the plate to distribute cells evenly.
8. Culture cells in a 37 °C incubator.

9. Change the medium daily by aspirating off the old medium and adding 2 mL of fresh E8 medium per well (*see* **Note 4**).
10. Human iPSCs are routinely tested to confirm the lack of mycoplasma contamination.

3.2 iPSC Freezing

1. Usually, when cells reach ~50%–70% confluency after 3–4 days of culture, the iPSCs are dissociated as described above (see **steps 3–6** in Subheading 3.1).
2. After EDTA dissociation, the iPSC freezing medium (instead of E8 medium) is used to detach iPSC colonies.
3. The iPSC colonies are transferred into cryogenic vials at 0.5 or 1 mL/vial (*see* **Note 5**). The vials are placed into a Mr. Frosty container, which is then placed at −80 °C overnight before transferring the cryogenic vials into a liquid nitrogen tank for long-term storage.

3.3 iPSC Thawing

1. Take out the frozen iPSCs from the liquid nitrogen tank, place them in a 37 °C water bath immediately, and incubate for several minutes until the cells are completely thawed.
2. Mix the cells once gently using a 1 mL pipettor before transferring them into a 15 mL conical tube with 1 mL of E8 medium containing 10 μM Y-27632 (*see* **Note 6**).
3. Spin down at 200 g for 2 min.
4. Aspirate off the supernatant.
5. Resuspend cells in 1 mL of E8 medium with 10 μM Y-27632 and seed 0.5 mL of iPSCs into one well of a Matrigel-coated six-well containing 2 mL of E8 medium with 10 μM Y-27632.
6. The next day, the medium is replaced by fresh E8 medium without 10 μM Y-27632.
7. Culture iPSCs with daily E8 medium change.
8. Culture iPSCs for at least two passages before using them for generating brain organoids.

3.4 Generation of Brain Organoids

Usually, four wells (*see* **Note 7**) of iPSCs cultured in a six-well plate with 70–80% confluency are used for generating brain organoids.

3.4.1 EB Formation

1. On day 0, wash cells once by aspirating off the old medium from iPSCs and adding 1 mL DPBS into each well.
2. Aspirate off DPBS and add 1 mL 0.5 mM EDTA solution to each well.
3. Incubate the plate at room temperature for 5–7 min.
4. Aspirate off the EDTA solution and detach iPSC colonies by gently blowing cells once using E8 medium with 5 μM Y-27632 (1 mL/well).

5. Transfer iPSC colonies into six wells of a suspension six-well plate containing E8 medium with 5 μM Y-27632 (3–4 mL/well) (*see* **Note 8**). Mix well (*see* **Note 9**) and culture the cells in a 37 °C incubator.
6. On day 1, the EB spheres with 100–300 μm diameter can be observed. *See* **Note 10**.
7. Blow the EBs gently using a 1 mL pipettor to detach all the EBs and dead cells. and then aspirate off most of the old medium to remove the dead cells and tiny EBs (< 50 μm diameter).
8. Add 3 mL of fresh E8 medium without Y-27632 into each well. Shake the plate to distribute EBs evenly before placing the plate back into the incubator.
9. On days 2–4, perform daily half-medium change by repeating **Steps 7–8** in Subheading 3.4.1 (*see* **Note 11**).

3.4.2 Neural Induction

1. On day 5, the EBs should exhibit a smooth surface under a microscope. Shake the plate gently to let all the EBs collect in the center of the well and then aspirate off all the old medium using vacuum (alternatively, transfer EBs to a 15 mL conical tube and then aspirate off the supernatant). Add 3 mL of freshly prepared NIM medium (which can be stored at 4 °C for up to 1 month) to each well. Mix well and put the plate back to the 37 °C incubator.
2. On days 6–7, perform half medium change each day (*see* **Note 12**).

3.4.3 Matrigel Embedding and Removal

1. On day 8, put Matrigel on ice for ~30 min to 1 h to completely thaw it (*see* **Note 1**). Mix the Matrigel with cold NIM medium to generate an embedding solution (25% Matrigel) and keep the solution on ice.
2. Blow the differentiated EBs gently using a 1 mL pipettor and then transfer the EBs from one well of a six-well plate into a 15 mL conical tube. Usually, the EBs in each conical tube can be embedded into one well of a 12-well plate.
3. Aspirate off all the medium and add 0.7 mL of cold embedding solution (25% Matrigel) into the tube using a 1 mL pipettor. Then, gently mix the EBs once using a 1 mL pipettor and immediately transfer them into one well of a 12-well plate. Shake the plate to distribute the EBs evenly. Put the plate back to the 37 °C incubator. Complete this step quickly (*see* **Note 13**).
4. Four to six hours later, slowly and gently add 2 mL of NIM per well using a 5 mL serological pipette. Do not detach the Matrigel and EBs (*see* **Note 14**).

5. On day 9, gently add an additional 2.5 mL of NIM into each well.
6. On days 10–11, perform daily half-medium change.
7. On day 12, detach the Matrigel and brain organoids using a 1 mL pipettor by blowing them gently (*see* **Note 15**).
8. Aspirate off half medium after organoids sink down and transfer the organoids to a six-well suspension plate with 2–3 mL of NIM medium. Shake the plate to distribute EBs evenly before putting the plate back in the incubator.
9. On days 13–19, perform daily half-medium change (*see* **Note 16**).

3.4.4 Screening and Maintaining

1. On day 20, transfer brain organoids into a suspension T25 flask with about 7 mL freshly prepared brain organoid differentiation medium (which can be stored at 4 °C for up to 1 month) and culture the brain organoids on a shaker at 50 rpm rotating speed in the incubator.
2. On days 21–40, change the medium every 3 days by aspirating off 5 mL of old medium and adding 5 mL of fresh brain organoid differentiation medium. Do not let the medium color turn yellow.
3. On days 21–40, screen brain organoids when high-quality neural rosettes that exhibit bright cycles under the phase-contrast microscopy are observed. Gently transfer brain organoids from a T25 flask to one well of a six-well plate using a 5 mL serological pipette. Remove the organoids without a clear neural rosette structure and only keep the organoids that are mainly composed of high-quality neural rosettes for further culture.
4. After screening, brain organoids (see an example in Fig. 1a) are cultured in brain organoid differentiation medium for up to 4 months on a shaker in the incubator with medium change every 3 days (*see* **Note 17**). We can get hundreds of organoids from four 6-well plates of iPSCs (Fig. 1b).
5. Alternatively, on days 50–70, one brain organoid can be sliced into two halves to reduce the inner cell death and extend the culture time up to 5 months.

3.4.5 Immunostaining of Brain Organoids

1. Transfer brain organoids into a 1.5 mL tube (*see* **Note 18**). Aspirate off the medium and wash the organoids with phosphate-buffered saline (PBS).
2. Aspirate off the PBS and add 1 mL of 4% paraformaldehyde (PFA) followed by incubation at 4 °C overnight.
3. Aspirate off the PFA solution and wash brain organoids with PBS twice.

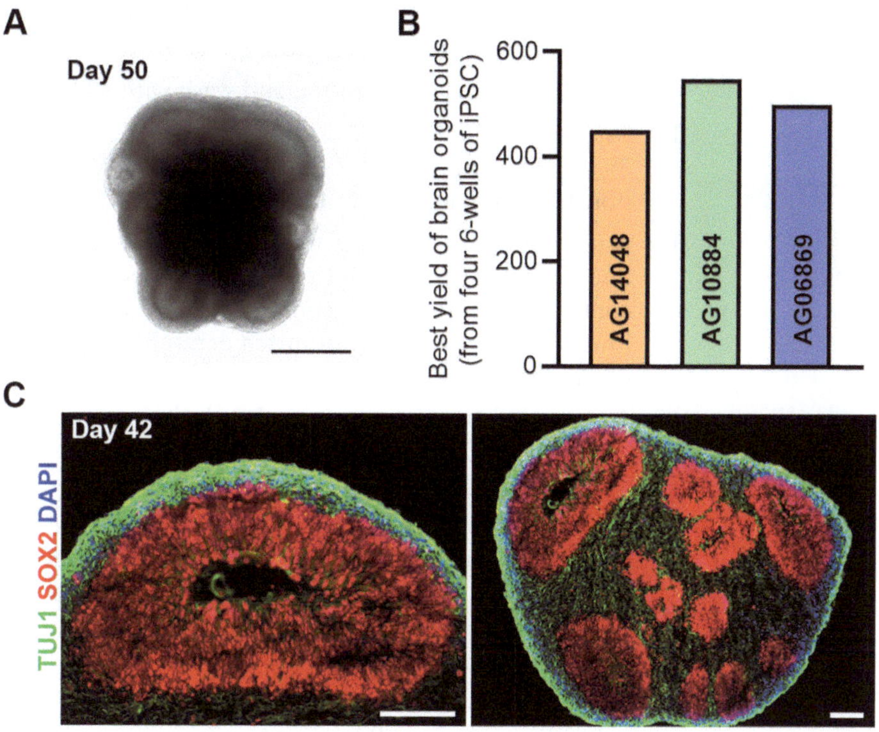

Fig. 1 (**a**) A representative bright field image of AG06869 brain organoids at day 50. Scale bar, 100 μm. (**b**) The highest yield of brain organoids after screening from four wells of iPSCs in a 6-well plate. (**c**) Representative images of SOX2 and TUJ1 immunostaining of AG06869 brain organoids at day 42. Scale bar, 100 μm. AG14048, AG06869, and AG10884 fibroblasts were purchased from Coriell and reprogrammed to iPSCs in our laboratory

4. Aspirate off the PBS and add 1 mL of 30% sucrose solution (in PBS) followed by incubation at 4 °C for 2–4 h or overnight.
5. Transfer the brain organoids into a cryomold and aspirate off the sucrose solution. Let the organoids sit at the center of the cryomold.
6. Add the O.C.T. compound into the cryomold to cover the organoids and place the cryomold on dry ice or at −80 °C for a while to form a solid matrix. The embedded organoids can be kept at −80 °C for long-term storage.
7. Perform cryo-sectioning to generate brain organoid sections with 14–20 μm thickness and transfer them onto microscope slides. Store the slides at −20 °C.
8. Take out the slides from the freezer and dry them at room temperature.
9. Circle the organoid tissues using a hydrophobic slide marker.
10. Wash the organoid sections with PBS twice.

11. Aspirate off the PBS solution and add 4% PFA followed by incubation at room temperature for 15 min.
12. Aspirate off the PFA solution and wash with PBS twice.
13. Aspirate off the PBS and add the blocking solution (PBS with 3% donkey serum and 0.2% Triton X-100) followed by incubation at room temperature for 1–2 h.
14. Aspirate off the blocking solution and add the primary antibody solution (primary antibodies are diluted in the PBS buffer containing 3% donkey serum and 0.01% Triton X-100) followed by incubation at room temperature for 2 h or at 4 °C overnight.
15. Aspirate off the primary antibody solution and wash with PBS containing 0.01% Triton X-100 once.
16. Aspirate off the washing buffer and add the secondary antibody solution (secondary antibodies labeled with Alexa Fluor dyes are diluted in the PBS buffer containing 3% donkey serum and 0.01% Triton X-100) followed by incubation for 1 hour at room temperature. Protect the slides from light from this point forward.
17. Aspirate off the secondary antibody solution and wash twice with PBS containing 0.01% Triton X-100.
18. Aspirate off the washing buffer and add the DAPI solution followed by incubation at room temperature for 5 min.
19. Aspirate off the DAPI solution and wash twice with PBS containing 0.01% Triton X-100.
20. Aspirate off the washing solution and mount the slide with a mounting solution.
21. Dry the slide in a dark place. Then store the slides at 4 °C protected from light.
22. Image acquisition is performed using a fluorescence microscope imaging system (See an example in Fig. 1c).

4 Notes

1. Always keep Matrigel, coating solution, and embedding solution on ice because Matrigel is temperature sensitive. To thaw Matrigel faster, blow Matrigel several times using 0.5 mL of cold medium with a 1 mL pipettor until Matrigel is completely thawed and then make coating solution or embedding solution by mixing the thawed Matrigel with cold medium.

2. It is convenient to prepare the plates for iPSC culture in advance because the coated plates can be kept for several days in the incubator.
3. Usually, iPSCs are ready for passaging after 3–4 days of culture. It is not suggested to culture iPSCs for more than 4 days before passaging. iPSCs can be passaged even if they have not reached 50% confluence after 3–4 days of culture. In this case, enough iPSCs should be seeded during passaging.
4. When changing the medium for iPSCs, the medium volume should be adjusted based on the cell density to avoid the color of the medium turning yellow.
5. High iPSC cell density is recommended for making cell stocks, which is good for maintaining cell status and getting enough iPSCs for experiments in a short period of time after thawing.
6. The ROCK inhibitor Y-27632 is necessary when thawing iPSCs, which helps the survival of iPSCs.
7. Starting with more iPSCs will increase the yield of brain organoids for large-scale organoid differentiation.
8. If 5 μM Y-27632 is not sufficient to support iPSC survival, increasing the concentration to up to 10 μM can help.
9. When seeding iPSCs to make EBs, make sure the iPSC colonies are mixed well before putting them back in the incubator. Otherwise, EBs may merge to form big clusters (> 500 μm diameter), which can be too big to generate brain organoids.
10. Since Y27632 is added on the first day, it is normal to observe attached EBs, which should be easily detached by gently blowing with a 1 mL pipettor. It is also normal to observe dead cells, some of which can be on the EB surface.
11. In Subheading 3.4.1, adjust the medium volume according to the EB number to conserve reagents. Do not let the medium color turn yellow.
12. In Subheading 3.4.2, adjust the medium volume according to the EB number to conserve reagents. Do not let the medium color turn yellow. On day 8, the differentiated EBs should be bright and have a smooth surface.
13. Since Matrigel is temperature-sensitive, complete the organoid embedding (**Step 14** in Subheading 3.4.3) quickly. If more than 6 wells of differentiated EBs are prepared, multiple 12-well plates are recommended for embedding the differentiated EBs.
14. The neural rosette-like structure in the differentiated EBs should be observed at **step 15** in Subheading 3.4.3 4–6 h after embedding in Matrigel.

15. Clear neural rosette-like structure should be observed in organoids at **Step 18** in Subheading 3.4.3 4 days after embedding in Matrigel. Organoids with good quality usually do not have many cells migrating out. Organoids with low quality usually have a large number of cells that migrate out and limited number of clear neural rosettes.
16. Matrigel can be separated from the organoids by blowing organoids during daily half-medium change.
17. After screening, do not let the medium turn yellow when culturing organoids. When the size of the organoids increases, dividing them from one flask to multiple flasks can help maintain organoids healthy.
18. When transferring brain organoids, it is important not to break them. It is recommended to cut the 1 mL tip at the top before using it for transferring brain organoids.

Acknowledgments

The authors would like to thank Louise and Herbert Horvitz, the Christopher Family, the Judy and Bernard Briskin Fund, and the Sidell Kagan Foundation for their generosity. This work was supported by the National Institute of Aging of the National Institutes of Health under Grant Nos. R01 AG056305, RF1 AG061794, R01 AG072291, and RF1 AG079307 to Y.S.

References

1. Takahashi K et al (2007) Induction of pluripotent stem cells from adult human fibroblasts by defined factors. Cell 131(5):861–872
2. Yu J et al (2007) Induced pluripotent stem cell lines derived from human somatic cells. Science 318(5858):1917–1920
3. Shi Y (2009) Induced pluripotent stem cells, new tools for drug discovery and new Hope for stem cell therapies. Curr Mol Pharmacol 2(1): 15–18
4. Shi Y et al (2020) Induced pluripotent stem cell technology: venturing into the second decade. In: Lanza R et al (eds) Principles of tissue engineering. Elsevier
5. Shi Y et al (2017) Induced pluripotent stem cell technology: a decade of progress. Nat Rev Drug Discov 16(2):115–130
6. Li L, Chao J, Shi Y (2018) Modeling neurological diseases using iPSC-derived neural cells: iPSC modeling of neurological diseases. Cell Tissue Res 371(1):143–151
7. Lancaster MA, Knoblich JA (2014) Organogenesis in a dish: modeling development and disease using organoid technologies. Science 345(6194):1247125
8. Lancaster MA et al (2013) Cerebral organoids model human brain development and microcephaly. Nature 501(7467):373–379
9. Qian X et al (2016) Brain-region-specific organoids using mini-bioreactors for modeling ZIKV exposure. Cell 165(5):1238–1254
10. Pasca AM et al (2015) Functional cortical neurons and astrocytes from human pluripotent stem cells in 3D culture. Nat Methods 12(7): 671–678
11. Sun G et al (2020) Modeling human cytomegalovirus-induced microcephaly in human iPSC-derived brain organoids. Cell Rep Med 1(1):100002
12. Pasca SP (2018) The rise of three-dimensional human brain cultures. Nature 553(7689): 437–445

13. Kelava I, Lancaster MA (2016) Stem cell models of human brain development. Cell Stem Cell 18(6):736–748
14. Gordon A et al (2021) Long-term maturation of human cortical organoids matches key early postnatal transitions. Nat Neurosci 24(3): 331–342
15. Di Lullo E, Kriegstein AR (2017) The use of brain organoids to investigate neural development and disease. Nat Rev Neurosci 18(10): 573–584
16. Feng L et al (2023) Developing a human iPSC-derived three-dimensional myelin spheroid platform for modeling myelin diseases. iScience 26(11):108037
17. Cerneckis J, Shi Y (2023) Myelin organoids for the study of Alzheimer's disease. Front Neurosci 17:1283742
18. Chen X et al (2021) Modeling sporadic Alzheimer's disease in human brain organoids under serum exposure. Adv Sci (Weinh):e2101462
19. Huang S et al (2022) Chimeric cerebral organoids reveal the essentials of neuronal and astrocytic APOE4 for Alzheimer's tau pathology. Signal Transduct Target Ther 7(1):176
20. Lin YT et al (2018) APOE4 causes widespread molecular and cellular alterations associated with Alzheimer's disease phenotypes in human iPSC-derived brain cell types. Neuron 98(6):1294
21. Zhao J et al (2020) APOE4 exacerbates synapse loss and neurodegeneration in Alzheimer's disease patient iPSC-derived cerebral organoids. Nat Commun 11(1):5540
22. Cerneckis J, Bu G, Shi Y (2023) Pushing the boundaries of brain organoids to study Alzheimer's disease. Trends Mol Med 29(8): 659–672
23. Chen X et al (2021) Human induced pluripotent stem cell–based modeling of Alzheimer's disease, a glial perspective. In: Birbrair A (ed) iPSCs for modeling central nervous system disorders. Elsevier, pp 21–35
24. Qian X, Song H, Ming GL (2019) Brain organoids: advances, applications and challenges. Development 146(8)
25. Giandomenico SL et al (2019) Cerebral organoids at the air-liquid interface generate diverse nerve tracts with functional output. Nat Neurosci 22(4):669–679
26. Qian X et al (2020) Sliced human cortical organoids for modeling distinct cortical layer formation. Cell Stem Cell 26(5):766–781 e9
27. Xu R et al (2021) Developing human pluripotent stem cell-based cerebral organoids with a controllable microglia ratio for modeling brain development and pathology. Stem Cell Rep 16(8):1923–1937
28. Popova G et al (2021) Human microglia states are conserved across experimental models and regulate neural stem cell responses in chimeric organoids. Cell Stem Cell 28(12): 2153–2166 e6
29. Cerneckis J, Shi Y (2023) Modeling brain macrophage biology using human iPSC-derived neuroimmune organoids. Front Cell Neurosci 17:1198715
30. Schafer ST et al (2023) An in vivo neuroimmune organoid model to study human microglia phenotypes. Cell 186(10):2111–2126 e20
31. Cerneckis J, Shi Y (2023) Context matters: hPSC-derived microglia thrive in a humanized brain environment in vivo. Cell Stem Cell 30(7):909-910
32. Cakir B et al (2019) Engineering of human brain organoids with a functional vascular-like system. Nat Methods 16(11):1169–1175
33. Matsui TK et al (2021) Vascularization of human brain organoids. Stem Cells 39(8): 1017–1024
34. Mansour AA et al (2018) An in vivo model of functional and vascularized human brain organoids. Nat Biotechnol 36(5):432–441

Chapter 14

Primary Culture of Dissociated Neurons from the Embryonic Cerebral Cortex

Shiho Ito, Megumi Kawasaki, and Takeshi Kawauchi

Abstract

Primary neuronal culture is a valuable in vitro model for analyzing the molecular mechanisms underlying the development and function of neural circuits. In contrast to neurons in vivo, primary cultured neurons can easily be transfected with genes of interest or treated with chemicals such as agonists and inhibitors of a specific target molecule. Furthermore, time-dependent morphological changes, such as the acquisition of neuronal polarity, axon elongation, and dendrite branch formation, can be analyzed by using primary neuronal cultures. Here, we describe a method for preparing a primary culture of neurons from the developing cerebral cortex, together with a method for gene transfer to primary cultured cortical neurons.

Key words Neuron, Primary culture, Cerebral cortex, Embryonic brain, Mouse, Dissociated cell culture

1 Introduction

In brain tissues, neurons have a highly polarized, unique structure typically consisting of a cell body, dendrites which function to integrate synaptic inputs from other neurons, and an axon which serves to transmit information in functional neural circuits. To investigate the structure and function of mammalian brains, it is necessary to understand the mechanisms that regulate the dynamic morphological changes and electrical activities of neurons. For such studies, in vitro culture of neurons offers many advantages, including experimental convenience. In 1988, Dotti et al. established a culture method for isolated primary hippocampal neurons [1]. The cultured neurons form a rapidly elongating axon and branched dendrites. Synapse formation and electrophysiological responses to stimuli have also been observed in long-term cultured neurons [2, 3].

It is known that primary cultured neurons differ in part from neurons in vivo [4]. For example, immature cortical neurons

Koh-ichi Nagata (ed.), *Cerebral Cortex Development: Methods and Protocols*, Methods in Molecular Biology, vol. 2794, https://doi.org/10.1007/978-1-0716-3810-1_14,

exhibit multiple modes of migration during the development of the cerebral cortex, whereas migration hardly occurs in primary neuronal culture. However, the molecular pathways regulating cortical neuronal migration are largely conserved and activated in immature primary cortical neurons [5–7]. Thus, neuronal cultures can be used to analyze the molecular mechanisms underlying morphological changes and function of neurons during brain development.

To prepare primary neuronal cultures, cerebral cortices dissected from embryonic brains are dissociated with trypsin and the cells are cultured on dishes or plates. These cultured cells can be subjected to various biological analyses, such as immunostaining, immunoblotting, and time-lapse imaging. Furthermore, the transfection of plasmids into primary cultured neurons for knockdown or overexpression experiments is also available, which enables us to analyze the functions of specific genes in neural development. Here, we describe a method for preparing a primary culture of neurons from the developing cerebral cortex, together with a method for gene transfer to these neurons using plasmids.

2 Materials

2.1 Brain Dissection

1. Phosphate-buffered saline (PBS).
2. Forceps No. 2, 3, and 5.
3. Curved forceps.
4. Small scissors.
5. 35 mm dish.
6. 10 cm dish.
7. HEPES-HBSS: Hanks' balanced salt solution (HBSS) without calcium and magnesium, and 10 mM HEPES.

2.2 Coating Dishes

1. 0.1 mg/mL poly-D-lysine dissolved in sterile deionized distilled water (ddH_2O).
2. Sterile ddH_2O.
3. 6- or 12-well plates or 6 cm dishes for culture.

2.3 Gene Transfection and Culture

1. Trypsin-EDTA: 0.25% trypsin and 1 mM EDTA.
2. Wash medium: Dulbecco's modified Eagle's medium (DMEM, high-glucose), 10% fetal bovine serum, 100 unit/mL penicillin, and 0.1 mg/mL streptomycin.
3. DNA solution: plasmid DNA in TE buffer (10 mM Tris–HCl (pH 8.0), and 1 mM EDTA (pH 8.0)).
4. Gene transfer equipment; here we use Nucleofector 2b and the accompanying cuvettes and single-use pipettes (Lonza).

5. Gene transfer solution; here we use Amaxa mix (Nucleofector Solution for Mouse Neurons and the Supplement are mixed at the ratio of 4.5:1 according to the manufacturer's protocol) (Lonza, #VPG-1001).
6. NB medium: Neurobasal, B-27 (x50), 2 mM GlutaMAX supplement, 100 unit/mL penicillin, and 0.1 mg/mL streptomycin.

2.4 Obtaining the Cell Lysate

1. PBS.
2. Cell scraper.
3. Lysis buffer: 20 mM Tris–HCl (pH 7.5), 150 mM NaCl, 1% Triton X-100, protease inhibitors, and phosphatase inhibitors.
4. Handy sonicator (or ultrasonic generator/disrupter).

3 Methods

3.1 Coating Dishes and Preparation for Culture

1. Coat the culture dishes or plates with poly-D-lysine. Add 0.1 mg/mL of poly-D-lysine solution to the dishes/plates (4 mL per 6 cm dish; 2 mL per 1 well of 6-well plate; 1 mL per 1 well of 12-well plate) and incubate them at room temperature for 70–80 min (*see* **Note 1**). After the incubation, remove the poly-D-lysine solution, wash the dishes/plates with sterile ddH_2O 4 times, and allow them to dry under sterile conditions (*see* **Note 2**).
2. Add NB medium to the coated dishes/plates (5 mL per 6 cm dish; 2 mL per 1 well of 6-well plate; 1 mL per 1 well of 12-well plate) and incubate them at 37 °C in an atmosphere containing 5% CO_2.
3. Take two 1.5 mL microfuge tubes, one for dissociation of tissues with trypsin (**step 3.2.7**) and the other for suspending cells in NB medium (**step 3.3.2**). Put the 1.5 mL microfuge tubes with 500 μL of 0.25% trypsin and 1 mM EDTA (trypsin–EDTA) on ice and keep the 1.5 mL microfuge tubes with 1.2 mL of NB medium at 37 °C in an atmosphere containing 5% CO_2.

3.2 Neuron Isolation

1. After cervical dislocation, clean a pregnant ICR mouse carrying embryos at embryonic day 15 (E15) with 70% ethanol. Make an abdominal incision to excise the uterine horn containing the embryos, and transfer it into ice-cold PBS in a 10 cm dish (*see* **Note 3**).
2. Take out an embryo from the uterine horn and decapitate it with small scissors (Fig. 1a) (*see* **Note 4**). Fix the embryonic head around the medulla with forceps No.3 held in the

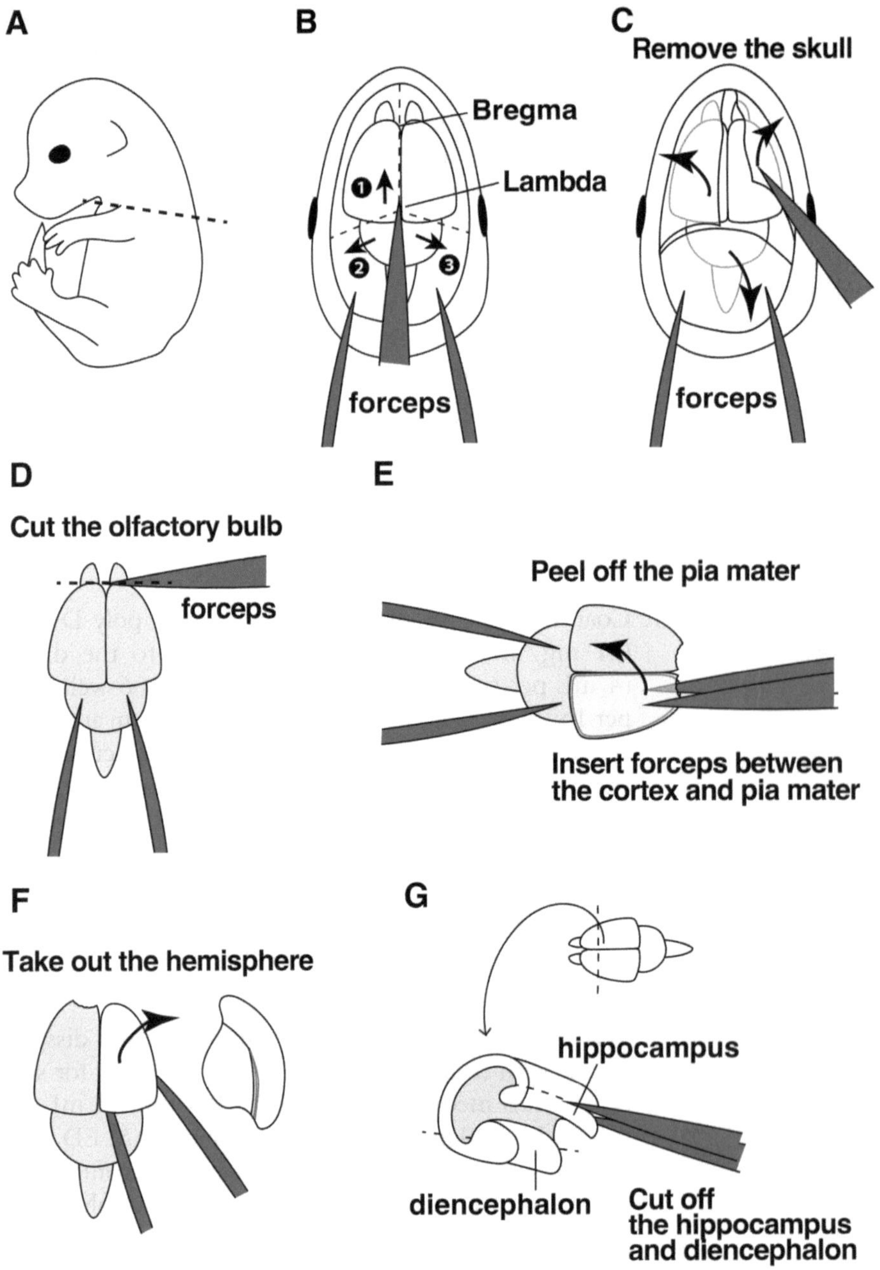

Fig. 1 Dissection of a mouse embryonic brain. (**a**) The head is removed along the dotted line. (**b**, **c**) The skull is removed with forceps. (**d**, **e**) After cutting the olfactory bulb, the pia matter is peeled off. (**f–g**) The hemisphere is taken out and the hippocampus and diencephalon are removed to obtain the cerebral cortex

nondominant hand. Remove the skin from the embryonic head with forceps No.5 held in the dominant hand (*see* **Note 5**).

3. While fixing the embryonic head with forceps No. 3 held in the nondominant hand, prick the skull at the lambda with one side

of forceps No. 5 and cut toward the bregma (Fig. 1b) (*see* **Note 6**).

4. Remove the skull with forceps No. 5 and take out the brain (Fig. 1c) (*see* **Note 7**). Transfer the brain to a 35 mm dish with HEPES-HBSS using curved forceps.
5. Cut off the olfactory bulbs, leaving a small hole in the pia mater (Fig. 1d). Insert one side of forceps No. 5 into the small hole between the cerebral cortex and pia mater and peel off the pia mater (Fig. 1e).
6. Take out the hemisphere with forceps No. 5 and cut off the hippocampus and diencephalon (Fig. 1f, g).
7. Transfer the dissected cerebral cortices (two hemispheres from an embryonic brain) to the prepared 1.5 mL microfuge tube with 500 μL of trypsin–EDTA (**step 3.1.3**) by using curved forceps and incubate them at 37 °C for 11–15 min.
8. Remove the trypsin–EDTA using a P1000 micropipette, add 1 mL of the Wash medium to the cerebral cortices, and remove them. Repeat 5 times of the washing step (*see* **Note 8**).
9. After the final washing step, remove the wash medium using a P1000 micropipette, and then thoroughly remove the remaining wash medium with a P200 micropipette (*see* **Note 9**).

3.3 DNA Transfection

1. Add 100 μL of Amaxa mix to the 1.5 mL microfuge tube containing the cerebral cortices and subsequently add the DNA solution (*see* **Note 10**). Gently pass the cerebral cortices through a P1000 micropipette tip 5–7 times to break up the tissues and subsequently pass the cortices through a P200 micropipette to dissociate them into single cells (*see* **Note 11**).
2. If the cortical neurons are to be cultured without gene transfection, the cerebral cortices should be dissociated in NB medium, instead of the Amaxa mix. Then move to **step 3.4.1**.
3. Transfer the single cell suspension containing plasmid DNA into the cuvette with a P200 micropipette. Insert the cuvette into Nucleofector 2b and apply the program "O-005." After finishing the program, take out the cuvette and transfer the transfected cell suspension immediately into the prepared 1.5 mL microfuge tube containing 1.2 mL of NB medium (**step 3.1.3**) by using the pipette (*see* **Note 12**).

3.4 Culture and Preparation of Cell Lysates

1. After pipetting, plate the transfected cells into the coated dishes/plates with NB media (**step 3.1.2**) (600 μL per 6 cm dish, 300 μL per 1 well of 6-well plate, 100 μL per 1 well of 12-well plate) and incubate them overnight at 37 °C in an atmosphere containing 5% CO_2.

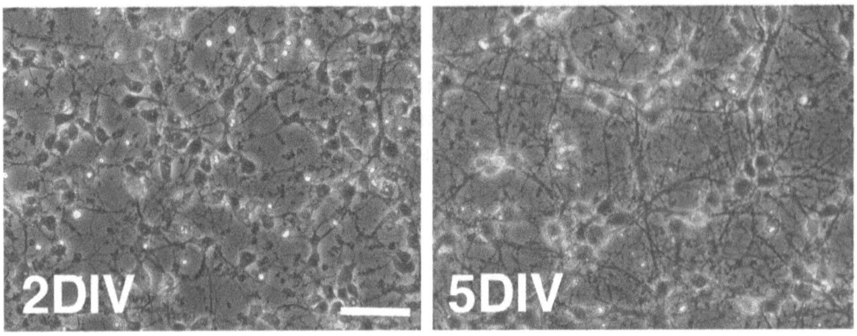

Fig. 2 Phase-contrast microscopy images of primary cultured cortical neurons. Left and right panels show neurons cultured in vitro for 2 days or 5 days (2DIV and 5DIV, respectively). Scale bar: 20 μm

2. After 12–20 h of incubation, replace the medium with fresh NB medium and further incubate the transfected cells for several days at 37 °C in an atmosphere containing 5% CO_2 (Fig. 2).
3. To prepare the cell lysates, wash the cultured cortical neurons with ice-cold PBS twice, treat them with ice-cold lysis buffer and harvest the roughly lysed cells with a cell scraper. After 1 h incubation on ice, sonicate the cell lysates with a handy sonicator and centrifuge them at 5000 G for 5 min at 4 °C to remove cell debris. The supernatants can be subjected to immunoblotting or immunoprecipitation (or other experiments) (*see* **Note 13**).

4 Notes

1. 0.1 mg/mL poly-D-lysine solution is stored at 4 °C and can be reused approximately 5 times.
2. If the dishes or plates are used soon after coating, they do not need to be dried. Without drying, add the NB medium to the dishes or plates and use them for neuronal culture.
3. The uterine horn containing embryos should quickly be transferred into ice-cold PBS. The embryos and dissected brains were also put in ice-cold PBS or HEPES-HBSS to minimize the loss of cell viability. Even if put on ice, all brains should be dissected and cultured within 4 h.
4. When decapitating, cut off the embryonic head diagonally from the neck to the upper back, which provides the place to prick for fixation of the head (Fig. 1a).
5. Before use, forceps were cleaned with 70% ethanol and irradiated with UV light. Autoclave sterilization was also applicable.

6. A deep incision beyond the tips of the olfactory bulb should be made and the skull should be carefully opened (Fig. 1c).
7. When taking out the brain, insert forceps below the brain. Note that the skull should be completely removed because the remains of a broken skull may damage the cerebral cortex.
8. Add the wash medium along the wall of the microfuge tube to raise the cortices in the medium and remove them gently after they have sunk to the bottom of the tube.
9. Do not pass the cerebral cortices through a micropipette to avoid breaking up of the tissues.
10. The amount of Amaxa mix used is 100 μL per cortex from one embryo. In co-transfection experiments, up to 3 μg of plasmid DNA(s) can be transfected.
11. Quick pipetting with a P200 micropipette provides the single cell suspension. Do not create air bubbles.
12. Perform the procedure quickly because long-term exposure to the Amaxa mix may damage the transfected cells.
13. Incubate them for 1 h on ice to allow the cells to be completely lysed.

References

1. Dotti CG, Sullivan CA, Banker GA (1988) The establishment of polarity by hippocampal neurons in culture. J Neurosci 8(4):1454–1468
2. Brachet A, Lario A, Fernández-Rodrigo A, Heisler FF, Gutiérrez Y, Lobo C, Kneussel M, Esteban JA (2021) A kinesin 1-protrudin complex mediates AMPA receptor synaptic removal during long-term depression. Cell Rep 36(5): 109499. https://doi.org/10.1016/j.celrep.2021.109499
3. Grabrucker A, Vaida B, Bockmann J, Boeckers TM (2009) Synaptogenesis of hippocampal neurons in primary cell culture. Cell Tissue Res 338(3):333–341. https://doi.org/10.1007/s00441-009-0881-z
4. Shikanai M, Nishimura YV, Sakurai M, Nabeshima YI, Yuzaki M, Kawauchi T (2018) Caveolin-1 promotes early neuronal maturation via caveolae-independent trafficking of N-cadherin and L1. iScience 7:53–67. https://doi.org/10.1016/j.isci.2018.08.014
5. Kawauchi T, Chihama K, Nabeshima Y, Hoshino M (2003) The in vivo roles of STEF/Tiam1, Rac1 and JNK in cortical neuronal migration. EMBO J 22(16):4190–4201
6. Kawauchi T, Chihama K, Nabeshima Y, Hoshino M (2006) Cdk5 phosphorylates and stabilizes p27kip1 contributing to actin organization and cortical neuronal migration. Nat Cell Biol 8(1):17–26
7. Shikanai M, Ito S, Nishimura YV, Akagawa R, Fukuda M, Yuzaki M, Nabeshima YI, Kawauchi T (2023) Rab21 regulates caveolin-1-mediated endocytic trafficking to promote immature neurite pruning. EMBO Rep 24(3):e54701. https://doi.org/10.15252/embr.202254701

Chapter 15

Immunocytochemistry of Primary Cultured Cerebral Cortical Neurons

Shiho Ito and Takeshi Kawauchi

Abstract

Immunocytochemistry combined with confocal or superresolution microscopy allows us to observe molecular localization and intracellular structures. However, it is challenging to analyze individual neurons in brain tissue, where neurons are densely packed. In contrast, we can easily observe structures such as the axonal growth cone and dendritic spines in dissociated individual neurons. Thus, the immunocytochemistry of primary cultured neurons is often used because it reflects the in vivo condition at least in part. Here, we describe a method for indirect fluorescence immunocytochemistry of primary cultured neurons from the embryonic cerebral cortex. This involves multiple steps including fixation, permeabilization, and antibody reaction, and in particular, we introduce an optimized protocol for permeabilization to enable the precise localization of target molecules.

Key words Immunocytochemistry, Primary culture, Neuron, Cerebral cortex, Brain, Membrane-associated protein, Cytosolic soluble protein, Digitonin, Triton X-100

1 Introduction

Immunocytochemistry employs specific antibodies to detect the expression and localization of cellular target molecules, including proteins and gangliosides, in combination with a fluorescent dye–conjugated secondary antibody that serves to visualize the primary antibody. Furthermore, multicolor immunofluorescence staining using different fluorescence dyes allows us to analyze the colocalization of two or more different molecules simultaneously.

Immunohistochemistry of brain sections is challenging because brain tissues contain a high density of neurons, which typically have a long, thin axon and complex dendritic arbors. On the other hand, immunocytochemistry using primary cultured neurons is a more effective approach to understanding a detailed localization of molecules. For example, the components of axonal growth cones, dendritic spines, and intracellular organelles and endosomes can be

Koh-ichi Nagata (ed.), *Cerebral Cortex Development: Methods and Protocols*, Methods in Molecular Biology, vol. 2794, https://doi.org/10.1007/978-1-0716-3810-1_15,

easily analyzed. Furthermore, unlike cell lines, such as PC12 and Neuro2a cells, primary cultured neurons form mature axons and dendrites and exhibit electrical activities via synaptic transmission, retaining many of the in vivo features [1, 2]. Thus, immunocytochemistry using primary cultured neurons is an important approach for the analysis of complicated molecular mechanisms such as those involved in organelle function, intracellular molecular interaction, and cell adhesion in the brain.

To detect the behavior of target molecules, not only advanced techniques of microscopy but also appropriate sample preparation methods are important. Briefly, the usual procedure of sample preparation is as follows. Firstly, cells are cultured on coverslips for microscopic analysis. Secondly, cells are fixed to retain cell morphology, usually with 4% paraformaldehyde, which cross-links intracellular proteins, although methanol is also used depending upon the antibodies. For intracellular targets, cells are permeabilized with detergents, such as 0.1–0.5% Triton X-100, which is generally used for the perforation of the plasma membrane and nuclear membrane, to enable the antibody to access the epitope. However, Triton X-100 dissolves membrane lipids and some of the cytosolic soluble proteins are lost [3]. In the case of staining membrane components or cytosolic soluble proteins, digitonin, a milder detergent, is recommended [4]. The cells are then blocked with serum from a host species different from the host species of the primary antibody to prevent nonspecific binding of the primary antibody. Next, the cells are incubated with primary antibodies that bind the target molecules and then incubated with secondary antibodies labeled with a fluorescent dye to bind the primary antibodies. Finally, coverslips are placed on the slide glass using mounting media to prevent the fading of the fluorescent dye.

For immunocytochemistry, the conditions of each procedure, such as fixation and permeabilization, should be optimized depending on the target molecules and the purpose of the experiments. Here, we describe a procedure for immunocytochemistry of primary cultured neurons, focusing especially on the optimization of permeabilization.

2 Materials

2.1 Coating Coverslips

1. 18 mm coverslip.
2. 0.1 mg/mL poly-D-lysine dissolved in sterile ddH_2O.
3. Sterile deionized distilled water (ddH_2O).
4. 70% ethanol in ddH_2O.
5. 100% ethanol.

2.2 Fixation and Permeabilization

1. Phosphate-buffered saline (PBS).
2. 4% Paraformaldehyde (PFA).
3. Goat serum.
4. Triton X-100.
5. Digitonin.
6. Permeabilization solutions.
   0.01% Digitonin solution: 0.01% digitonin and 10% goat serum in PBS on ice.
   0.15% Triton X-100 solution: 0.15% Triton X-100 and 10% goat serum in PBS.
   0.5% Triton X-100 solution: 0.5% Triton X-100 and 10% goat serum in PBS.

2.3 Blocking and Antibody Reaction

1. Blocking solution: 10% goat serum in PBS.
2. Antibody dilution buffer: 10% goat serum in PBS.
3. Primary antibody: Anti-GD3 antibody (abcam; ab117799), anti-caveolin-1 (Cell Signaling Technology; 3238), anti--β-tubulin (sigma; T5201), anti-calreticulin (Cell Signaling Technology; 12,238), and anti-γH2AX (Merck; 05-636) antibodies.
4. Secondary antibody: A fluorescence dye–conjugated anti-mouse IgG antibody and a fluorescence dye–conjugated anti-rabbit IgG antibody.

2.4 Mounting and Observation

1. Mounting medium with DAPI.
2. Confocal microscope.

3 Methods

3.1 Coating Coverslips and Culturing Neurons

1. Coat coverslips with poly-D-lysine in 12-well plates (*see* **Note 1**). Immerse coverslips in 100% ethanol and flame-dry them. Then, place the coverslips on a 12-well plate. Add 1 mL of 0.1 mg/mL poly-D-lysine solution to the 12-well plate with coverslips and incubate the plate at room temperature for 70–80 min (*see* **Note 2**).
2. After the incubation, remove the poly-D-lysine solution, wash the coverslips with 1 mL of sterile ddH_2O 4 times, and let them dry under sterile conditions (*see* **Note 3**).
3. Seed primary neurons from the embryonic cerebral cortex on the coated coverslips and incubate them at 37 °C with 5% CO_2 for the desired period (*see* **Note 4**).

3.2 Fix the Neurons

1. Remove the culture medium and wash the cells twice with cold PBS.
2. Fix the cells with 1 mL of 4% PFA per well at room temperature for 15–20 min.
3. Remove the 4% PFA and wash the coverslips with 1 mL of PBS twice.

3.3 Permeabilization

1. Prepare a permeabilization solution. Triton X-100 solution (0.15%) is generally used for permeabilization in immunostaining of most proteins, such as intracellular organelle proteins and cytoplasmic proteins. Digitonin solution (0.01%) is suitable for permeabilization for the immunostaining of molecules localized at the plasma membrane, while the Triton X-100 solution (0.5%) is suitable in the case of nuclear proteins (*see* **Note 5**). Before use, keep the Triton X-100 solution at room temperature and the digitonin solution on ice (*see* **Note 6**).
2. Add 1 mL of an appropriate permeabilization solution to the cells, and incubate under the following conditions. When using 0.15% Triton X-100 solution, incubate the cells at room temperature for 5 min. When using 0.5% Triton X-100 solution, incubate the cells at room temperature for 10 min. When using 0.01% digitonin solution, incubate the cells at room temperature for 1 min.
3. Remove the permeabilization solution and wash the cells on the coverslips with 1 mL of PBS twice.

3.4 Blocking and Antibody Reaction

1. Add 1 mL of the blocking solution and incubate the cells at room temperature for 1 h (*see* **Note 7**).
2. Prepare the primary antibody solution. Dilute an antibody directed against a protein of interest in 100 μL (per one coverslip) of the antibody dilution buffer.
3. Place 100 μL of the primary antibody solution on a sheet of parafilm and overlay the coverslips on the primary antibody solution with the cell-adherent side down by using forceps. Incubate them at 4 °C for 24–72 h in a humidified chamber (Fig. 1) (*see* **Notes 8–11**).
4. Put the coverslips on a sheet of parafilm with the cell-adherent side up and load 300 μL of PBS on the coverslips (Fig. 1). Incubate the coverslips at room temperature for 10 min with gentle shaking and remove the PBS. Repeat this washing step 3 times.
5. Prepare the secondary antibody solution. Dilute a secondary antibody conjugated with a fluorescent dye in 100 μL (per one coverslip) of the antibody dilution buffer (1:400) (*see* **Note 12**).

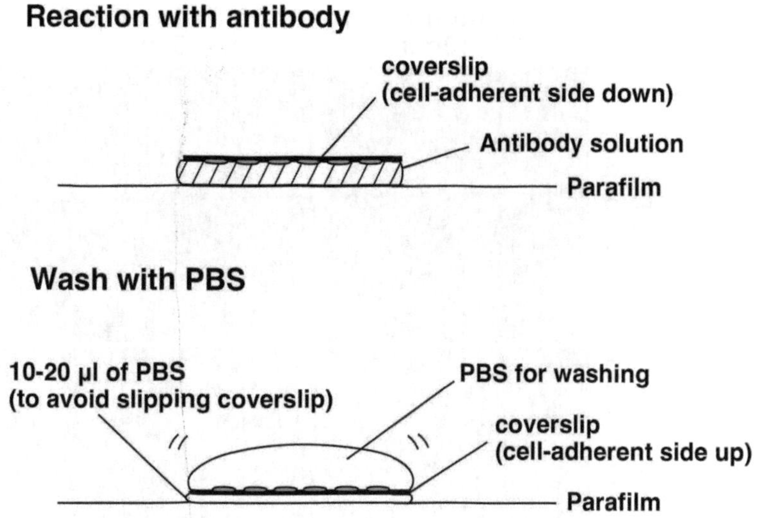

Fig. 1 Schematic illustration of the reaction of cells on coverslips with antibodies and washing with PBS on parafilm. For the reaction with antibodies, place the antibody solution (diagonal pattern) on a sheet of parafilm, and overlay the coverslip on the antibody solution with the cell-adherent side down. For washing with PBS, place 10–20 μL of PBS on the sheet of parafilm and overlay the coverslip on it with the cell-adherent side up. Then, place 300 μL of PBS for washing on the coverslip and shake it gently

6. Put 100 μL of the secondary antibody solution on a sheet of parafilm and overlay the coverslips on the secondary antibody solution with the cell-adherent side down by using forceps, followed by incubation at room temperature for 1 h in the dark.
7. Put the coverslips on a sheet of parafilm with the cell-adherent side up and load 300 μL of PBS on the coverslips. Incubate the coverslips at room temperature for 10 min with gentle shaking in the dark and remove the PBS. Repeat this washing step 3 times. After the final wash step, immerse the coverslips in ddH_2O to wash out the salts derived from the PBS.

3.5 Fixing Coverslips on the Slide Glass and Analyzing

1. Place the mounting medium containing DAPI on the slide glass and overlay the coverslips gently on the mounting medium with the cell-adherent side down by using forceps (*see* **Note 13**).
2. After the mounting medium has been solidified, observe the cells using a confocal microscope or another fluorescence microscope (Fig. 2). As shown Fig. 2, the signal of target molecules is influenced by the permeabilization solution used. The fluorescence signals of ganglioside GD3, a glycolipid

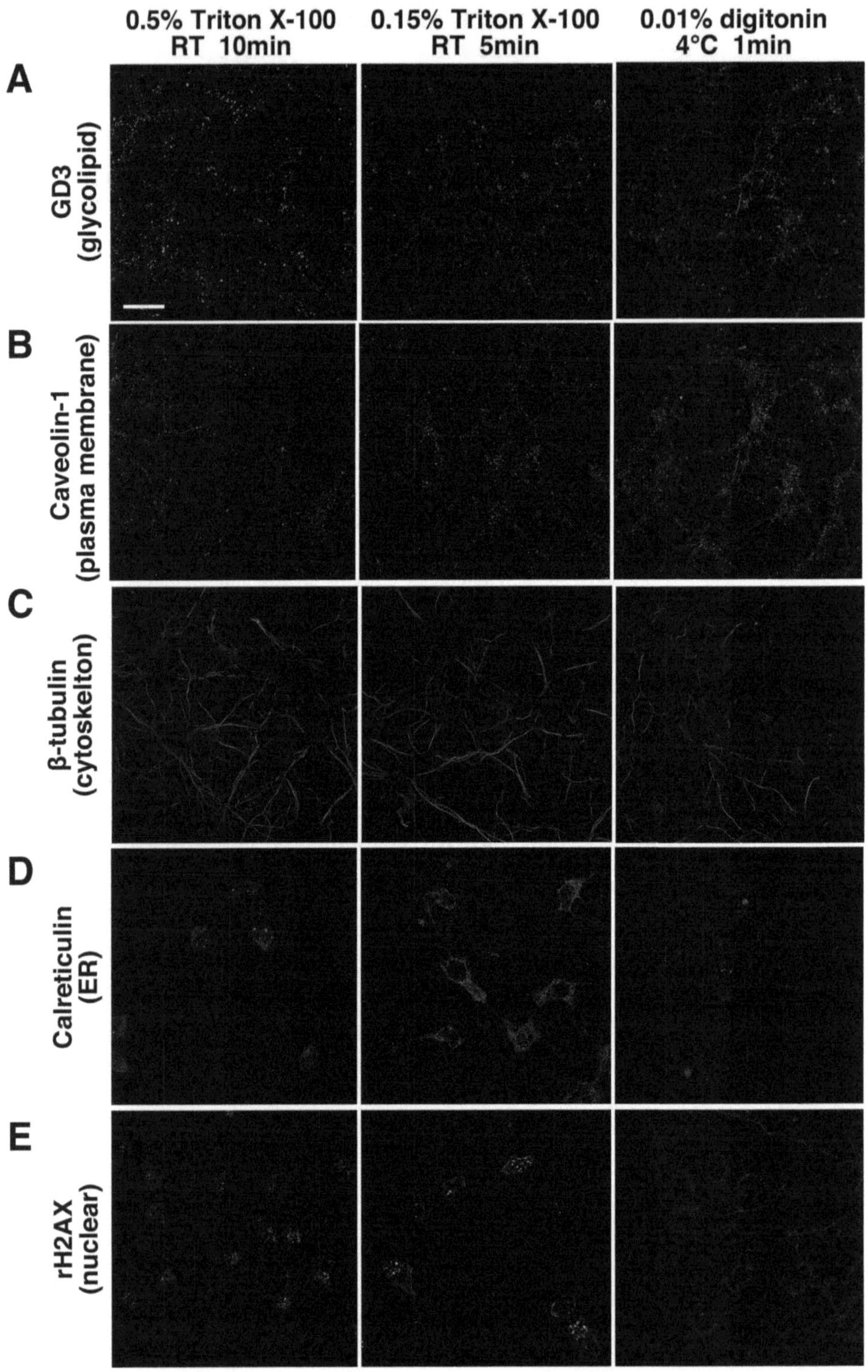

Fig. 2 Immunofluorescence images of organelle markers after using different detergents for permeabilization. Cells (left) were permeabilized using 0.5% Triton X-100 and incubated for 10 min at room temparature. Cells

localized at the plasma membrane [5], and caveolin-1, a membrane-associated protein [6], were detected more clearly, especially near the plasma membrane, when digitonin was used compared with Triton X-100 (Fig. 2a, b). The staining signals of β-tubulin, a major component of the microtubule cytoskeleton [7], showed little difference among the three permeabilization conditions (Fig. 2c). The signal of calreticulin, an ER marker [8, 9], was observed at the expected localization only when 0.15% Triton X-100 was used (Fig. 2d). γH2AX, a nuclear protein [10], was clearly localized in the nucleus when 0.5% Triton X-100 was used, but no nuclear localization was detected when digitonin was used (Fig. 2e) (*see* **Note 14**).

4 Notes

1. Appropriate culture dishes, such as coverslips, a glass-bottomed dishes, or chamber slides, and appropriate types of coating for dishes, such as PDL or laminin, should be selected according to the cell type because cell adhesion and morphology are influenced by the type of culture dish and coating. Here, we describe an immunostaining method using coverslips.
2. 18 mm coverslips can be stored in 70% ethanol. Coverslips are flame-dried briefly to prevent breakage.
3. If the coverslips are to be used soon after the coating step, they need not be dried. Without drying, add 1 mL of the culture medium to the 12-well plate with the coverslips and move to the procedure for cell culture.
4. The cell density on the coverslips should be optimized to ensure proper cell morphology and molecular localization. We recommend that 30–50 μL of cell suspension is plated in one well of a 12-well plate, when a total of 1.2 mL of cell suspension is prepared from two hemispheres (one brain) of an E15 embryo. When performing gene transfection, the amount of the plating cell suspension should be increased.
5. The optimal detergent type in the permeabilization solution should be selected depending on the purpose of the experiments. The detergent concentration and reaction temperature should also be optimized. The first choice may be 0.15% Triton X-100 because it is suitable for immunostaining studies of most proteins. However, Triton X-100 is not recommended for

Fig. 2 (continued) (middle) were permeabilized using 0.15% Triton X-100 and incubated for 5 min at room temperature. Cells (right) were permeabilized using cold 0.01% digitonin and incubated for 1 min. Organelle markers were stained with the indicated antibodies. The images were obtained with a confocal microscope

plasma membrane proteins because it solubilizes membrane components.

6. Brief incubation (1 min) with cold 0.01% digitonin solution is suitable for immunostaining studies of membrane proteins.
7. When coverslips are not used immediately for the primary antibody reaction, they can be kept for a few hours in the blocking solution at 4 °C or for 2–3 days in PBS at 4 °C after blocking.
8. Antibodies diluted in the antibody dilution buffer should be mixed thoroughly because antibody solution frequently contains glycerol, which tends to sink in a water-based solution.
9. During the primary antibody reaction on the coverslips, the humidified chamber should be gently shaken.
10. To conserve antibodies, the coverslips can be put onto a drop of the antibody solution (50–100 μL) on parafilm. The primary antibodies are diluted at a ratio of 1:100 in general, but check the product information sheet for the optimal concentration.
11. The humidified chamber can be created by using a box with a tightly fitting lid and seating parafilm on the bottom of the box. Putting wet paper at the side of the box can humidify the box.
12. After the secondary antibody reaction, perform all subsequent procedures in the dark.
13. Put the edge of the coverslip onto the mounting medium by using forceps and slope it gently until the coverslip is placed completely on the slides. Avoid creating air bubbles under the coverslip.
14. If probes for live cells, such as LysoTracker (marker of lysosomes), MitoTracker (marker of mitochondria), and a fluorescence dye–conjugated WGA (marker of plasma membrane), are used, the cells should be fixed after staining with the probes. However, the signals of Mito Tracker and WGA are lost during permeabilization using a strong detergent such as 0.5% Triton X-100 (Fig. 3), so permeabilization should be optimized according to the target molecules and purpose of the experiments.

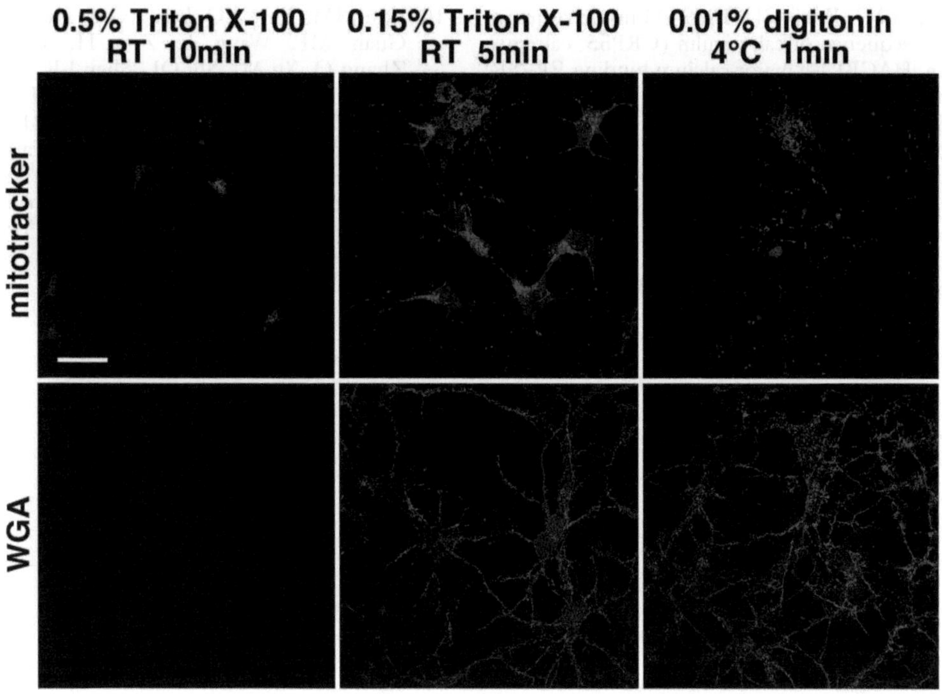

Fig. 3 Fluorescence images of organelle markers stained in live cells. After staining, cells were fixed and permeabilized with the indicated detergents. Cells (left) were permeabilized using 0.5% Triton X-100 and incubated for 10 min at room temperature (RT). Cells (middle) were permeabilized using 0.15% Triton X-100 and incubated for 5 min at RT. Cells (right) were permeabilized using ice-cold 0.01% digitonin and incubated for 1 min. Organelle markers were stained with (upper) Mito Tracker (Thermo; M22426) and (lower) Alexa647-conjugated WGA (Thermo; W32466) according to the manufacturer's instructions. Images were acquired with a confocal microscope. Scale bar: 20 μm

References

1. Dotti CG, Sullivan CA, Banker GA (1988) The establishment of polarity by hippocampal neurons in culture. J Neurosci 8(4):1454–1468
2. Evans RJ, Derkach V, Surprenant A (1992) ATP mediates fast synaptic transmission in mammalian neurons. Nature 357(6378): 503–505. https://doi.org/10.1038/357503a0
3. Kureishi Y, Kobayashi S, Amano M, Kimura K, Kanaide H, Nakano T, Kaibuchi K, Ito M (1997) Rho-associated kinase directly induces smooth muscle contraction through myosin light chain phosphorylation. J Biol Chem 272(19):12257–12260. https://doi.org/10.1074/jbc.272.19.12257
4. Ohsaki Y, Cheng J, Suzuki M, Fujita A, Fujimoto T (2008) Lipid droplets are arrested in the ER membrane by tight binding of lipidated apolipoprotein B-100. J Cell Sci 121(Pt 14): 2415–2422. https://doi.org/10.1242/jcs.025452
5. Shikanai M, Nishimura YV, Sakurai M, Nabeshima YI, Yuzaki M, Kawauchi T (2018) Caveolin-1 promotes early neuronal maturation via Caveolae-independent trafficking of N-cadherin and L1. iScience 7:53–67. https://doi.org/10.1016/j.isci.2018.08.014
6. Shikanai M, Ito S, Nishimura YV, Akagawa R, Fukuda M, Yuzaki M, Nabeshima YI, Kawauchi T (2023) Rab21 regulates caveolin-1-mediated endocytic trafficking to promote immature neurite pruning. EMBO Rep 24(3): e54701. https://doi.org/10.15252/embr.202254701
7. Kawauchi T, Chihama K, Nabeshima Y, Hoshino M (2003) The in vivo roles of STEF/Tiam1, Rac1 and JNK in cortical neuronal migration. EMBO J 22(16):4190–4201

8. Smith MJ, Koch GL (1989) Multiple zones in the sequence of calreticulin (CRP55, calregulin, HACBP), a major calcium binding ER/SR protein. EMBO J 8(12):3581–3586. https://doi.org/10.1002/j.1460-2075.1989.tb08530.x
9. Wang B, Zhao Z, Xiong M, Yan R, Xu K (2022) The endoplasmic reticulum adopts two distinct tubule forms. Proc Natl Acad Sci USA 119(18):e2117559119. https://doi.org/10.1073/pnas.2117559119
10. Lou MM, Tang XQ, Wang GM, He J, Luo F, Guan MF, Wang F, Zou H, Wang JY, Zhang Q, Xu MJ, Shi QL, Shen LB, Ma GM, Wu Y, Zhang YY, Liang AB, Wang TH, Xiong LL, Wang J, Xu J, Wang WY (2021) Long noncoding RNA BS-DRL1 modulates the DNA damage response and genome stability by interacting with HMGB1 in neurons. Nat Commun 12(1):4075. https://doi.org/10.1038/s41467-021-24236-z

Chapter 16

Brain Sample Preparation After Performing in Utero Electroporation to Proliferating and Differentiated Cells in the Developing Mouse Neocortex

Minkyung Shin and Yoichi Kosodo

Abstract

In utero electroporation (IUE) enables labeling and manipulating specific types of cells by introducing DNA plasmids with desired promoters. After the surgery, mouse brains are fixed at any stage and analyzed after staining using specific antibodies. Here, we describe the flow of the IUE experiment from the preparation to microscopic observations.

Key words In utero electroporation, DNA plasmid, Microcapillary, Electroporator, Brain dissection, Cryostat, Immunohistochemistry

1 Introduction

The mammalian neocortex comprises a 6-laminar structure required for higher brain functions [1]. The neuroepithelial cells on the ventricle side first increase in number through self-renewal and then change their characteristics to radial glial cells (RGCs), which give rise to the progeny of the central nervous system (CNS), including neurons, astrocytes, oligodendrocytes, and adult neural stem cells [2]. The RGCs generate neurons and glial cells sequentially, referred to as neurogenesis and gliogenesis [3, 4]. During neurogenesis, the newly generated immature neurons from RGCs and intermediate progenitors find their positions in the laminar structure in a "birthdate" dependent and "inside-out" manner, the early-born neurons locate more deeply while late-born neurons locate more superficially [5–7]. With a combination of intrinsic and extrinsic factors, RGCs switch to produce glial cells. During gliogenesis, RGCs mainly generate astrocytes and oligodendrocytes [8].

In utero electroporation (IUE) is a method to introduce plasmids into the cells aligning at the ventricular surface of the cortex

Koh-ichi Nagata (ed.), *Cerebral Cortex Development: Methods and Protocols*, Methods in Molecular Biology, vol. 2794, https://doi.org/10.1007/978-1-0716-3810-1_16,

by injecting the plasmids into the ventricles and applying positive electrodes toward the desired orientation [9, 10]. This method is based on the principle that the DNA with a negative charge is incorporated into the cells by a series of electric pulses with appropriate electric power [11, 12]. By introducing a plasmid-expressing fluorescent protein, such as EGFP and RFP, the IUE enables the visualization of neuronal migration in a subset population of glutamatergic pyramidal cortical neurons during brain development [13, 14]. With the development of the technique of IUE, it is possible to target specific brain regions, including the prefrontal cortex, motor cortex, hippocampus, and cerebellum, by adjusting the position and angle of the electrode [15–18].

The coding sequence on a plasmid is translated into protein within the neural progenitors in a few hours. By regulating the promoter of plasmids, the production level of protein can be controlled in target neurons and glial cells [19, 20]. Furthermore, genetic manipulations in both gain-of-function and loss-of-function are also possible. Combinations with RNA interference [21] and CRISPR/Cas9 genome editing [22] are commonly utilized in the latter case. Analyses of fluorescent-labeled cells, such as for the morphology of dendrites, leading processes, axons, and cellular locations among layers, can be combined with genetic modifications. Here, we describe a typical method from IUE to prepare brain samples for immunostaining, which will guide the researchers to obtain similar results (Fig. 1). In addition, we introduce a labeling method of proliferative cells using thymidine analogs in vivo, which can be combined with IUE [23].

2 Materials

2.1 Plasmid Preparation

1. The recommended promotor of plasmid DNA: Cytomegalovirus (CMV) enhancer fused to the chicken beta-actin promoter (CAG) [24, 25].
2. Resolving plasmid buffer: 0.2 μm filtered EB buffer or nuclease-free water:
   - Prepare plasmids using the Midi-prep or Maxi-prep Kit.
   - Stock of plasmid should be above the final concentration.
3. Fast-green stock solution: 0.1% Fast-green solution dissolved in 0.2 μm of filtered EB buffer.
4. The final concentration of plasmid for surgery: 0.2–2 μg/mL, depending on the requirements (*see* **Note 1**):
   - 0.01% Fast-green solution.
   - For fluorescent plasmids, 0.2–1 μg/mL is recommended.
   - Dilute using 0.2 μm of filtered EB buffer or nuclease-free water.

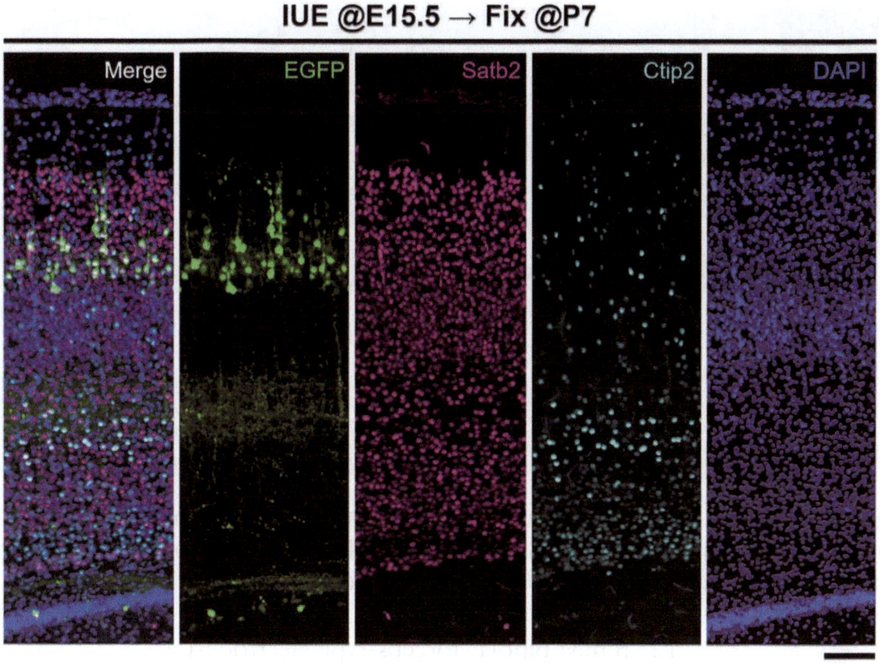

Fig. 1 IHC staining after IUE to the mouse somatosensory cortex. The brain was obtained at P7 after IUE of pCAG-EGFP-N1 + pCAGGS at E15.5. After antigen retrieval, the staining was performed with an anti-Satb2 antibody (magenta), an anti-Ctip2 antibody (cyan), and DAPI (blue). Secondary antibodies: Alexa 555 for Ctip2, Alexa 647 for Satb2. Scale bar: 100 μm

2.2 Microcapillary Preparation

1. Glass pipette puller for making glass microcapillaries.
2. Break the capillaries using fine-tip forceps (*see* **Note 2**).

2.3 Surgery Preparation

1. A broken glass microcapillary connected to the needle holder with a filter, and 1 mL syringe for mouth suction (Fig. 2a).
2. One fine-tip forceps (Fig. 2b).
3. One Graefe forceps (Fig. 2b).
4. Two scissors (Fig. 2b).
5. Tungsten carbide jaws (Fig. 2b).
6. Medical tape (Fig. 2b).
7. Autoclaved PBS in a 50 mL syringe with a 0.2 μm filter (Fig. 2b).
8. Medial threads (nylon and silk) (Fig. 2b).
9. A 5 mm diameter platinum electrode (Fig. 2c).
10. Isoflurane anesthesia machine (Fig. 2e).
11. Electroporator (Fig. 2f).

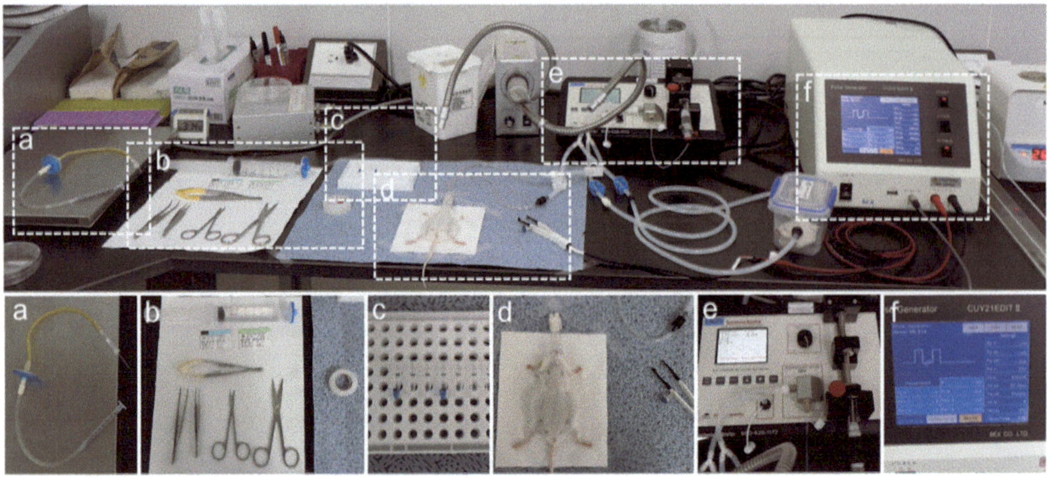

Fig. 2 Surgery preparation for IUE. (**a**) Broken glass microcapillary connected to the needle holder with a filter and 1 mL syringe for mouth suction. (**b**) Surgical tools. (**c**) Plasmid solution with 0.01% Fast-green. (**d**) Anesthetized mouse. (**e**) Isoflurane anesthesia machine. (**f**) Electroporator

12. 5 mM bipolar forceps-type electrodes.
13. Sterilizer.

2.4 Animal Preparation

1. Theoretically, all pregnant mice can be the subject of surgery.
2. This protocol uses an ICR mouse because of many embryos (>10).
3. It is recommended to perform the surgery from the embryonic day (E)12.5 to E16.5.
4. Electroporation at E12.5 will mainly label deeper layer neurons (layers IV–V), while E15.5 will mainly label superficial layer neurons (layers II–III).
5. If the surgery is performed later than E16.5 (too close to the birth date-E19), the pregnant mouse may give a premature birth.

2.5 Dissection for Fixation of Brains

1. Autoclaved PBS (store at 4 °C).
2. 4% paraformaldehyde (PFA) (store at 4 °C).
3. Two fine-tip forceps.
4. Scissors.
5. 15 mL or 50 mL conical tube.
6. Laboratory rocker (store at 4 °C).
7. Ice.
8. Anesthesia machine (if necessary).

2.6 Brain Sample Preparation for Cryostat

1. 20% sucrose in autoclaved PBS.
2. 30% sucrose in autoclaved PBS.
3. OCT compound:30% sucrose in autoclaved PBS = 3:1.
4. One fine-tip forceps.
5. Two knives.
6. 10 cm Petri dish.
7. Cryomold.
8. Liquid nitrogen or dry ice.
9. Metal block.
10. Deep freezer (−80 °C).

2.7 Cryostat for Generating Brain Slices

1. Cryostat.
2. Specimen disk.
3. Cryomold with brains (store at −80 °C).
4. A round-type brush.
5. OCT compound.
6. Slide glasses.
7. Slide holder.

2.8 Immunohistochemistry (IHC)

1. 0.05% Triton X-100 in PBS (PBSTx).
2. Glass or plastic jar.
3. 2% BSA in 0.05% PBSTx.
4. Primary antibodies.
5. Secondary antibodies.
6. 4′,6-Diamidino-2-phenylindole (DAPI).
7. Liquid blocker.
8. Cover glasses.
9. Shaker.

(*Optional*) Thymidine-analogs for labeling proliferative cells

1. 5-Ethynyl-2′-deoxyuridine (EdU): Prepare a stock solution of 0.5 mg/mL in PBS.
2. 5-Bromo-2′-deoxyuridine (BrdU): Prepare a stock solution of 2 mg/mL in PBS.

For both EdU and BrdU solutions, aliquot 500 μL into sterile tubes (a suitable volume for injecting into a 50 g pregnant mouse). Store the tubes at −20 °C. After thawing the solution, warm it to room temperature (RT) before injecting into a mouse.

3 Methods

3.1 Preparation

1. Prepare plasmid solution according to the experimental purpose (*see* Subheading 2.1):
   - Example of the preparation: The fluorescent reporter construct (pCAG-EGFP-N1) with the empty vector (pCAGGS).

	pCAG-EGFP-N1	pCAGGS	EB	0.1% FGS
Stock	3.2 μg/μL	10 μg/μL		
Dilution	0.5/3.2	1/10		
Final	0.5 μg/μL	1 μg/μL		
33 μL	4.7 μL	3 μL	22.3 μL	3 μL

2. Prepare the microcapillary and electroporator (*see* Subheadings 2.2 and 2.3):
   - Turn on the electroporator and connect it to the forceps-type electrodes.
   - Connect the foot pedals to the electroporator.
   - Break the microcapillary using sterilized or 70% ethanol-cleaned fine-tip forceps and connect it to the needle holder with a filter and a 1 mL syringe for mouth suction (*see* **Note 3**).
   - Example of electroporator setting at E15.5.

Pulse voltage	35 V
Programmed pulses	4
Pulse width	50 ms
Space between pulses	950 ms

3. Prepare surgical instruments (*see* Subheading 2.3):
   - Clean with 70% ethanol just before using it.
4. Prepare a pregnant mouse (*see* Subheading 2.4).

3.2 Anesthesia

1. Fill enough isoflurane for the surgery's duration and connect it to the anesthesia machine. Turn on the device for use.
2. Put a pregnant mouse in the chamber and close it until anesthetized.
3. Connect the face of the mouse to the nose mask.
4. Place the abdominal side to be exposed and tide feet using medical tapes.

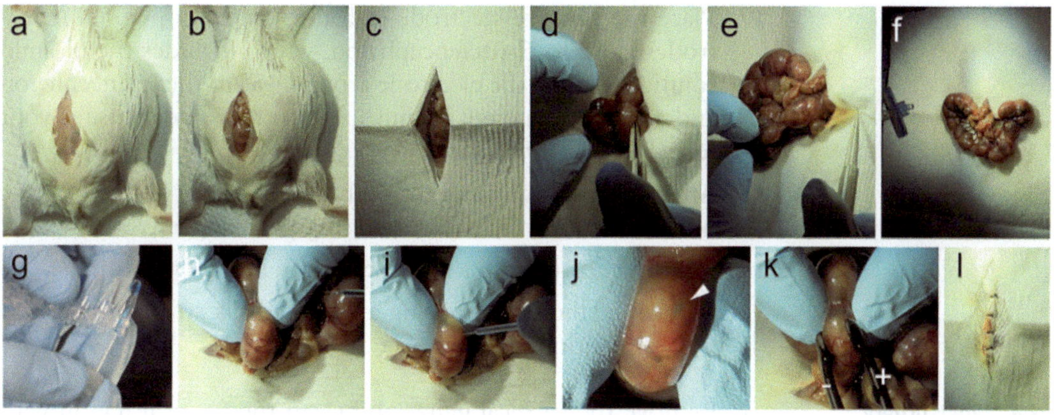

Fig. 3 Surgery procedure. (**a**–**l**) Details are described in Subheading 3.3. (**j**) Plasmid solution with Fast-green spreads adequately in the right side of the lateral ventricle (arrowhead)

3.3 Surgery

1. Replace the medical gloves or clean the gloves with 70% ethanol thoroughly before using surgical instruments.
2. Clean the surgical region with 70% ethanol.
3. Raise the skin using Graefe forceps and make a 1.5–2 cm vertical incision in the middle of the abdomen to expose abdominal muscles (Fig. 3a).
4. Raise the abdominal muscles using fine-tip forceps and make a 1.5–2 cm vertical incision according to the white line (Fig. 3b).
5. Cut the surgical gauze into a rhombus shape and put the gauze on the mouse's abdomen to expose only the surgical incision (Fig. 3c).
6. Drop the autoclaved PBS into the exposed region.
7. Pull out one side of the abdominal muscle using the fine-tip forceps to expose the uterine horn, and carefully pull out the right and left uterine horns using your fingers (Fig. 3d) (*see* **Note 4**).
8. Drop the autoclaved PBS to the uterine horns, put back one of the uterine horns, and retain one of the uterine horns for surgery (Fig. 3f) (*see* **Note 5**).
9. Fill up the microcapillary with the plasmid solution (Fig. 3g).
10. Orient and grab the embryo carefully, using your fingers to visualize the telencephalon (Fig. 3h).
11. Gently poke the microcapillary with the plasmid solution into the lateral ventricle and immediately inject the plasmid solution into the ventricle (Fig. 3i). If the injection is performed appropriately, the Fast-green solution will be visualized like an arrowhead (Fig. 3j) (*see* **Notes 6** and 7).

12. Place the positive electrode of 5 mm diameter platinum electrodes close to the telencephalon, which includes the plasmid solution. Touch the electrode around 45° to the telencephalon and apply the voltage by pushing the foot pedals using your feet (Fig. 3k) (*see* **Note 8**).
13. After electroporation of the embryos, carefully put the uterine horns back into the abdomen using your fingers.
14. Start to suture using a sterilized nylon string (*see* **Note 9**).
15. Suture the mouse skin using a sterilized silk string in the same way as mentioned above (Fig. 3l).

3.4 Postsurgery

1. Remove the surgical gauze and clean the surgical region with 70% ethanol.
2. Remove the medical tapes and the nose mask.
3. Wrap the mouse using tissue softly and place it in the cage with its abdominal side exposed.
4. Allow the mouse to recover in a warm environment.
5. The mouse wakes up during the recovery.
6. Place it back in the cage and observe whether it behaves normally.
7. Sterilize the surgery instruments using a sterilizer.

3.5 Dissection for Fixation of Brains

(*Optional*) Perform an intra-peritoneal (IP) injection of the solution of thymidine analogs (EdU or BrdU) at the preferred time before dissection for fixation. For instance, 30 min is suitable for labeling S-phase cells in the cell cycle, while 2 h is recommended for labeling S/G2/M-phase cells. Dual labeling using both EdU and BrdU with different time points is possible.

1. If the plasmids that express fluorescent proteins are included, they can be visualized under the fluorescent microscope after birth. For the selection of electroporated mice, it is recommended to check the fluorescence before the postnatal day (P) 2 (*see* **Note 10**).
2. Prepare the required materials before the dissection (*see* Subheading 2.5) (Fig. 4a):
   - Autoclaved PBS and 4% PFA on the ice.
   - Prepare 30 mL of 4% PFA in a 50 mL conical tube.
3. At the expected stage, dissect the brains:
   - Pregnant stage: Sacrifice the pregnant mouse and immediately pull out of the embryos. Cut the right and left uterine horns and collect embryos in cold PBS on the ice.

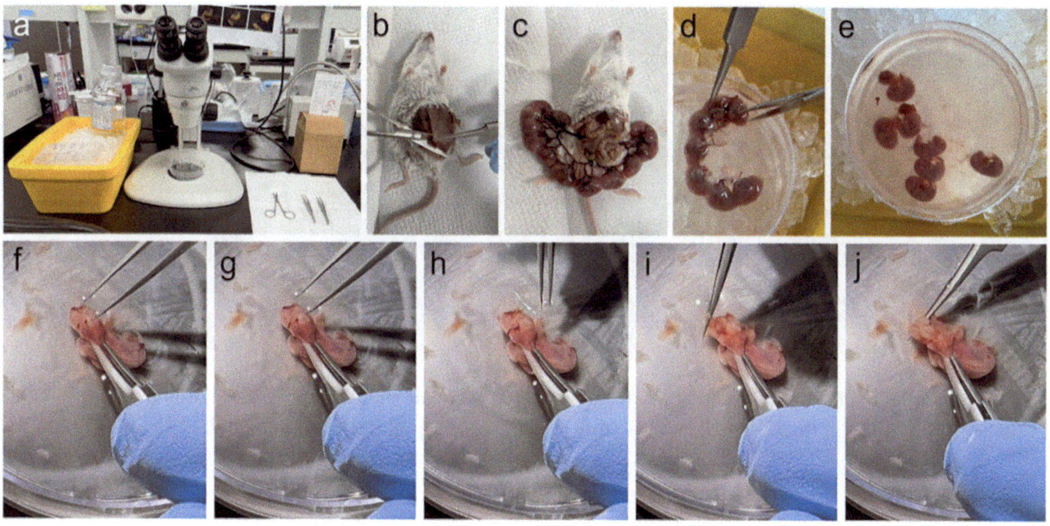

Fig. 4 Dissection procedure. (**a**) Preparation for brain dissection at E16.5 after IUE of pCAG-EGFP-N1 + pCAGGS at E14.5. (**b–j**) Details are described in Subheading 3.5

- Postnatal stage: Proceed with gas anesthesia and then perfuse the mice transcardially under the fume hood (*see* **Note 11**).

4. Raise the skin using forceps and make a 3–4 cm vertical incision in the middle of the abdomen to expose abdominal muscles.
5. Raise the abdominal muscles using forceps and make a 3–4 cm vertical incision to expose the embryos (Fig. 4b).
6. Carefully pull out the right and left uterine horns using fingers (Fig. 4c).
7. Transfer the cut uterine horn to PBS on ice and remove the amniotic membrane (Fig. 4d, e).
8. Secure the nape of the embryo with fine-tip forceps, then use another pair of forceps to remove the skin surrounding the brain (Fig. 4f).
9. Make a puncture with fine-tip forceps between the cerebellum and cerebral fissure and drag it forward to expose the brain (Fig. 4g).
10. Grasp one side of the skull with fine-tip forceps and hold it open. Perform the same procedure on the other side of the skull to fully expose the brain (Fig. 4h, i).
11. Insert the fine-tip forceps under the cerebellum to lift and separate the brain (Fig. 4j).
12. Transfer the brain into a 50 mL conical tube filled with 4% PFA using the fine-tip forceps or a spoon.
13. Put the 50 mL conical tube onto the laboratory rocker at 4 °C overnight.

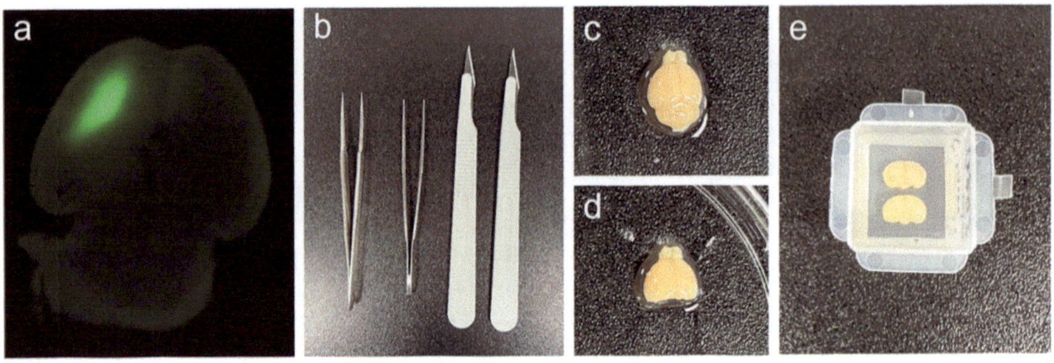

Fig. 5 Brain sample preparation procedure. (**a**) One of the examples of EGFP-expressing brains. The brain was obtained at P7 after IUE of pCAG-EGFP-N1 + pCAGGS at E14.5. (**b–e**) Details are described in Subheading 3.6

3.6 Brain Sample Preparation for Cryostat

1. Next day, replace 4% PFA with 20% sucrose solution for cryoprotection at 4 °C overnight (*see* **Note 12**).
2. Replace 20% sucrose solution with 30% sucrose solution at 4 °C overnight (*see* **Note 13**).
3. Place the brain in the Petri dish with 30% sucrose solution.
4. Prepare required materials (*see* Subheading 2.6) (Fig. 5a).
5. Put the brains in the plate using the fine-tip forceps and drop the OCT compound to block the brains from drying up (Fig. 5b).
6. Remove the cerebellar cortex using two knives by crossing it (*see* **Note 14**) (Fig. 5c).
7. Transfer the brain to the cryomold and place it perpendicular to the surface.
8. Fill with the OCT compound in 30% sucrose solution until the mouse brain is submerged (Fig. 5d).
9. Put the metal block in the container and fill it with the liquid nitrogen (*see* **Note 15**).
10. Put the brain samples on the metal block until the OCT compound freezes (*see* **Note 16**).
11. Store the brain samples in a deep freezer at −80 °C.

3.7 Cryostat for Generating Brain Slices

1. Place the brain block in the cryostat for 1–2 h to equilibrate the temperature of the brain sample to the cryostat.
2. Place the specimen disk in the cryostat, apply the OCT compound, and immediately put the brain block on the disk (Fig. 6a).
3. Section and collect the brain slices to 10–30 μm thickness using a round-type brush (Fig. 6b–f).
4. After drying up, store brain slices in the slide holder at −20 °C until performing IHC.

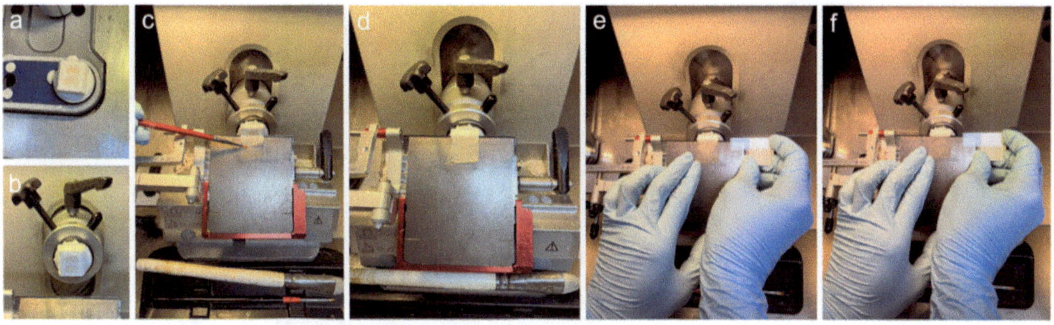

Fig. 6 Cryostat procedure. (**a**–**f**) Details are described in Subheading 3.7

3.8 IHC

1. Antigen retrieval can be performed before staining (*see* **Note 17**).

   (*Optional*) To stain with BrdU antibody, wash 3 times with PBS for 5 min each, and then incubate the brain slices with 2 N HCl for 1 h.
2. Wash the brain slices using PBSTx 3 times for 10 min each.
3. Blocking: Incubate the brains in 2% BSA in PBSTx for 1 h at RT.
4. Primary antibody solution: Add the required amount of primary antibodies in 2% BSA in PBSTx and incubate the brains at 4 °C overnight.
5. Wash the brain slices using PBSTx 3 times for 10 min each.

   (*Optional*) Perform EdU detection for 30 min. The reaction solution should be prepared according to the manufacturer's instructions. Then wash 3 times with PBSTx for 5 min each.
6. Secondary antibody and DAPI solution: Add the required amount of secondary antibodies and DAPI in 2% BSA in PBSTx and incubate the brains for 1 h at RT.
7. Wash the brain slices using PBSTx 2 times for 15 min each.
8. Mount the brain slices in a mounting medium (Fig. 7) (*see* **Note 18**).
9. Store the specimen at 4 °C before imaging.

4 Notes

1. The required volume of the final plasmid mixture is prepared just before the surgery. The total volume can be adjusted according to the number of embryos for surgery. About 50 μL of the plasmid mixture covers more than 10 embryos. Less than 20 μL of the mix is not recommended because it will

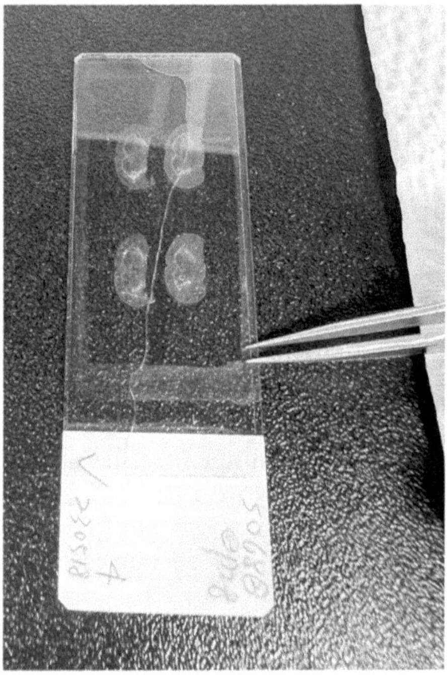

Fig. 7 Mounting procedure

increase the chance of breaking the glass microcapillaries while replenishing the mixture.

2. Microcapillaries should be prepared using a glass pipette puller and then break them into diagonal shapes using 70% ethanol-cleaned fine-tip forceps. Ensure broken microcapillaries are sharp enough to poke the brains but not damage them [26].
3. A 1 mL syringe is for mouth suction.
4. Do not pull out too strong, which may break the uterine capillaries.
5. Check how many embryos you can proceed with for the surgery.
6. The lateral ventricle is not far from the skull. Once the microcapillary pokes the skull, it may reach the lateral ventricle. If the plasmid solution with Fast-green spreads well, the microcapillary reaches the ventricle appropriately.
7. It is better to inject plasmids into all embryos first and then apply electroporation next. It can reduce the total surgery time, which is also critical to the survival of the embryos. Recommended total surgery time is less than 30 min.
8. The embryos move their feet when the voltages are applied.
9. Four to five passes are acceptable.

10. The head skin of pups becomes thickened, which hinders observing fluorescence.
11. The neonatal stage of mouse pups is too small to perform transcardiac perfusion. Cutting the head immediately after anesthesia can be a way to collect the brains.
12. Overnight incubation is fine. Do not incubate brains in the 4% PFA solution too long. It will reduce the reactivity to the antigen–antibody reaction.
13. The brain should have sunk after overnight incubation with 20% sucrose solution.
14. This method is for making coronal sections of the brain.
15. Do not put the brains into the liquid nitrogen directly.
16. Just before putting the brains on the metal block, adjust it again if it is not perpendicular to the surface.
17. Antigen retrieval will increase the reactivity of antigen–antibody, which will increase the fluorescent signals.
18. Slowly put the cover glass from one side to the other side so as not to include bubbles.

Acknowledgments

We are grateful to all the members of Kosodo laboratory and Dr. Kazunori Nakajima for their valuable discussions. This project was supported by KBRI basic research program through the Korea Brain Research Institute funded by Korea government Ministry of Science and ICT (MSIT) (23-BR-01-03 and 23-BR-03-01 to Y.K.) and by the National Research Foundation of Korea (NRF) grant funded by Korea government MSIT (No.2022R1A2C1002728 to Y.K. and No.2022R1F1A1074446 to M.S.).

Permission to Animal Experiments The animal protocol M3-IACUC-19-00031 has been approved by the IACUC (KBRI).

References

1. Rakic P (2009) Evolution of the neocortex: a perspective from developmental biology. Nat Rev Neurosci 10(10):724–735
2. Rowitch DH, Kriegstein AR (2010) Developmental genetics of vertebrate glial-cell specification. Nature 468(7321):214–222
3. Miller FD, Gauthier AS (2007) Timing is everything: making neurons versus glia in the developing cortex. Neuron 54(3):357–369
4. Yamaguchi M et al (2016) Neural stem cells and neuro/gliogenesis in the central nervous system: understanding the structural and functional plasticity of the developing, mature, and diseased brain. J Physiol Sci 66(3):197–206
5. Angevine JB Jr, Sidman RL (1961) Autoradiographic study of cell migration during histogenesis of cerebral cortex in the mouse. Nature 192:766–768

6. Rakic P (1974) Neurons in rhesus monkey visual cortex: systematic relation between time of origin and eventual disposition. Science 183(4123):425–427
7. Greig LC et al (2013) Molecular logic of neocortical projection neuron specification, development and diversity. Nat Rev Neurosci 14(11):755–769
8. Malatesta P, Appolloni I, Calzolari F (2008) Radial glia and neural stem cells. Cell Tissue Res 331(1):165–178
9. Briz CG et al (2017) In utero electroporation approaches to study the excitability of neuronal subpopulations and single-cell connectivity. J Vis Exp 120:55139
10. dal Maschio M et al (2012) High-performance and site-directed in utero electroporation by a triple-electrode probe. Nat Commun 3:960
11. Yamashiro K, Ikegaya Y, Matsumoto N (2022) In utero electroporation for manipulation of specific neuronal populations. Membranes (Basel) 12(5):513
12. Zhang L, Getz SA, Bordey A (2021) Dual in utero electroporation in mice to manipulate two specific neuronal populations in the developing cortex. Front Bioeng Biotechnol 9: 814638
13. Tabata H, Nakajima K (2001) Efficient in utero gene transfer system to the developing mouse brain using electroporation: visualization of neuronal migration in the developing cortex. Neuroscience 103(4):865–872
14. Saito T, Nakatsuji N (2001) Efficient gene transfer into the embryonic mouse brain using in vivo electroporation. Dev Biol 240(1):237–246
15. Kitazawa A et al (2014) Hippocampal pyramidal neurons switch from a multipolar migration mode to a novel "climbing" migration mode during development. J Neurosci 34(4):1115–1126
16. Kanatani S et al (2015) The COUP-TFII/Neuropilin-2 is a molecular switch steering diencephalon-derived GABAergic neurons in the developing mouse brain. Proc Natl Acad Sci USA 112(36):E4985–E4994
17. Baumgart J, Baumgart N (2016) Cortex-, hippocampus-, thalamus-, hypothalamus-, lateral septal nucleus- and striatum-specific in utero electroporation in the C57BL/6 mouse. J Vis Exp 107:e53303
18. Hazra D et al (2022) Rhythmic activation of excitatory neurons in the mouse frontal cortex improves the prefrontal cortex-mediated cognitive function. Cereb Cortex 32(23):5243–5258
19. Hamabe-Horiike T et al (2021) Glial cell type-specific gene expression in the mouse cerebrum using the piggyBac system and in utero electroporation. Sci Rep 11(1):4864
20. Tabata H et al (2022) Erratic and blood vessel-guided migration of astrocyte progenitors in the cerebral cortex. Nat Commun 13(1):6571
21. LoTurco J, Manent JB, Sidiqi F (2009) New and improved tools for in utero electroporation studies of developing cerebral cortex. Cereb Cortex 19(Suppl 1):i120–i125
22. Tsunekawa Y, Terhune RK, Matsuzaki F (2021) Protocol for de novo gene targeting via in utero electroporation. Methods Mol Biol 2312:309–320
23. Etienne O et al (2014) Assessing cell cycle progression of neural stem and progenitor cells in the mouse developing brain after genotoxic stress. J Vis Exp 87:51209
24. Miyazaki J et al (1989) Expression vector system based on the chicken beta-actin promoter directs efficient production of interleukin-5. Gene 79(2):269–277
25. Dou Y et al (2021) The CAG promoter maintains high-level transgene expression in HEK293 cells. FEBS Open Bio 11(1):95–104
26. Meyer-Dilhet G, Courchet J (2020) In utero cortical electroporation of plasmids in the mouse embryo. STAR Protoc 1(1):100027

Chapter 17

Live Imaging of Migrating Neurons and Glial Progenitors Visualized by in Utero Electroporation

Masashi Nishikawa, Koh-ichi Nagata, and Hidenori Tabata

Abstract

During cortical development, both neurons and glial cells are generated in the germinal zone near the lateral ventricle, migrate in the correct direction, and settle in their appropriate locations. This developmental process can be clearly visualized by introducing fluorescent protein-expression vectors via in utero electroporation. In this chapter, we describe labeling methods for migrating neurons and glial progenitors, as well as methods for slice culture, and time-lapse imaging.

Key words In utero electroporation, Cerebral cortex, Neurodevelopmental disorder, Slice culture, Time-lapse imaging

1 Introduction

Neuronal cells, including neurons and glial cells, are located within specific brain regions and perform roles in neuronal networks and physiological functions. A characteristic feature of brain development is that the neural cells are generated from the neural stem cells in the ventricular zone (VZ) lining the lateral ventricle or from secondary progenitor cells in the subventricular zone adjacent to the VZ and migrate toward their predestined positions where they perform physiological functions [1, 2]. The migration processes are tightly regulated during development, and their disturbance caused by external environmental factors or genetic abnormalities can lead to psychiatric disorders with structural brain abnormalities including band heterotopia, lissencephaly, and polymicrogyria [3, 4]. In utero electroporation is a method to introduce plasmid DNA into neural stem cells in the VZ [5, 6]. Owing to its high transfection efficiency, introducing green fluorescent protein (GFP) or other fluorescent protein expression vectors via in utero electroporation allows clear visualization of cells derived from the VZ with a high signal-to-noise ratio. Organotypic slice cultures of the brain

Koh-ichi Nagata (ed.), *Cerebral Cortex Development: Methods and Protocols*, Methods in Molecular Biology, vol. 2794, https://doi.org/10.1007/978-1-0716-3810-1_17,

recapitulate normal development in vivo for at least 24 h [7]. Time-lapse observation of fluorescent protein-labeled cells in slice cultures reveals the movement of individual cells, including their migration speed and direction [8, 9]. Using in utero electroporation, two or more plasmids can be easily cotransfected by mixing them, allowing the effects of overexpression or knockdown on cell behavior to be visualized with fluorescent proteins.

Here, we describe a method for the time-lapse observation of migrating neurons and glial progenitors labeled with fluorescent proteins using in utero electroporation. Please refer to another chapter for detailed information about in utero electroporation (Chap. 16). We provide labeling methods not only for neurons but also for glial cells. The slice culture method described here is applicable for the observation of both neurons and glial cells.

2 Materials

2.1 In Utero Electroporation

1. Electric square pulse generator for in vivo electroporation (*see* **Note 1**).
2. Tweezers-type electrode (5 mM diameter) (*see* **Note 2**).
3. An aspirator tube assembly composed of a mouthpiece, silicon tube, and capillary holder (*see* **Note 3**).
4. Plate warmer: 23 × 16 cm or larger. Set the temperature to 38 °C.
5. Surgical instruments: fine forceps × 2, surgery scissors × 2, ring forceps (8 × 10-mM rings), surgical suture, surgical needle, and needle holder (for suturing).
6. Micropipettes for injection: Make from 1-mM diameter glass capillary tubes with a puller (*see* **Note 4**). Cut the tip of the micropipette obliquely using fine forceps under a dissecting microscope and attach it to the holder of the aspirator tube assembly.
7. HEPES-buffered saline (HBS): HEPES 5 g/L, NaCl 8 g/L, KCl 0.37 g/L, $Na_2HPO_4 \cdot 12H_2O$ 0.251 g/L, and glucose 1 g/L in water. Alternatively, purchase 2× concentrate solution (Sigma) and dilute in autoclaved water in a 1:1 ratio.
8. MMB anesthetic: A mixture of medetomidine (30 μg/mL), midazolam (0.4 mg/mL), and butorphanol (0.5 mg/mL) in autoclaved water (*see* **Note 5**).
9. Atipamezole solution: An antagonist of MMB anesthesia. Prepare a 150 μg/mL solution in autoclaved water.
10. Plasmids for labeling neurons: Conventional CAG promoter-driven expression vectors. Plasmids are dissolved in HBS. To monitor the location of the injected DNA solution in the brain,

add 1/10 volume of 0.2% Fast Green solution (w/v in distilled water) to the DNA solution.

11. Plasmids for labeling glial progenitors: The transposon vector system consists of a CAG promoter-driven transposase expression vector and a donor plasmid to be integrated into the host genome (*see* **Note 6**). Plasmid solutions are prepared as described for neuron labeling.

2.2 Slice Culture

1. Hanks' Balanced Salt Solution (HBSS) (+): Purchase 500 mL of HBSS (−) and add 5 mL each of autoclaved $CaCl_2$ (14 g/L) and $MgSO_4$-$7H_2O$ (20 g/L) (*see* **Note 7**). Before use, the solution is chilled on ice and bubbled through a mixture of 95% O_2 and 5% CO_2 for 10 min to achieve oxygen saturation (*see* **Note 8**). The solution is stored at 4 °C.
2. Culture medium: a mixture of 10% fetal bovine serum (FBS), 2% B27, 0.5 mM L-glutamine, and approximately 50 units/mL penicillin-streptomycin in the neurobasal medium (*see* **Note 9**).
3. After filling a 35-mM glass-bottom dish with the culture medium, place the culture insert membrane on it carefully without introducing any bubbles. Subsequently, add the medium onto the membrane (*see* **Note 10**). Place the dish on ice.
4. Put 20 mL of phosphate-buffered saline (PBS) into two 100-mM glass dishes and place them on ice.
5. Put 20 mL of HBSS (+) into a 100-mM glass dish and place it on ice.
6. Prepare a solution of 3% low melting point agarose in HBSS. Heat the solution using autoclaving or microwaving until the agarose is completely dissolved. Incubate the solution in a temperature-controlled water bath at 55 °C prior to use.
7. Turn on the confocal laser microscope and set the incubator chamber to 37 °C, 95% humidity, 5% CO_2, and 40% O_2 (*see* **Note 11**).

3 Methods

3.1 In Utero Electroporation

1. Anesthetize a pregnant mouse at embryonic day (E) 14 (for labeling neurons) or E16 (for labeling glial cells) by intraperitoneally injecting MMB anesthetic at a dose of 10 μL/g body weight.
2. After 10 min, place the mouse on its back on a plate warmer with a temperature maintained at 38 °C (*see* **Note 12**).

3. Make an approximately 2 cm skin incision along the midline using one set of forceps and scissors.
4. Then, make a 2 cm midline incision in the abdominal wall along the linea alba using another set of forceps and scissors.
5. Gently draw the uterine horns out using ring forceps.
6. After observing the orientation of the embryos through the wall of the uterine horn, insert a micropipette into one side of the lateral ventricle and inject 1 μL of the plasmid solution via expiratory pressure using an aspirator tube assembly (*see* **Note 13**).
7. Pinch the head of the embryo with a tweezer-type electrode, and apply electronic pulses using an electroporator (35 V, 50 ms, 5 times with 450 ms intervals).
8. After electroporation, place the uterine horn back in the abdominal cavity.
9. Pour 1–2 mL of sterile PBS into the abdominal cavity (*see* **Note 14**).
10. Apply atipamezole solution into the abdominal cavity at a dose of 10 μL/g body weight.
11. Close the abdominal wall and skin with surgical sutures. Alternatively, the skin can be closed using an autoclip.
12. Warm the mice on a warmer plate for approximately 30 min after surgery (*see* **Note 12**).

3.2 Slice Preparation

1. Sacrifice a pregnant mouse at E16 (electroporated at E14 to label neurons) or E17 (electroporated at E16 to label glial cells) and dissect out the uterus. Transfer the uterus to a 100-mM glass dish containing chilled PBS.
2. Collect the embryos from the uterus and transfer them to another 100-mM glass dish filled with chilled PBS to wash off any blood (Fig. 1a).
3. Harvest brains from the embryos and transfer them to a 10 cm glass dish filled with chilled HBSS (+) (*see* **Note 15**).
4. Embed the brains into 3% low melting point agarose dissolved in HBSS (+) with the dorsal side of the brain facing upward (*see* **Note 16**) (Fig. 1b).
5. After the agarose is solidified, cut it into blocks.
6. Glue the agarose-embedded brain tissue blocks to the slicer stage with the olfactory bulb at the top and the cerebellum at the bottom to make coronal sections (*see* **Note 17**) (Fig. 1c).
7. Immerse the brain tissue blocks on the slicer stage in chilled HBSS (+) (*see* **Note 18**) (Fig. 1c).

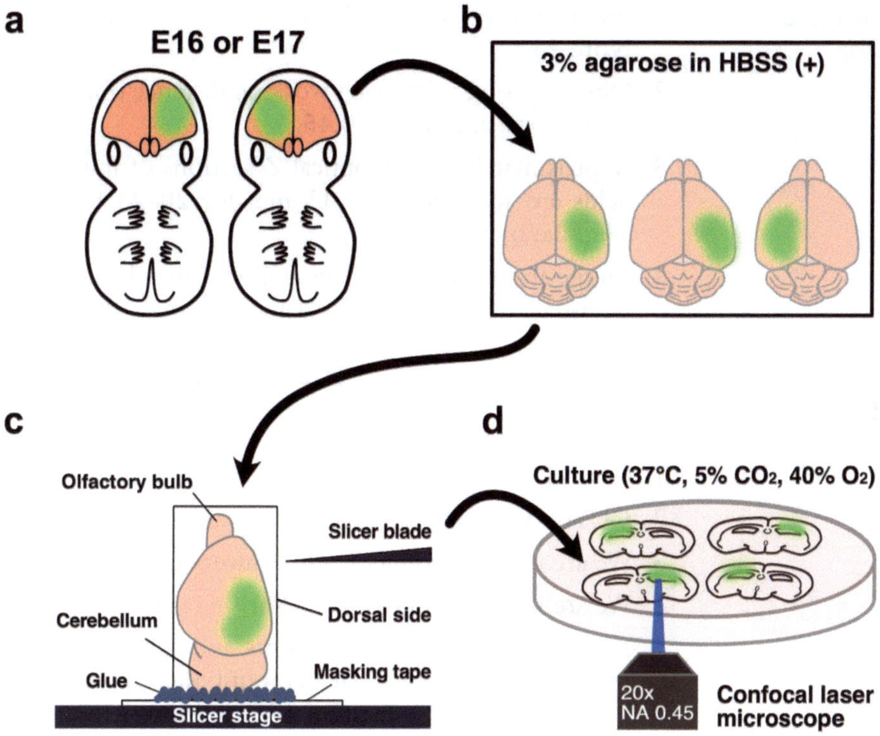

Fig. 1 Experimental procedures for time-lapse imaging of cells migrating out from the cortical ventricular zone. (**a**) Embryos electroporated at E14 (for labeling neurons) or E16 (for labeling glial cells) are sampled at E16 or E17, respectively. (**b**) Brains are dissected out and embedded in 3% low melting point agarose gel. (**c**) Brains are set on a stage and sliced coronally using a vibrating microtome. (**d**) Slices are placed on a filer membrane, and labeled cells are observed using a confocal laser microscope

8. Remove the rust preventive coating from the slicer blade using acetone (*see* **Note 19**).
9. Make coronal slices (250 μm thickness) from the interventricular foramen using a microtome.
10. Place the slices on the insert membrane in a 35-mM glass bottom dish containing the culture medium (*see* **Note 20**) (Fig. 1d). Keep the dish on ice.
11. To stabilize the adhesion between the brain slices and the insert membrane, carefully remove any excess culture medium (*see* **Note 21**).
12. Immerse the brain slices attached to the membrane surface by adding the culture medium (*see* **Note 22**).
13. Mount the dish on the incubator chamber of a confocal laser microscope, maintained at 5% CO_2, 40% O_2, and 37 °C (*see* **Note 23**) (Fig. 1d).

3.3 Time-Lapse Observation

1. Incubate the slices for 30 min to allow them to recover (*see* **Note 24**).
2. Observe using a 20× long working distance objective lens with a numerical aperture of 0.45.
3. Approximately 10–15 optical Z sections of multiple objective fields are acquired every 15 min for 20–24 h (*see* **Note 25**) (Fig. 2a, b).
4. Merge around 10 focal planes (approximately 50-μm thickness) to visualize the entire cellular morphology (*see* **Note 26**).

4 Notes

1. CUY21E, NEPA21 (NEPA gene) or equivalent.
2. CUY650P5 (NEPA gene) or equivalent.
3. We are using a product from Drummond.
4. We are using a product from Narishige (GD-1).
5. Purchase the three components individually and prepare a mixture in an appropriate amount (approximately 25 mL). The mixture can be stored at 4 °C in the dark for a few months.
6. We are using the PiggyBac transposon vector system [10], a gift from the Wellcome Trust Sanger Institute.
7. We are using HBSS(−) (Gibco).
8. If the oxygen concentration decreases after approximately 2 weeks, the bubbling process must be repeated.
9. We are using Neurobasal Medium (Gibco).
10. After adding 1 mL of the culture medium to the 35-mM dish, place the culture insert membrane carefully, ensuring that there are no bubbles. Subsequently, add 700 μL of the medium onto the membrane.
11. In our apparatus, the O_2 and CO_2 gas mixture passes through a preheated water bath at 37 °C before entering the incubation chamber.
12. MMB significantly lowered the body temperature of the mice. Warming during and after surgery is necessary.
13. Alternatively, the plasmid solution can be injected using a gas microinjector. If the injectant fills the lateral ventricle, the whole shape of the lateral ventricle will emerge with Fast Green color.
14. This is to avoid dehydration.
15. Under a stereomicroscope, the brain is extracted from the embryo using thin, fine forceps. It is recommended to maintain a low temperature during operation.

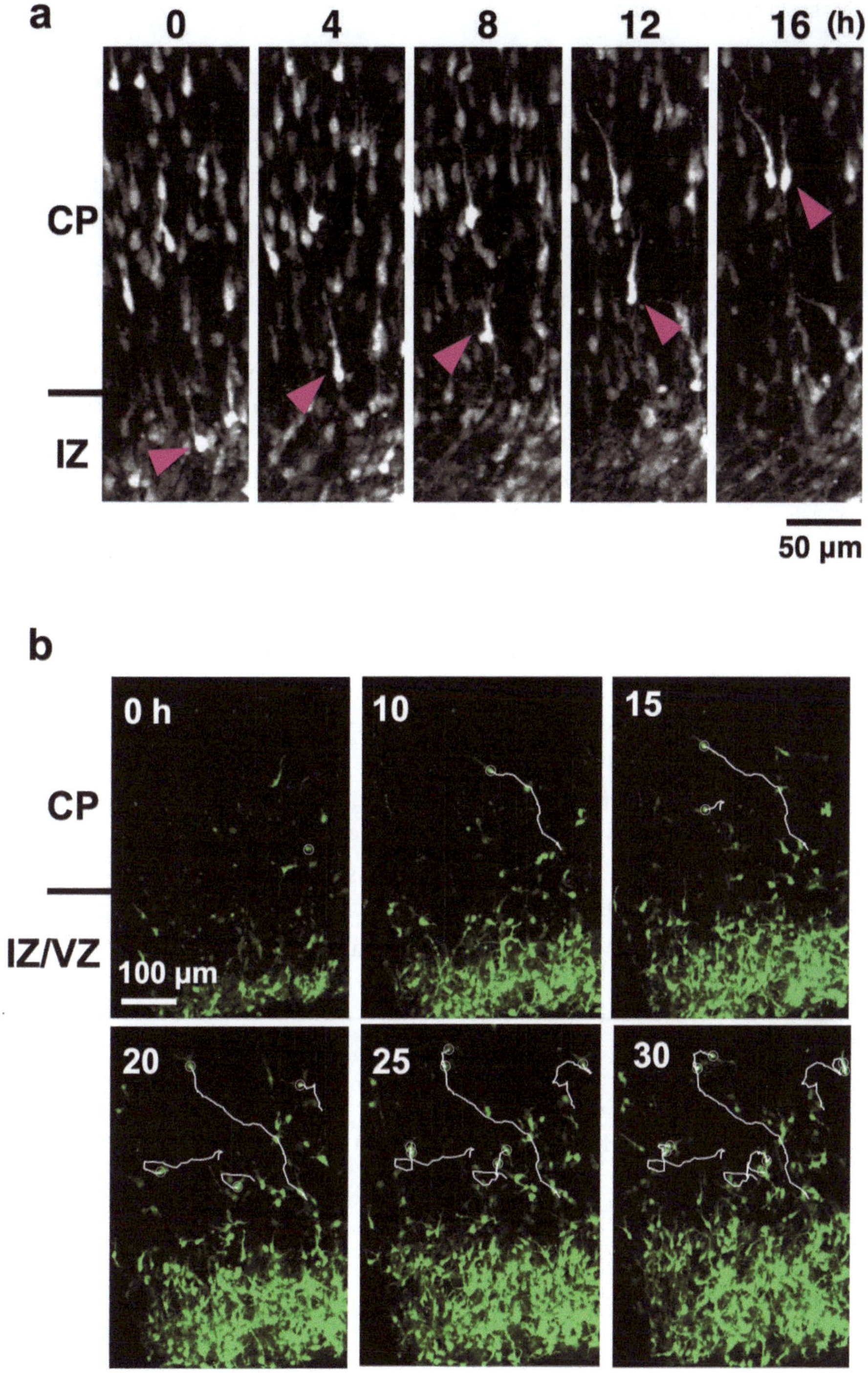

Fig. 2 Time-lapse observations of neuronal and glial migration in cortical slices. (**a**) Time-lapse imaging was performed to observe the migration of cortical neurons. The pCAG-EGFP (0.5 μg) was electroporated into the ventricular zone progenitor cells at E14. Coronal sections were then prepared at E16. Neurons expressing EGFP (arrowheads) were traced from the upper intermediate zone (IZ) to the cortical plate (CP), indicated by arrowheads. Scale bar, 50 μm. (**b**) Migration of astrocyte progenitors. E16 mouse embryos were electroporated with pCAG-hyPBase (0.5 μg) and pPB-CAG-EGFP (1 μg). Coronal slices were prepared 1 day later (E17) and subjected to time-lapse observation. Densely labeled neurons were found in the VZ to IZ (VZ/IZ). Glia progenitors (in this case, astrocyte progenitors) entered the CP faster than neurons, and migrated with characteristic zigzag movement, erratic migration. The migration trajectories of four astrocyte progenitors were traced (white lines). Scale bar, 100 μm

16. Take care to avoid the introduction of bubbles between the agarose and brain during embedding. The brain should be gently rolled in the solution several times.
17. It is recommended to align the dorsal side of the brain toward the slicer blade.
18. Place ice around the slicer stage to maintain a low temperature.
19. Perform this step immediately before cutting to prevent rapid rusting of the blade.
20. We are using a 5 mM square metal spatula to collect brain slices. The agarose attached to the slices does not require removal.
21. Remove the excess culture medium from the dish to ensure that the brain slices are slightly above the level of the medium. Keep for 2 min.
22. To prevent brain slices from moving, it is recommended not to add an excessive amount of culture medium. Adding approximately 400 μL of medium is typically sufficient.
23. The entire process, from **steps** 1 to **13**, should be completed within 90 min. The key to the success of this experiment is using "fresh slices."
24. If the bottom glass of the dish becomes fogged or condensed, wipe it off gently before proceeding.
25. Electroporated neurons are expected to be located near the intermediate zone of the lateral ventricle.
26. Acquired images were analyzed with ImageJ.

Acknowledgments

This work was supported by JSPS KAKENHI (Grant Number JP21K07309) to H. Tabata.

References

1. Buchsbaum IY, Cappello S (2019) Neuronal migration in the CNS during development and disease: insights from in vivo and in vitro models. Development 146:dev163766. https://doi.org/10.1242/dev.163766
2. Kriegstein A, Alvarez-Buylla A (2009) The glial nature of embryonic and adult neural stem cells. Annu Rev Neurosci 32:149–184. https://doi.org/10.1146/annurev.neuro.051508.135600
3. Severino M, Geraldo AF, Utz N et al (2020) Definitions and classification of malformations of cortical development: practical guidelines. Brain 143:2874–2894. https://doi.org/10.1093/brain/awaa174
4. Klingler E, Francis F, Jabaudon D, Cappello S (2021) Mapping the molecular and cellular complexity of cortical malformations. Science 371:eaba4517. https://doi.org/10.1126/science.aba4517
5. Saito T, Nakatsuji N (2001) Efficient gene transfer into the embryonic mouse brain using in vivo electroporation. Dev Biol 240:237–246. https://doi.org/10.1006/dbio.2001.0439
6. Tabata H, Nakajima K (2001) Efficient in utero gene transfer system to the developing mouse brain using electroporation: visualization of neuronal migration in the developing cortex. Neuroscience 103:865–872. https://doi.org/10.1016/s0306-4522(01)00016-1
7. Tabata H, Nakajima K (2003) Multipolar migration: the third mode of radial neuronal migration in the developing cerebral cortex. J Neurosci 23:9996–10001. https://doi.org/10.1523/jneurosci.23-31-09996.2003
8. Scala M, Nishikawa M, Ito H et al (2022) Variant-specific changes in RAC3 function disrupt corticogenesis in neurodevelopmental phenotypes. Brain 145:3308–3327. https://doi.org/10.1093/brain/awac106
9. Tabata H, Sasaki M, Agetsuma M et al (2022) Erratic and blood vessel-guided migration of astrocyte progenitors in the cerebral cortex. Nat Commun 13:6571. https://doi.org/10.1038/s41467-022-34184-x
10. Yusa K, Rad R, Takeda J, Bradley A (2009) Generation of transgene-free induced pluripotent mouse stem cells by the piggyBac transposon. Nat Methods 6:363–369. https://doi.org/10.1038/nmeth.1323

Chapter 18

Ca^{2+} Imaging in Immature Cortical Neurons

Shin-ichiro Horigane and Sayaka Takemoto-Kimura

Abstract

Ca^{2+} signaling plays a central role in various neurodevelopmental steps, and immature neurons exhibit spontaneous Ca^{2+} activity. To analyze Ca^{2+} dynamics in migrating immature neurons, we developed a method for Ca^{2+} imaging and offline analysis of Ca^{2+} dynamics.

Key words Ca^{2+} imaging, Genetically encoded Ca^{2+} indicator, Confocal laser microscopy, Immature neurons, In utero electroporation

1 Introduction

Ca^{2+} is an essential second messenger regulating multiple cellular events involved in neural circuit formation, including neuronal migration [1]. Migrating neurons generate spontaneous Ca^{2+} transients, which are key regulators of migratory movements [2–6]. Here, we describe a method for Ca^{2+} imaging and quantitative analysis of Ca^{2+} dynamics in radially migrating neurons of the developing cortex. Acute slice cultures are prepared after in utero electroporation (IUE) of plasmids carrying a genetically encoded Ca^{2+} indicator [7, 8], GCaMP6s [9]. Ca^{2+} imaging is performed using confocal laser microscopy for tens of minutes to several hours, and Ca^{2+} dynamics (e.g., event frequency, amplitude, and duration) are analyzed offline.

2 Materials

Prepare all solutions using ultrapure water and analytical-grade reagents. Store all reagents at room temperature unless otherwise indicated. Diligently follow all waste disposal regulations when discarding waste materials. Perform recombinant DNA and animal

Koh-ichi Nagata (ed.), *Cerebral Cortex Development: Methods and Protocols*, Methods in Molecular Biology, vol. 2794, https://doi.org/10.1007/978-1-0716-3810-1_18,

experiments following regulations and guidelines for the care and use of experimental animals of your institution committees.

2.1 Brain Slice Cutting Solution

1. Prepare 1 M stock solutions of KCl, $MgSO_4$, and $CaCl_2$.
2. Prepare 500 mL brain cutting solution: 38.01 g of sucrose, 1.25 mL of 1 M KCl, 0.109 g of $NaH_2PO_4{\cdot}2H_2O$, 0.050 g of L-ascorbic acid, 3.5 mL of 1 M $MgSO_4$, 0.5 mL of 1 M $CaCl_2$, and 1.134 g of $NaHCO_3$. Dilute reagents and stock solutions with water and adjust to 500 mL in a graduated cylinder. Final concentrations in brain slice cutting solution: 27 mM $NaHCO_3$, 1.4 mM NaH_2PO_4, 2.5 mM KCl, 0.5 mM ascorbic acid sodium salt, 7.0 mM $MgSO_4$, 1.0 mM $CaCl_2$, and 222.1 mM sucrose [10]. Prepare fresh brain slice cutting solution on the day of use.

2.2 Slice Culture Medium

1. Inactivate horse serum for 30 min in a 56 °C water bath (*see* **Note 1**).
2. Prepare 100 mM stock solution of $CaCl_2$ and $MgSO_4$ using water.
3. Mix the solution and reagents on a clean bench. Prepare 100 mL of minimum essential media (MEM) without glutamine and phenol red supplemented with 50 mL of Earle's balanced salt solution (EBSS) without calcium, magnesium, and phenol red, 50 mL horse serum, 4 mL serum-free B-27 supplement, 4 mL of 32.5% D-glucose, 2.5 mL of 1 M HEPES, 1 mL of 200 mM GlutaMAX supplement, 0.9 mL of 100 mM $CaCl_2$, and 0.4 mL of 100 mM $MgSO_4$ (*see* **Note 2**). Final composition of slice culture medium: MEM without glutamine and phenol red, 23.5% EBSS without calcium, magnesium, and phenol red, 23.5% horse serum, 1.9% serum-free B-27 supplement, 0.61% D-glucose, 11.75 mM HEPES, 0.94 mM GlutaMAX, 0.42 mM $CaCl_2$, and 0.19 mM $MgSO_4$ [5].
4. Sterilize the mixture with a 0.2-μm pore size nylon membrane vacuum filter system.
5. Store slice culture medium at 4 °C.

2.3 Plasmids and Anesthetics Used for IUE

1. Prepare high-copy plasmids carrying a CAG promoter and sequences encoding Ca^{2+} indicator and cytosolic marker. We have constructed a pCAG-TagRFP-P2A-GCaMP6s plasmid where TagRFP encodes the red fluorescence protein and GCaMP6s encodes a green fluorescent protein (GFP)-based Ca^{2+} indicator [9, 11]. Both coding sequences are fused through the self-cleaving peptide P2A [12] (*see* **Note 3**).
2. Plasmids are amplified using *E. coli* DH5α competent cells and purified with an endotoxin-free plasmid extraction kit (*see*

Note 4). Plasmid concentration is adjusted to 4 g/L and stored at 4 °C (for weeks) or − 30 °C (for months to years).

3. Three types of anesthetic agents are mixed as follows: 1.875 mL of 1 g/L medetomidine hydrochloride, 2 mL of 5 g/L midazolam, and 2.5 mL of 5 g/L butorphanol tartrate. The mixture is diluted with 18.625 mL saline and stored at 4 ° C. The final concentrations in the stock solution are 0.075 g/L medetomidine hydrochloride, 0.4 g/L midazolam, and 0.5 g/ L butorphanol tartrate.

3 Methods

3.1 IUE and Slice Preparation

1. Inject the anesthetic mixture (0.1 mL per 10 g of mouse body weight) into the abdominal cavity of a mouse pregnant with embryos at embryonic days E14–15. Make an incision of the abdominal midline and gently pull out the uterine horns.
2. Inject 1.0 μL of plasmid solution (2.0 g/L of pCAG-TagRFP-P2A-GCaMP6s) into the right or left lateral ventricle of the mouse embryos using a glass capillary (1.5 mm outer diameter and 1.17 mm inner diameter). The glass capillary is pulled using a micropipette puller and cut at 40–50 μm in diameter using surgical scissors. The glass capillary tip is beveled using a microgrinder.
3. Perform electroporation (5 square pulses of 45 V, 50 ms duration with 950 ms intervals) using an electroporator and a tweezer-type electrode. Place the uterine horns back in their original location and suture the abdominal muscles and skin. Cover animals with a heating pad.
4. Seventy-two hours after IUE, embryonic brains are quickly removed and placed into an ice-cold slice cutting solution bubbled with a gas mixture of 95% O_2 and 5% CO_2. Then, brains are embedded into 3% low-melting temperature agarose diluted in slice cutting solution.
5. Prepare coronal slices (200–300 μm) using a vibratome.
6. Mount brain slices on cell culture plate inserts (30 mm diameter, hydrophilic polytetrafluoroethylene [PTFE], 0.4 μm pore size) (*see* **Note 5**) placed into 35-mm glass bottom dishes filled with slice culture medium (*see* **Note 6**).
7. Incubate brain slices in a CO_2 incubator at 37 °C for at least 1 h.

3.2 Ca^{2+} Imaging

1. After incubation in a CO_2 incubator at 37 °C, the brain slice cultures are transferred to a stage CO_2 incubator equipped with an inverted confocal laser microscope (Fig. 1a).

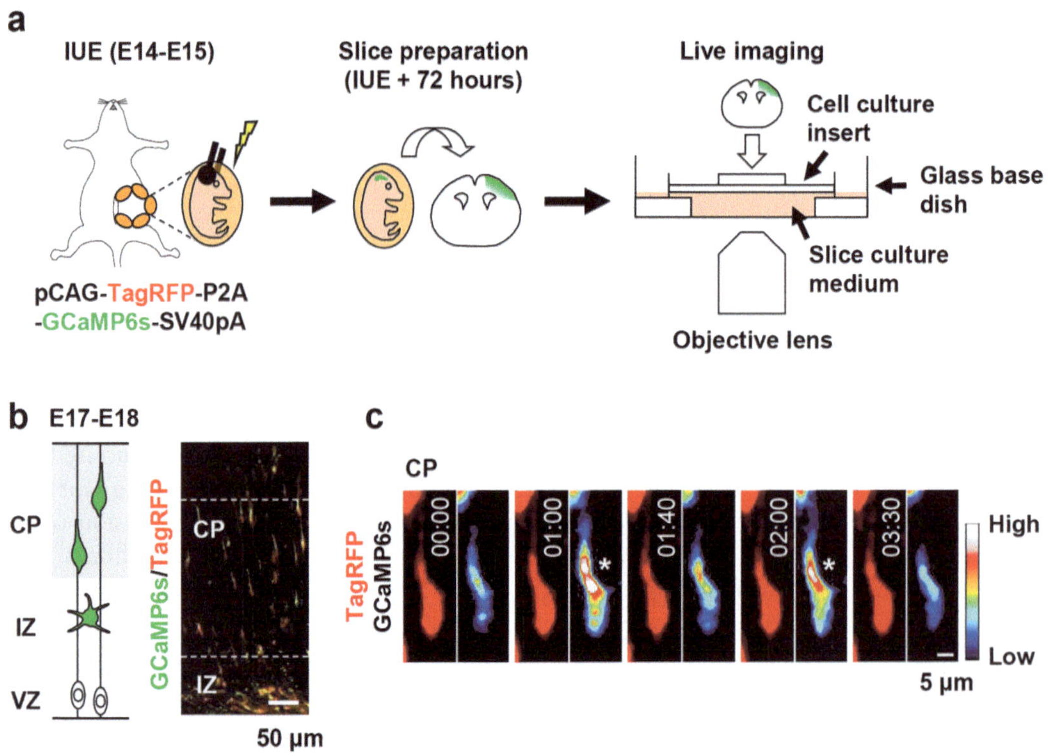

Fig. 1 Ca^{2+} imaging of cortical immature neurons. (**a**) Schematic illustration of IUE, slice preparation for brain slice culture, and live imaging of brain slices. Ca^{2+} imaging is performed using an inverted confocal laser microscope. Brain slices are incubated in a stage CO_2 incubator for time-lapse imaging. (**b**) Seventy-two hours after IUE of E14–15 embryos, neurons labeled with fluorescent proteins in the CP and IZ are observed. (**c**) Spontaneous Ca^{2+} transients (asterisks) in immature migrating neurons in the CP. Time stamps indicate minutes and seconds

2. Acquire GCaMP6s and TagRFP 512 × 512-pixel images sequentially at 0.1–1 Hz in the intermediate zone (IZ) or cortical plate (CP) (*see* **Note 7**) (Fig. 1b). Excite GCaMP6s and TagRFP using Argon (488 nm) and HeNe (561 nm) laser lines, respectively (*see* **Note 8**). Select emission signals using a photomultiplier tube spectral detector.
3. Acquire a series of time-lapse images for tens of minutes to hours (*see* **Note 9**). Keep the inside of the stage CO_2 incubator moist (*see* **Note 10**) (Fig. 1c).
4. After live imaging, fix brain slices in 4% paraformaldehyde (PFA) dissolved in phosphate-buffered saline overnight and stain the fixed slices with Hoechst 33342 (*see* **Note 11**).
5. Capture Hoechst 33342 images and identify the cortical layers, ventricular zone (VZ), IZ, and CP areas investigated by live imaging.

3.3 Quantitative Analysis of Ca^{2+} Transients

1. To extract regions of interest (ROIs) from a series of TagRFP images, subtract the background using a 30-pixel radius rolling ball from the acquired 512 × 512-pixel images (*see* **Note 12**).
2. Apply a median filter with a 1-pixel radius for the TagRFP images after background subtraction.
3. Split the processed TagRFP time-lapse images into fragments of the XY axis containing a single migrating neuron.
4. Extract serial TagRFP ROIs for each migrating neuron. To identify cell bodies, test as many intensity threshold values or thresholding algorithms as possible and apply the best ones to target cells.
5. Acquire a series of GCaMP6s and TagRFP signal intensity values for the TagRFP ROIs. Define ROIs on unprocessed original images.
6. After background subtraction from GCaMP6s and TagRFP signals (*see* **Note 13**), normalize GCaMP6s signal intensities with a series of TagRFP signal intensities to correct for motion and focusing artifacts (Fig. 2a, b).
7. Further normalize the GCaMP6s signal intensity using moving median smoothing (tens of minutes for each median window) to exclude baseline fluctuations (*see* **Note 14**).

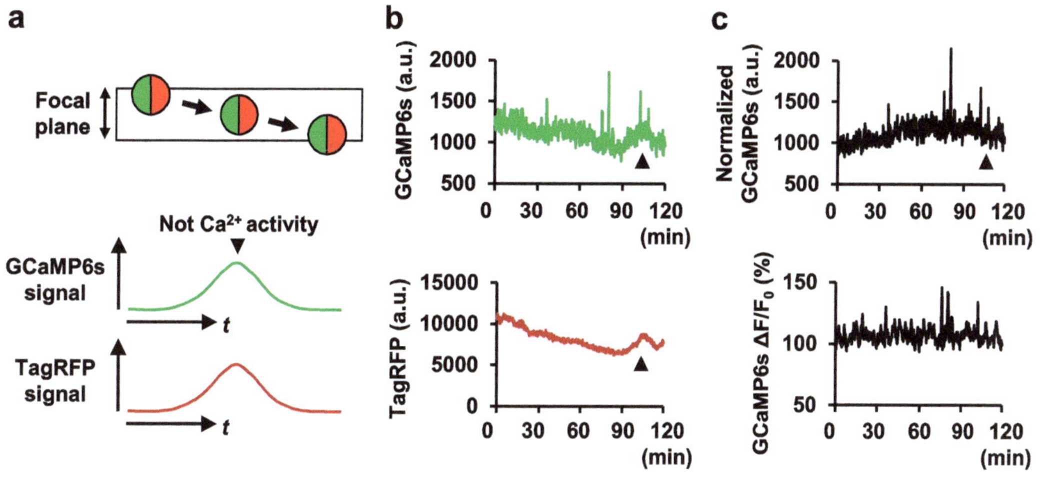

Fig. 2 Processing of GCaMP6s numerical data. (**a**) During long-term Ca^{2+} imaging, target cells expressing both GCaMP6s and TagRFP move in the Z-direction of the focal plane, resulting in changes in GCaMP6s signal intensity. Since these changes in GCaMP6s signal intensity are not derived from Ca^{2+} activity, we remove the signal fluctuation using TagRFP signal intensities that vary with GCaMP6s signal intensity. (**b**) Serial measurement of GCaMP6s and TagRFP signal intensities. The baseline fluctuation (arrowheads) is observed for both GCaMP6s and TagRFP signal intensities. (**c**) Baseline fluctuation of GCaMP6s is normalized with a series of TagRFP intensities (arrowhead). After assessing the moving median and moving average smoothing, GCaMP6s $\Delta F/F_0$ is calculated. Baseline and noise-like fluctuations are removed in these processes

8. Apply rolling average smoothing to exclude noise-like fluctuations (tens of seconds for each average window) (*see* **Note 15**).
9. After these processing steps, calculate GCaMP6s $\Delta F/F_0$ as $(F - F_0)/F_0$, where F is the processed GCaMP6s signal intensity and F_0 is the baseline intensity defined as the mean of values below 10–25% of the processed GCaMP6s signal intensity (*see* **Note 16**) (Fig. 2c).
10. Define Ca^{2+} transients as events of processed GCaMP6s signal with suprathreshold intensities. Set the threshold value based on the standard deviation (SD) of values below 10–25% of processed GCaMP6s signal intensity (*see* **Note 17**).
11. Calculate parameters such as event frequency, amplitude, duration, and area under the curve to analyze the Ca^{2+} dynamics of target cells.

4 Notes

1. We purchase 500 mL horse serum and inactivate it. Then, we prepare 50-mL aliquots of house serum and stored them at −30 °C.
2. Ca^{2+} imaging is performed using an inverted confocal microscope. Excitation and emission light pass through the slice culture medium as the objectives are below the slice cultures. MEM without phenol red is used to reduce the background signal. B-27 and GlutaMAX supplements are purchased from Thermo Fisher Scientific.
3. We use GCaMP6s and TagRFP for Ca^{2+} imaging and moving ROI extraction. Previously, we performed co-electroporation of two plasmids; however, this resulted in different plasmid copy numbers in each cell. Thus, the quantification of signal intensities was difficult (e.g., although GCaMP6s signal is strong enough to be measured, TagRFP signal is weak and, consequently, ROI extraction is not optimal). Douglas Kim & GENIE Project provided pGP-CMV-GCaMP6s plasmid (Addgene plasmid # 40753), and TagRFP-C DNA sequence was purchased from Evrogen.
4. Plasmids purified using a plasmid extraction kit, such as one without endotoxin-free grade, may harm the embryonic brain after IUE and induce hydrocephalus. We have tested several plasmid DNA extraction kits and use NucleoBond Xtra Midi/Maxi EF (Macherey-Nagel).
5. Embryonic brain slices are soft and easily misshapen. We leave the agarose around the brain slices to preserve the slice shape for tens of minutes to hours of live imaging. This is achieved by gently scooping brain slices surrounded by agarose from the

vibratome buffer tray using a metal spatula and transferring them to a cell culture plate insert filled with the slice cutting solution. Slice cutting solution is removed after mounting brain slices on the cell culture plate insert. We recommend not using syringes to transfer the brain slices as agarose may detach from the slices. We mount 4–6 brain slices per cell culture plate insert.

6. To visualize a wide field with the inverted confocal microscope, we use dishes with a large bottom glass (glass 27φ). We add 1.9 mL slice culture medium into each dish, but the optimal volume of the slice culture medium depends on the shape of the cell culture inserts and dishes. Therefore, you need to determine the appropriate medium volume for your imaging system.
7. Seventy-two hours after the IUE of E14–15 embryos, radially migrating neurons start arriving at their final destination near the pial surface. We usually capture migrating neurons in the IZ and CP. In these cortical layers, migrating neurons produce Ca^{2+} transients for over 30–60 s in our imaging system using GCaMP6s; therefore, we set the sampling rate as 0.1–1 Hz. Determine the appropriate sampling rate for your imaging system and target cells.
8. A high-power excitation laser may cause photobleaching of fluorescent proteins. Reduce the power of the excitation laser as much as possible.
9. We use an LSM710 laser confocal microscope (Zeiss) with a 10×/NA-0.45 objective (Plan-Apochromat, Zeiss) or a 20×/NA-0.7 objective (UCPLFLN, Olympus) that has a long working distance of 0.80–1.80 mm. We capture 12 or 16-bit images.
10. In a dry environment in the stage CO_2 incubator, the volume of the slice culture medium will decrease during live imaging. This volume reduction affects the distance between the objective lens and the brain slice culture, thus resulting in target cells being out of focus.
11. Do not peel away the brain slices from the cell culture insert during PFA fixation to prevent brain slice deformation. We replace the slice culture medium below the cell culture insert with PFA solution and pour PFA solution on top of the brain slices.
12. We process and analyze acquired time-lapse images with Fiji [13] using self-made Macro and Microsoft Excel with visual basic analysis.
13. We set a background ROI in a region of brain slice culture without neurons and acquire background values from a series of GCaMP6s and TagRFP images.

14. For 2 h time-lapse, images were captured using a 0.1–1 Hz frame rate. We apply a moving median for 30 min for each median window and determine the window size according to the fluctuation of the data.
15. We apply moving average smoothing for 30 s for each average window to time-lapse images as described above.
16. We average all values below 10–25% of processed GCaMP6s signal intensity to determine F_0. We have set the percentage value by taking into consideration the Ca^{2+} event frequency. If target cells show high Ca^{2+} event frequency, we set a small percentage value to exclude Ca^{2+} event time point(s) from F_0.
17. We have set 5–10 SD as the Ca^{2+} event threshold by taking into consideration the baseline fluctuation and Ca^{2+} event amplitude.

Acknowledgments

This work was supported in part by grants from the KAKENHI grants from JSPS (S.T.-K., S.H.), the Takeda Science Foundation (S.H.), Toyoaki Scholarship Foundation (S.H.), and the Nakatomi Foundation (S.H.)

References

1. Rosenberg SS, Spitzer NC (2011) Calcium signaling in neuronal development. Cold Spring Harb Perspect Biol 3:a004259. https://doi.org/10.1101/cshperspect.a004259
2. Komuro H, Rakic P (1992) Selective role of N-type calcium channels in neuronal migration. Science 257(80):806–809. https://doi.org/10.1126/science.1323145
3. Bortone D, Polleux F (2009) KCC2 expression promotes the termination of cortical interneuron migration in a voltage-sensitive calcium-dependent manner. Neuron 62:53–71. https://doi.org/10.1016/j.neuron.2009.01.034
4. Kamijo S, Ishii Y, Horigane S-I et al (2018) A critical neurodevelopmental role for L-type voltage-gated calcium channels in neurite extension and radial migration. J Neurosci 38:5551–5566. https://doi.org/10.1523/JNEUROSCI.2357-17.2018
5. Sichiro H, Hamada S, Kamijo S et al (2021) Development of an L-type Ca2+ channel-dependent Ca2+ transient during the radial migration of cortical excitatory neurons. Neurosci Res 169:17–26. https://doi.org/10.1016/J.NEURES.2020.06.003
6. Horigane S, Ozawa Y, Yamada H, Takemoto-Kimura S (2019) Calcium signalling: a key regulator of neuronal migration. J Biochem 165:401–409. https://doi.org/10.1093/jb/mvz012
7. Lo Turco J, Manent JB, Sidiqi F (2009) New and improved tools for in utero electroporation studies of developing cerebral cortex. Cereb Cortex 19:i120–i125. https://doi.org/10.1093/CERCOR/BHP033
8. Kolk SM, De Mooij-Van Malsen AJ, Martens GJM (2011) Spatiotemporal molecular approach of in utero electroporation to functionally decipher endophenotypes in neurodevelopmental disorders. Front Mol Neurosci 4:37. https://doi.org/10.3389/FNMOL.2011.00037
9. Chen T-W, Wardill TJ, Sun Y et al (2013) Ultrasensitive fluorescent proteins for imaging neuronal activity. Nature 499:295–300. https://doi.org/10.1038/nature12354
10. Namiki S, Sasaki T, Matsuki N, Ikegaya Y (2009) Regional difference in stainability with calcium-sensitive acetoxymethyl-ester probes

in mouse brain slices. Int J Neurosci 119:214–226

11. Merzlyak EM, Goedhart J, Shcherbo D et al (2007) Bright monomeric red fluorescent protein with an extended fluorescence lifetime. Nat Methods 47(4):555–557. https://doi.org/10.1038/nmeth1062
12. Kim JH, Lee S-R, Li L-H et al (2011) High cleavage efficiency of a 2A peptide derived from porcine teschovirus-1 in human cell lines, zebrafish and mice. PLoS One 6:e18556. https://doi.org/10.1371/journal.pone.0018556
13. Schindelin J, Arganda-Carreras I, Frise E et al (2012) Fiji: an open-source platform for biological-image analysis. Nat Methods 9:676–682. https://doi.org/10.1038/nmeth.2019

Chapter 19

In Vitro Patch-Clamp

Toshihiro Nomura

Abstract

The patch-clamp technique is one of the most useful tools to analyze the function of electrically active cells such as neurons. This technique allows for the analysis of proteins (ion channels and receptors), cells (neurons), and synapses that are the building blocks of neuronal networks. Cortical development involves coordinated changes in functional measures at each of these levels of analysis that reflect both cellular and circuit maturation. This chapter explains the technical and theoretical basis of patch-clamp methodology and introduces several examples of how this technique can be applied in the context of cortical development.

Key words Patch-clamp, Electrophysiology, Action potential, Ion channel, Synapse, Development, Critical period

1 Introduction

The patch-clamp technique is the universally recognized gold standard of single-cell resolution electrophysiological analysis, enabling numerous historical discoveries. Erwin Neher and Bert Sakmann established the technique in the 1970s–1980s, for which they were awarded the Nobel Prize in Physiology or Medicine in 1991 [1–4]. Today, 40 years after this seminal innovation, it still stands as one of the most powerful and comprehensive tools to analyze the function of electrically active cells such as neurons. The patch-clamp technique allows for the analysis of intrinsic cellular properties such as ion channel activity as well as circuit properties such as synaptic activity [3, 5–8]. In neuroscience research, non-neuronal heterologous cells transfected to express recombinant neuronal ion channels and receptors, cultured neurons, acutely prepared brain slices, cultured brain slices, or even whole animal, in vivo, preparations can be analyzed using the patch-clamp technique. In this chapter, I will primarily discuss methods for in vitro patch-clamp recording from rodent acute brain slices. Cortical development is associated with coordinated changes in the functional properties of

Koh-ichi Nagata (ed.), *Cerebral Cortex Development: Methods and Protocols*, Methods in Molecular Biology, vol. 2794,
https://doi.org/10.1007/978-1-0716-3810-1_19,

both cortical neurons and circuits. Therefore, the patch-clamp technique is an ideal tool for interrogating this process. I will provide some practical technical tips for experiments and also discuss the logic underlying the technique, understanding of which is critical to accurately acquiring and interpreting data. Finally, I will present some examples of how the patch-clamp technique can be applied to the study of cortical development.

2 Materials

This section summarizes materials required for patch-clamp recordings from rodent acute brain slices.

2.1 Extracellular Solutions

The extracellular solution used in the patch-clamp technique, "artificial" cerebrospinal fluid (ACSF), mimics the ionic composition of physiological extracellular cerebrospinal fluid (CSF). The main cationic constituents are Na^+ or a suitable substitute. The author typically prepares three types of extracellular solutions for experiments. These are (1) slicing, (2) incubating, and (3) recording solutions. Examples of each solution are shown below. Intracellular solutions (pipette solutions) will be explained later. For best results, the incubating and recording extracellular solutions should be prepared daily. However, if needed, a working concentration (1×) of slicing solution can be stored for a few days at 4 °C, and 10× concentrated solutions can be stored for up to a few weeks (*see* **Note 1**):

1. Slicing solution: 85 mM NaCl, 2.5 mM KCl, 1.25 mM NaH_2PO_4, 25 mM $NaHCO_3$, 25 mM glucose, 75 mM sucrose, 0.5 mM $CaCl_2$ and 4 mM $MgCl_2$, 10 μM DL-APV (DL-2-amino-5-phosphonopentanoic acid), and 100 μM kynurenate (*see* **Note 2**).
2. Incubating solution: 125 mM NaCl, 2.4 mM KCl, 1.2 mM NaH_2PO_4, 25 mM $NaHCO_3$, 25 mM glucose, 1 mM $CaCl_2$, and 2 mM $MgCl_2$ with 10 μM DL-APV, and 100 μM kynurenate (*see* **Note 3**).
3. Recording solution: 125 mM NaCl, 2.4 mM KCl, 1.2 mM NaH_2PO_4, 25 mM $NaHCO_3$, 25 mM glucose, 2 mM $CaCl_2$, and 1 mM $MgCl_2$.
4. To maintain proper pH and to keep the tissue oxygenated, these extracellular solutions must be equilibrated and continually perfused with 95% O_2 and 5% CO_2 (carbogen) throughout experiments.

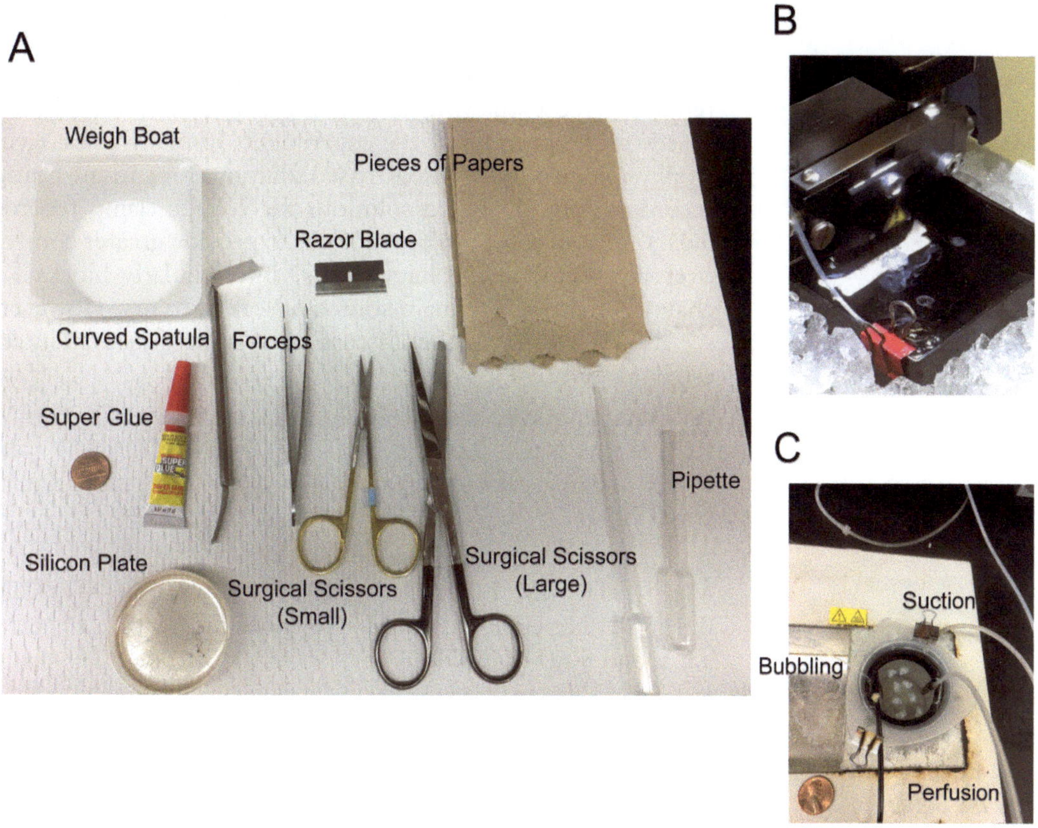

Fig. 1 Slicing. (**a**) Picture of tools for preparing slices. 1 ¢ coin is shown as a reference. Weigh boat, for decapitated animal body; Pieces of papers, for absorbing remaining solution/glue; Razor blade, for dissecting the brain; Curved spatula, for transferring the whole and dissected brain; Forceps, for peeling off the scull; Surgical scissors, for decapitating the animal and cutting the scull; Silicon plate, for dissecting the brain; and Pipette, for transferring the brain slices. (**b**) Dissected brain is cut using a vibrating blade slicer. (**c**) Brain slices in an incubating chamber. Tubes for bubbling (95% 0_2/5% CO_2), perfusion (incubating solution), and suction are indicated. 1 ¢ coin as a reference

2.2 Slicing

1. Vibrating blade tissue slicer.
2. Surgical tools for craniotomy and brain removal/dissection (Fig. 1a).
3. Incubating chamber containing nylon mesh to allow for solution circulation.
4. Heated chamber for incubating slices.

2.3 Electrodes and Intracellular Solutions

1. Borosilicate hollow glass capillary for glass electrodes.
2. Microelectrode puller.
3. Intracellular (pipette) solution: In contrast to extracellular solutions shown above, intracellular solution should mimic the physiological intracellular environment. The main cation constituent is K^+ or an appropriate substitute (e.g., Cs^+).

Intracellular solutions should be aliquoted to a small volume for daily use (typically 50–100 μL for the author) and stored in a freezer. Aliquoted solutions should be thawed daily to avoid repeated freeze–thawing. Intracellular solutions are chosen based on the purpose of the experiment. In many cases, electrophysiologists like to use K-based solutions for current-clamp recordings and Cs-based solutions for voltage-clamp recordings. Cs-based solutions are thought to provide greater control over the membrane voltage as Cs^+ intracellularly blocks K^+ channels that are the main source of leak currents. However, see also [9, 10] for further discussion or controversies, if interested. (*see* **Note 4**).

4. An example of a K-based solution: 125 mM $KMeSO_4$, 5 mM KCl, 5 mM NaCl, 0.02 mM EGTA, 11 mM HEPES, 1 mM $MgCl_2$, 10 mM phosphocreatine, 4 mM Mg-ATP, and 0.3 mM Na-GTP.
5. An example of a Cs-based solution: 135 mM $CsMeSO_3$, 1 mM $MgCl_2$, 0.2 mM EGTA, 10 mM HEPES, 2 mM Mg-ATP, 0.3 mM Na-GTP, 10 mM Na_2-phosphocreatine,10 mM TEA, and 5 mM QX-314.
6. Centrifuge filter (*see* **Note 5**).

2.4 Electrophysiology Rig SetUp

1. Faraday cage: A Faraday cage is made from wire mesh that surrounds the microscope, head stage, micromanipulators, and other equipment on the air table (see below). This helps isolate the rig from external electrical noise. Several pre-made Faraday cages for electrophysiological experiments are commercially available. When the environmental noise is well controlled and the signal-to-noise ratio is acceptably high, this may be not necessary.
2. Perfusion system: Either a gravity-driven perfusion or peristaltic pump can be used. When a peristaltic pump is used, steps should be taken to reduce or eliminate electrical noise from the pump.
3. Microscope: A microscope is required to visualize the cells to patch. The single-cell resolution allows for appropriate cells to be located. Additionally, contrast-enhancing techniques such as differential interference contrast (DIC), Dodt gradient contrast, or oblique illumination can be helpful in improving visibility.
4. Micromanipulator: Patch-clamp recordings require precise manipulation of electrode movement with μm resolution. Conventional oil pressure motorized or modern electrical motor motorized manipulators can be used for patch-clamp recordings. Electrical noise originating from electrically controlled manipulators should be reduced or eliminated.

5. Air table: Physical vibration should be reduced as much as possible. An air table minimizes vibrations via a floating table top that isolates the equipment on the table from the ground.
6. Amplifier: A key device to conduct patch-clamp recordings. In combination with the head stage (see below), an amplifier magnifies the small neuronal currents measured (function referred to as "gain") to allow for accurate measurement. Typically, an amplifier also contains circuitry that allows for removing components of recorded signal in a frequency-dependent manner in order to remove electrical noise or other unwanted components in the frequency domain (function referred to as "filter"). Proper adjustment of both the gain and filter settings is essential to successful recording. With proper settings, measurement of electrical signals in neuronal cell membranes with pA or mV resolution is possible.
7. Head stage: A device to hold the glass recording electrodes and transmit the electrical signals from the cells to the amplifier. A head stage is also equipped with a built-in electrical circuit that can reduce electrical noise.
8. Digitizer: This converts the physiological "analog" (A) signals to a "digital" (D) signal for analysis by computer software. It is also called an "A–D converter."
9. Computer: Data acquisition and analyses are performed with computer software in modern patch-clamp setups.
10. Data acquisition and analysis software: Most commercially available amplifiers and digitizers provide their own software for data acquisition and analysis. In addition, there are a variety of open-source software and codes available which enable patch-clamp analyses.

3 Methods

This section is a step-by-step guide to producing a patch-clamp recording. The key concept of the patch-clamp technique is to make a tight "patch" between the tip of the glass recording electrode and the cell membrane. Unlike in conventional sharp-electrode recordings, the electrode does not penetrate the cell membrane. In patch-clamp recordings, electrical access to the cell is made by rupturing the patch, a small portion of the membrane which forms a tight contact between the recording electrode. This enables precise control of the membrane voltage (voltage-clamp) and data acquisition with a better signal-to-noise ratio by generating an electrical circuit with low resistance between the recording electrode and the cell interior, and by making the resistance of a leak path several orders of magnitude higher (giga-seal).

3.1 Acute Brain Slice Preparation Using a Tissue Slicer

Preparing healthy brain slices is the first step to conducting successful patch-clamp recordings. For experiments using pre-made samples such as cultured immortal cells (e.g., HEK293 cells), cultured neurons, or cultured brain slices, this step can be skipped:

1. Heat incubating chamber filled with ~100 mL slicing solution to ~30 °C.
2. Deeply anesthetize animals by inhaled anesthetics such as isoflurane and/or injected anesthetics such as ketamine and xylazine (typically administered intraperitoneally in rodents). For neonatal animals which are resistant to these anesthetics, this procedure might be skipped or alternative methods (e.g., hypothermia) can be used. Check your regional and institutional guidelines.
3. Decapitate animals using surgical scissors or rodent guillotine.
4. Gently remove the scull using surgical scissors and forceps, avoiding damaging or touching the brain tissue as little as possible. Transfer the brain gently into an ice-cold slicing solution, again being careful not to damage the brain tissue with the skull or surgical instruments. This step is important, especially for making recordings from superficial regions such as the cortex as the tissue can be easily physically damaged during this process.
5. Dissect the brain in an ice-cold slicing solution (*see* **Note 6**).
6. Gently transfer the dissected brain to the slicing chamber of a vibrating blade slicer.
7. Slice the dissected brain in an ice-cold slicing solution using a vibrating blade slicer (Fig. 1b). Parameters for slicing can vary, but the author typically uses the setting as follows. Speed: 0.16 mm/s, Amplitude: 1 mm, Thickness: 250–400 μm (*see* **Note 7**).
8. Transfer brain slices to the incubating chamber. Incubate slices at 30 °C for 15–30 min.
9. Gradually exchange the slicing solution in the incubating chamber with the incubating solution at room temperature (Fig. 1c). The author typically prefills the chamber with ~100 mL of slicing solution, and after slicing, a flow regulator is used to slowly exchange this solution with incubating solution at a rate of ~2 mL/min.
10. After ~1.5 h of incubation, the slices are ready for electrophysiological recordings.

3.2 Manufacturing Electrodes

1. Manufacture glass recording electrodes using a microelectrode puller. The properties of the electrodes can vary depending on the purpose, but electrodes with tip resistances of 3–4 MΩ are

ideal for whole-cell recordings from many types of neurons. Pipettes with tip resistances of >10 MΩ are not suitable for whole-cell recordings.

2. Prepare electrodes for extracellular electrical stimulation if necessary. Either glass or tungsten electrodes can be used. Glass electrodes can be manufactured using the same electrode puller, but typically, the tip resistance is lower than that of recording electrodes (~1 MΩ or less). To make lower-resistance glass electrodes, the author sometimes makes a tiny break at the tip with metal wire under a microscope.
3. Filter thawed intracellular pipette solutions.
4. Load recording pipette electrodes with intracellular solution from the back. Stimulating electrodes are normally filled with ACSF or normal saline (NaCl).

3.3 Patching Cells

This section describes the most critical technical part of patch-clamp recording. Substantial practice is required until reliable and high-quality data can be obtained using the patch-clamp technique. While the procedure itself is technically challenging, understanding the theory underlying each step is also critical to monitoring the quality of recordings, correctly interpreting the data, and troubleshooting. Therefore, I will also discuss the logic of how electrophysiological signals are measured in patch-clamp recordings and the parameters to be monitored during the experiment to ensure a successful recording:

1. Transfer brain slices to the recording chamber on the rig and perfuse slices with recording solution, equilibrated with 95/5% of O_2/CO_2 (carbogen). Recording temperature can vary depending on the purpose, but the author normally keeps the solution temperature at ~30 °C.
2. Visualize cells under the recording microscope.
3. It is important to understand the electrical equivalent circuit, which dynamically changes during patch-clamp experiments. The circuits are illustrated in Fig. 2a, b, and each apparatus is summarized as follows. $R_{pipette}$ (R_p): Pipette resistance. Resistance of the tip of the pipette filled with an intracellular solution. R_{access} (R_a): Access resistance. Resistance of the cell membrane in contact with the tip of the glass recording electrode, i.e., "patch." R_{series} (R_s): Series resistance (not shown as an icon in Figure). A sum of R_p and R_a which are connected in "series." $R_{membrane}$ (R_m) and $C_{membrane}$ (C_m): Membrane resistance and membrane capacitance. These are the intrinsic resistance and capacitance of the cell membrane, which are connected in parallel to each other, and in series with R_s. R_{leak} (or R_{seal}): Leak (or seal) resistance. Resistance of the seal

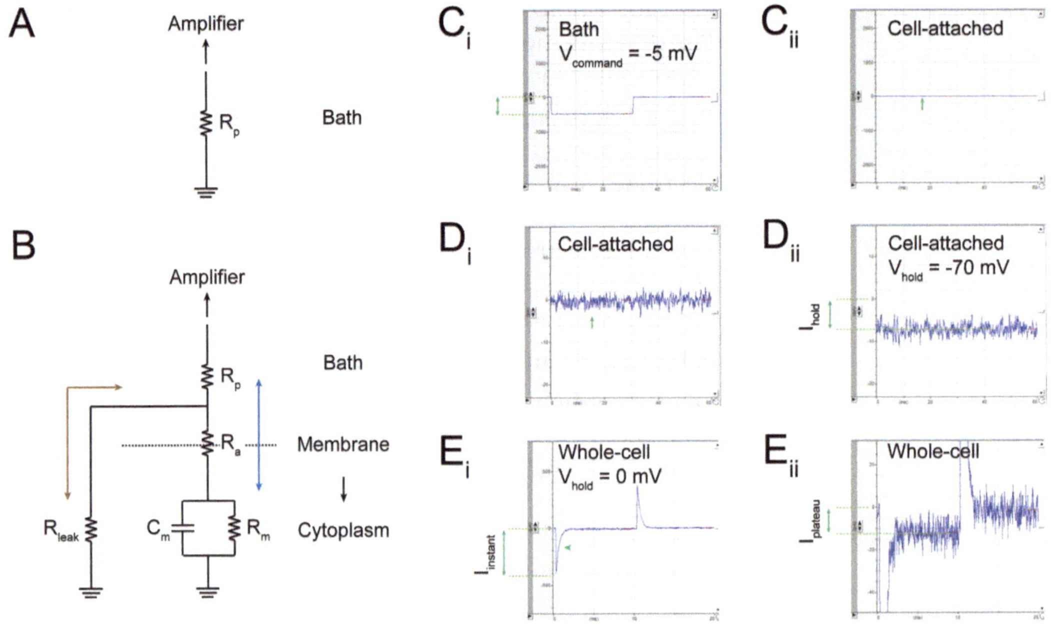

Fig. 2 Equivalent circuits and current responses to the test pulse. The locations of the bath, cell membrane, and cytoplasm are indicated. (**a**) Equivalent circuit in bath configuration. (**b**) Equivalent circuits in cell-attached and whole-cell configurations. R_p and R_a are connected in series forming R_s. The cell membrane is considered to be a resistor (R_m) and a capacitor (C_m) connected in parallel. When a tight seal is not formed or maintained, currents can leak through R_{leak} connected in parallel with R_m. Current pathways, which are dominant in cell-attached (brown) and whole-cell (blue) configurations, are indicated by two-sided arrows. (**c**–**e**) Examples of current responses to −5 mV hyperpolarizing square test pulse (30 ms width) in each configuration. These are responses using a model cell, not a real cell, containing an artificial electrical circuit composed of as follows, R_p (= R_s) 10 MΩ, R_{leak} 10 GΩ, R_m 500 MΩ, and C_m 33 pF. (**c**$_i$) Bath configuration. Note the square current response to the test pulse (two-sided arrow). Also, note the current amplitude of ~500 pA is consistent with the equation of $V_{command}$ (5 [mV]) = I (500 [pA]) × R_p (10 [MΩ]). (**c**$_{ii}$) Cell-attached configuration with giga-seal (10 GΩ). Note the current shown in **c**$_i$ becomes invisible (arrow). The current amplitude is theoretically 0.5 pA based on the equation of $V_{command}$ (5 [mV]) = I (0.5 [pA]) × R_{leak} (10 [GΩ]). (**d**$_i$) Same current response as shown in **c**$_{ii}$ in a different scale. The current is practically invisible (same level or smaller relative to the background noise) so long as giga-seal is being maintained (arrow). (**d**$_{ii}$) V_{hold} of −70 mV is applied in the same condition as shown in **d**$_i$. Note the small difference in I_{hold} with in **d**$_i$. A small I_{hold} with an amplitude of 7 pA (two-sided arrow) is apparent. The amplitude of I_{hold} is again consistent with the equation of V_{hold} (70 [mV]) = I (7 [pA]) × R_{leak} (10 [GΩ]). (**e**$_{i,ii}$) Whole-cell configuration. V_{hold} is set to 0 mV again for simplification. In this mode, the actual read of R_s = ~12 MΩ (while the apparatus is 10 MΩ). (**e**$_i$) Note the instantaneous current ($I_{instant}$) (two-sided arrow) which attenuates with exponential decaying kinetics (arrow-head). The amplitude of $I_{instant}$ is given by the equation of $V_{command}$ (5 [mV]) = $I_{instant}$ (417 [pA]) × R_s (12 [MΩ]). (**e**$_{ii}$) Magnified trace shown in **e**$_i$. Note the steady-state current which stands throughout the 30 ms test pulse application (two-sided arrow). The amplitude of the steady-state current ($I_{plateau}$) is given by the equation of $V_{command}$ (5 [mV]) = $I_{plateau}$ (10 [pA]) × R_m (500 [MΩ]). Note that the measurement of $I_{instant}$ and $I_{plateau}$ amplitudes is consistent/similar to the values given from the equations

between the pipette rim and the cell membrane that forms a parallel circuit to R_s and R_m/C_m.

4. Put the recording electrode in the recording chamber perfused with an extracellular solution. Apply a square electrical pulse through the recording electrode to measure the pipette resistance. Most commercially available patch-clamp amplifiers provide dedicated circuitry to generate this square pulse (test pulse) at high frequency to allow for calculating pipette resistance in real time (function referred to as "seal test" or "membrane test" in some amplifiers). In this configuration, the only electrical apparatus in the circuit is the pipette. Therefore, the equivalent electrical circuit is composed solely of a resistor with the resistance of R_p (Fig. 2a). The current–voltage relationship is subject to Ohm's law $V = IR$, and theoretically, the current response to the square test voltage pulse is also square (Fig. 2c$_i$). An example of typical parameters is as follows V (5 [mV]) = I (1 [nA]) × R (5 [MΩ]). This state is sometimes called "bath" configuration (electrode in extracellular solution "bath," not touching cells) (*see* **Note 8**).
5. Apply positive pressure to the pipette tip. This positive pressure is critical for successful experiments by blowing out debris and deposits accumulating on the cell surface, in order to generate a high-resistance connection between the pipette and the cell membrane (*see* **Note 9**).
6. Carefully bring the pipette close to the cell until the tip is pressed up against a patch of the cell membrane. The formation of the patch can be detected by (1) a visible deflection, or dimple, on the cell membrane and (2) a reduction in the amplitude of the square current response to the test pulse and an apparent increase in pipette resistance. This reduction in the amplitude of the current, and increase in resistance, is due to an addition of R_a and R_{leak} to the circuit (Fig. 2b). This state is called the "cell-attached" configuration. In this condition, the cell membrane is not ruptured yet and R_a is extremely high relative to R_{leak} as the patch is composed of the nonconductive lipid bilayer that forms the neuronal cell membrane. Therefore, nearly all current from the electrode passes through R_{leak} to the ground electrode, and the intracellular environment is still electrically isolated from the pipette. As a result, the cell membrane components (C_m and R_m) are negligible at this state, and the current–voltage relationship is still, for practical purposes, exclusively subject to Ohm's law (*see* **Note 10**).
7. When these signs (dimple formation and/or test current reduction) are detected, release the positive pressure and/or apply a small amount of negative pressure, if necessary, in order to help the membrane seal tightly to the pipette tip. The

current amplitude will decrease gradually or sometimes abruptly, and the measured resistance will increase (*see* **Note 11**). This occurs due to an increase in the magnitude of R_{leak} as the membrane is sealing around the pipette rim. When R_{leak} is high enough (typically >1GΩ; "giga-seal"), the square test current will be practically invisible (Fig. 2c_{ii}). To illustrate this, in Ohm's equation, when a 5 mV test pulse is applied to the circuit whose resistance (= R_p + R_{leak} in series) is 1 GΩ, the current response will be 5 pA in amplitude, which is a comparable level of the baseline noise in typical patch-clamp recordings; V (5 [mV]) = I (5 [pA]) × R (1 [GΩ]).

8. During seal formation (**step** 7 above), the holding current (I_{hold}), which is the current the head stage and the amplifier produce to generate the desired voltage drop to the ground electrode, can also be monitored. When R_{leak} is high enough (e.g., ~ 100 MΩ), applying a holding voltage (V_{hold}) of near physiological membrane potential (e.g., −70 mV) results in a jump of I_{hold} due to a large current leak from the seal between the pipette rim and the cell membrane (R_{leak}). (*see* **Note 12**). When the seal becomes tighter, as R_{leak} increases, I_{hold} becomes smaller. As an example, when the R_{leak} is 1 GΩ (giga-seal) and if the V_{hold} is −70 mV, the I_{hold} will be −70 pA, again based on Ohm's law; V (−70 [mV]) = I (−70 [pA]) × R (1 [GΩ]) (Fig. 2$d_{i,\ ii}$).
9. When the resistance (R_{leak}) reaches >1 GΩ, the setup is ready for the rupture of the cell membrane (often called "break-in").
10. Apply negative pressure and/or short electrical pulses (in some amplifiers, this electrical pulse is called "zap") to rupture the cell membrane until the shape of the current response to the test pulse changes drastically (see below). I_{hold} can also show a subtle change depending on the cell type, pipette solution, and V_{hold} (see below) (*see* **Note 13**).
11. When the membrane is ruptured, the intracellular environment is electrically connected to the recording devices (head stage and amplifier) through the pipette electrode. This state is called the "whole-cell" configuration (Fig. 2b). While R_a decreases drastically, some incompletely ruptured membranes do produce a resistive element between the electrode and the intracellular space (*see* **Note 14**). At this stage, if the giga-seal is well maintained, R_{leak} remains high, and if the cell membrane does not re-seal, R_s will be many orders of magnitude lower than R_{leak} (e.g., $R_s = R_a + R_p$ [several tens of MΩ] vs R_{leak} [GΩ]). In this configuration, experimenters can assume that practically all currents pass through R_s (= $R_p + R_a$), currents passing through R_{leak} (I_{leak}) are negligible, and currents originating from the cell membrane can be accurately measured by the recording

electrode (*see* **Note 15**). Disruption of giga-seal (a reduction in the magnitude of R_{leak}) can be monitored as an increase in I_{hold}. Intuitively, the increase in I_{hold} magnitude means that the amplifier is required to generate a large amount of current to maintain the membrane voltage at V_{hold}, which is typically set at near physiological resting membrane potential (e.g., −70 mV). When R_{leak} is high, as less current is leaking to the ground electrode, less current is required to maintain the membrane voltage at V_{hold} (Fig. 2b) (*see* **Notes 16** and **17**).

12. In addition to a change in the I_{hold}, the current response to the test pulse shows a drastic change in its shape following the rupture of the membrane (Fig. $2e_{i,ii}$). This results from the addition of the capacitor component of the membrane (C_m) to the electrical circuit. In an equivalent circuit, C_m is connected in parallel with the membrane resistance (R_m), and R_m and C_m are connected with R_s in series forming a resistor–capacitor (RC) circuit (Fig. 2b). The current–voltage relationship to the test pulse is no longer simply subject to Ohm's law but contains two distinct types of components which are (1) an instantaneous current which decays exponentially ($I_{instant}$) and (2) a steady plateau current ($I_{plateau}$) (Fig. $2e_{i,\ ii}$). By measuring these two types of currents, the intrinsic cellular membrane properties (R_m and C_m) as well as the quality of the whole-cell recording configuration (R_s) can be estimated (*see* **Note 18**). In contrast to R_{leak}, R_s MUST be as low and stable as possible. Large R_s also results in a considerable experimental error in voltage-clamping (e.g., a large R_s connected in series with R_m, and C_m means that a significant fraction of the holding voltage applied to the circuit produces a voltage drop across R_s relative to R_m) (*see* **Note 19**).

13. Once the whole-cell configuration is formed, and acceptably high R_{leak} and low R_s are confirmed, patch-clamp experiments are ready to get started. A variety of electrophysiological measures can be recorded using either voltage-clamp or current-clamp modes. In the next sections, examples of actual experiments related to cortical development will be presented.

3.4 Recording Action Potentials

This section and the following sections (Subheadings 3.5 and 3.6) detail example procedures of electrophysiological experiments. While discussing all possible types of experiments is clearly beyond the scope of this chapter, the examples presented here are among the most common patch-clamp experiments and ones most readers likely need to learn. In this section, procedures for how to record action potentials will be explained:

1. The ability to generate action potentials is a unique feature of certain electrically active cells such as neurons and distinguishes

them from other cell types in the brain (e.g., glial cells) and elsewhere. Action potentials can be measured, or recorded, in several ways using the patch-clamp technique.

2. In neurons that spontaneously generate action potentials such as cerebellar Purkinje cells, action potentials can be easily measured in the cell-attached configuration [11, 12].
3. Alternatively, current-clamp or voltage-clamp configurations can be used to measure the action potential or the action "current" (the current underlying the action potential), respectively. Due to their high amplitudes and "spike" like characteristic shapes, action potentials or action currents can be easily distinguished from the background signal.
4. Action potentials can also be "evoked" in several ways.
5. One of the simplest ways to evoke action potentials is by directly injecting a depolarizing current into the cell soma through the patch electrode. In whole-cell current-clamp configuration, neurons can be incrementally depolarized by injecting gradually increasing amounts of current until the membrane potential exceeds the threshold for action potential generation, and then the cell starts to "fire" [13] (Fig. 3a) (*see* **Note 20**).
6. Numerous pieces of information can be extracted from analyses of action potentials. For instance, the minimal amount of injected current required to evoke action potential, known as the "rheobase" current, can be measured in the experiment described above. The rheobase current reflects "intrinsic" cellular excitability, produced by the combination of a variety of factors including resting membrane potential, membrane resistance, or action potential threshold. When the number of action potentials evoked is plotted against the amount of currents injected, an input–output (I–O) curve of action potentials can be generated. This I–O curve is a direct measure of how excitable the neuron is and in some cases is even used as a synonym for "neuronal excitability." These measures developmentally change during cortical development, and, consequently, neuronal excitability is often used as a measure of neuronal maturation [13] (Fig. 3b) (*see* **Note 21**).

3.5 Recording Ion Channel Activity

A cell membrane is an incomplete electrical insulator composed of a relatively nonconductive lipid bilayer and conductive ion channels that permit electrical current to flow in the form of ion flux across the cell membrane. Many ion channels possess open, conductive, and closed, nonconductive states that are influenced by membrane voltage and are, therefore, referred to as voltage-dependent channels. Measuring the currents that pass through these channels at specific voltages is an ideal experiment for the voltage-clamp

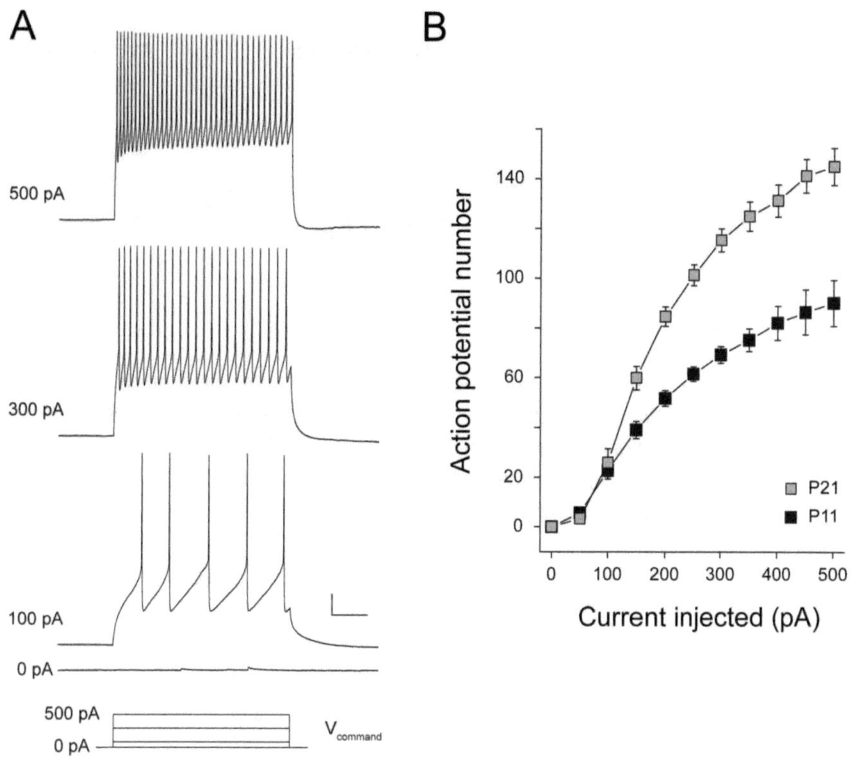

Fig. 3 Action potential analysis. (**a**) Example of action potentials evoked by somatic square current injection. The amount of injected currents (100 pA, 300 pA, and 500 pA) is shown next to each trace and at the bottom. Calibration: 100 ms, 10 mV. (**b**) Input–output (I–O) curve of evoked action potential recorded from cortical GABAergic interneurons during development at postnatal day 11 (P11) and 21 (P21) (P11, 22 cells and P21, 12 cells; each data point indicates averages in corresponding currents, and error bars indicate SEM). The number of action potentials is plotted against the amount of currents injected. Note that neurons from P21 mice fire more action potentials than those from P11 mice (I–O curve shows a leftward shift) demonstrating that the action potential firing property shows a developmental change. Modified from [13]

configuration. In this section, I will explain an example of recording of ion channel activity using whole-cell voltage-clamp. The patch-clamp technique also allows for the analysis of single-channel activity (currents passing through a single ion channel), but it will not be discussed here:

1. Most modern patch-clamp amplifiers can generate arbitrary voltage outputs on demand that can include even very complex patterns (Fig. 4). These protocols can be used to activate voltage-dependent channels.
2. There are several ways to isolate or distinguish the types of channels that mediate a given current. Certain types of current can be easily distinguished from others simply by their measured shapes, amplitude, and activation and deactivation kinetics (Fig. 5).

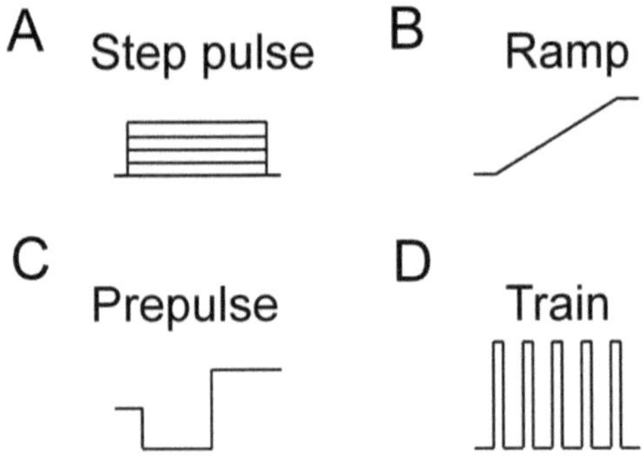

Fig. 4 Examples of command voltages. (**a**) Step pulse protocol: Incremental stepwise square depolarizations from baseline. (**b**) Ramp protocol: Gradual depolarization from baseline. (**c**) Prepulse protocol: Hyperpolarization (prepulse) followed by depolarization. (**d**) Train protocol: Repeated short depolarizations

3. Figure 5a is an example of applying incremental depolarizing voltage steps (−70 mV → +30 mV) to measure the current responses generated by the resulting activation of voltage-gated Na^+ channels and K^+ channels (Fig. 5a).
4. When the membrane voltage is depolarized above −40 mV, a very large inward current is observed with rapid onset and offset that is clearly distinct from the background current (Fig. 5b). This is produced by the opening of voltage-gated Na^+ channels, and the current recorded is due to the passage of largely Na^+ ions flowing through these channels (I_{Na}) into the cell down the concentration gradient. The activity of voltage-gated Na^+ channels is a main source of the depolarizing current during the initial rising phase of the action potential.
5. Following the Na^+ current, a slowly activating and sustained outward current is apparent (Fig. 5a). This current is mainly mediated by voltage-gated K^+ channels, and due to its current kinetics, it is called a delayed rectifier K^+ current (I_K). I_K plays a critical role in membrane repolarization following the action potential (*see* **Note 22**).
6. As discussed above, the shape of measured ion channel currents is one of the most useful parameters to determine which channels mediate the recorded currents. Several additional types of manipulations can be made to further characterize channel identity. In many cases, pharmacological compounds are available to selectively block certain types of channels. For instance, the application of the Na^+ channel blocker tetrodotoxin (TTX) in extracellular recording solution is known to block I_{Na} specifically while leaving I_K unaffected. Additionally, manipulating the ionic constituents of the recording or pipette solutions can also help isolate or block currents mediated by the channels of

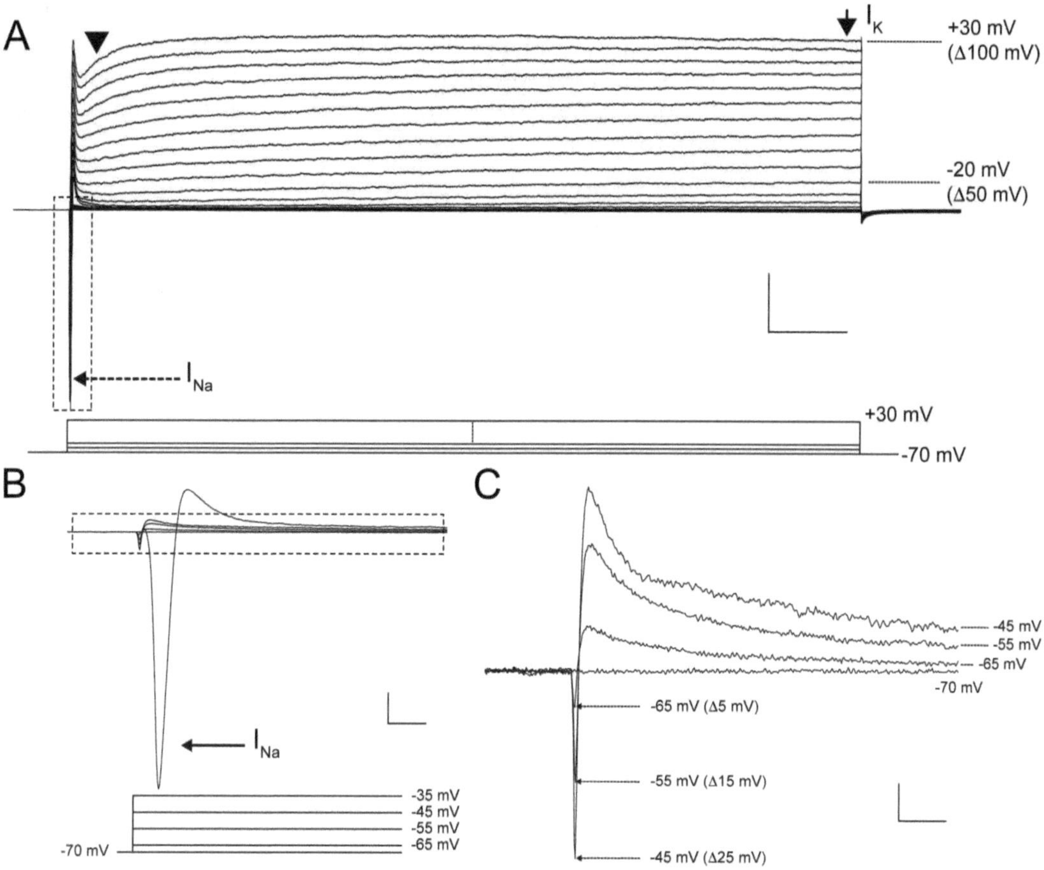

Fig. 5 Ion channel activity analysis. (**a**) Example of current responses evoked by somatic depolarization from −70 mV to +30 mV (5 mV increments). I_{Na} and I_K are indicated by a horizontal dashed line and a vertical solid line. Note the clear difference in activation and deactivation time scales between I_{Na} and I_K. I_{Na} is activated and deactivated very quickly (< a few ms) while I_K is activated relatively slowly (arrowhead) and sustains throughout the long (500 ms) depolarization period. I_K at −20 mV and + 30 mV of membrane voltages are indicated by horizontal dashed lines, which are nonproportional to the magnitude of depolarization (Δ50 mV and Δ100 mV). Note the clear difference from the currents shown in (**c**), which are proportional to each depolarization (Δ5 mV, Δ15 mV, and Δ25 mV). The dashed rectangle shows the portion of the trace magnified in (**b**). Calibration: 50 ms, 5 nA. (**b**) Magnified portion of the same traces shown in (**a**). Current responses at baseline (−70 mV) and four different depolarized membrane voltages (−65 mV, −55 mV, −45 mV, and −35 mV) are shown. I_{Na} is indicated by the solid line. Note that a huge (> 10 nA) inward I_{Na} abruptly appears when the membrane voltage depolarizes at a certain level (−35 mV). The dashed rectangle shows the portion of the trace magnified in (**c**). Calibration: 2 ms, 2 nA. (**c**) Magnified portion of the same traces shown in (**b**). The trace containing I_{Na} (−35 mV in **b**) is omitted. Note sharp inward currents in each trace approximately proportional to each depolarization (≈ injected current) indicated by dashed lines. This suggests that they are artificial responses (note the difference with I_K shown in **a**) (*see* **Note 22**). Calibration: 2 ms, 200 pA

interest. For instance, removing Ca^{2+} from the extracellular recording solution can block Ca^{2+} channel-mediated and Ca^{2+}-activated currents. Recordings from transgenic animals or cell lines can also help to dissect specifically which channels or component proteins are required for producing a certain current. For instance, preventing the expression of the genes encoding the protein components of certain ion channels, or "knocking out" these proteins, can disrupt the ability of channels to generate their characteristic currents.

7. During cortical development, ion channel activity undergoes developmental alterations that can reflect developmental changes in the expression level and subunit composition of constituent proteins as well as their interaction with modulatory molecules.

3.6 Recording Synaptic Activity

Neuronal tissue consists of a very complex interconnected network. Neurons are connected to each other through synapses where a chemical signal, neurotransmitters released by a presynaptic cell, is converted to an electrical signal produced by ions flowing through postsynaptic neurotransmitter receptors. The analysis of synaptic activity is, therefore, another good example where the patch-clamp technique can be used:

1. Synaptic activity is typically recorded in the whole-cell configuration. Either current-clamp or voltage-clamp configurations can be used. Immediately after a whole-cell configuration is formed, spontaneous synaptic activity (spontaneous postsynaptic current; sPSC in voltage-clamp, or spontaneous postsynaptic potential; sPSP in current-clamp) can be observed. Synaptic responses can also be evoked by external stimuli such as electrical stimulation of afferent connections through an extracellular stimulating electrode.
2. Synaptic transmission is roughly divided into two kinds, excitatory transmission, which generates depolarizing currents, and inhibitory transmission, which generates hyperpolarizing currents. In the brain, excitatory and inhibitory transmission are mediated mainly by glutamate and GABA, respectively.
3. In electrophysiological experiments, excitatory and inhibitory postsynaptic currents/potentials (EPSC and IPSC/EPSP and IPSP) can be pharmacologically isolated by the inclusion of blockers of either GABA receptors or glutamate receptors, respectively, to the external recording solution.
4. In cortical slices, particularly from adult mice, treatment of slices with GABA receptor blockers may cause contamination of polysynaptic burst-like activity that makes accurate measurement of individual synaptic responses difficult (Fig. 6a). Therefore, to isolate inhibitory and excitatory postsynaptic currents, voltage-clamping the neuron at the reversal potential for glutamatergic and GABAergic synaptic transmissions in ACSF without added GABA blockers may be recommended (Fig. 6b) (*see* **Note 23**).

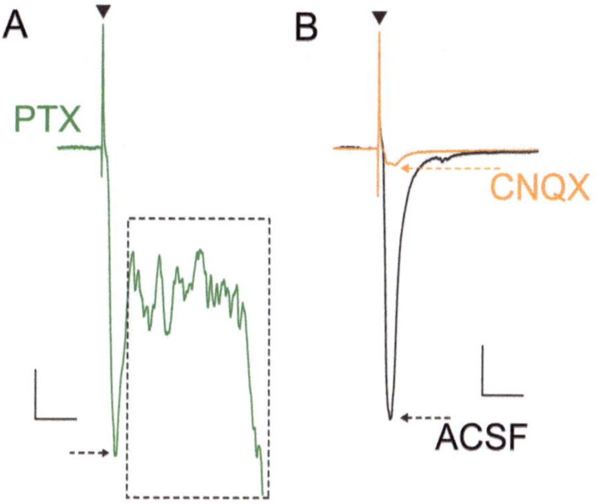

Fig. 6 Isolating excitatory and inhibitory synaptic responses. (**a**) Example of evoked synaptic response recorded from a cortical excitatory neuron in the presence of GABA receptor blocker picrotoxin (PTX). The arrowhead denotes stimulation artifact. Note burst-like responses (highlighted by a dashed box) following the first peak (highlighted by a dashed line). Calibration: 10 ms, 200 pA. (**b**) Example of synaptic responses in plain ACSF (black) and in the presence of AMPA type glutamate receptor blocker 6-Cyano-7-nitroquinoxaline-2,3-dione disodium (CNQX) (orange) recorded from the same cortical excitatory neuron. The arrowhead denotes stimulation artifact. The peaks in each condition are indicated by black- and orange-dashed lines. The neuron was voltage-clamped at −70 mV, which is close to the calculated reversal potential of Cl^-, the main ionic source of GABAergic synaptic transmission. Note that a clean synaptic response with a clear single peak was evoked, and > 90% of this is blocked with the CNQX treatment, suggesting that this response is a pure glutamatergic response (EPSC). Calibration: 10 ms, 200 pA

5. Synaptic activity shows dynamic alterations during brain development and maturation. For instance, neurons start to connect with each other, and functional synapses are formed in the early postnatal weeks in rodents. This process is called synaptogenesis [14–16]. Reflecting synaptogenesis, the frequency of spontaneous postsynaptic currents (sPSC) increases during these developmental stages (Figs. 7a, b) [13] (*see* **Note 24**).
6. In rodent cortices, network refinement such as synaptogenesis is most drastic during early postnatal weeks and disruption of the network refinement during this period is known to cause long-lasting consequences, as this period is commonly called the "critical period" [17–19] (*see* **Note 25**). Synaptic plasticity refers to activity-dependent increases (long-term potentiation; LTP) or decreases (long-term depression; LTD) in the functional strength of synapses. Synaptic plasticity is closely

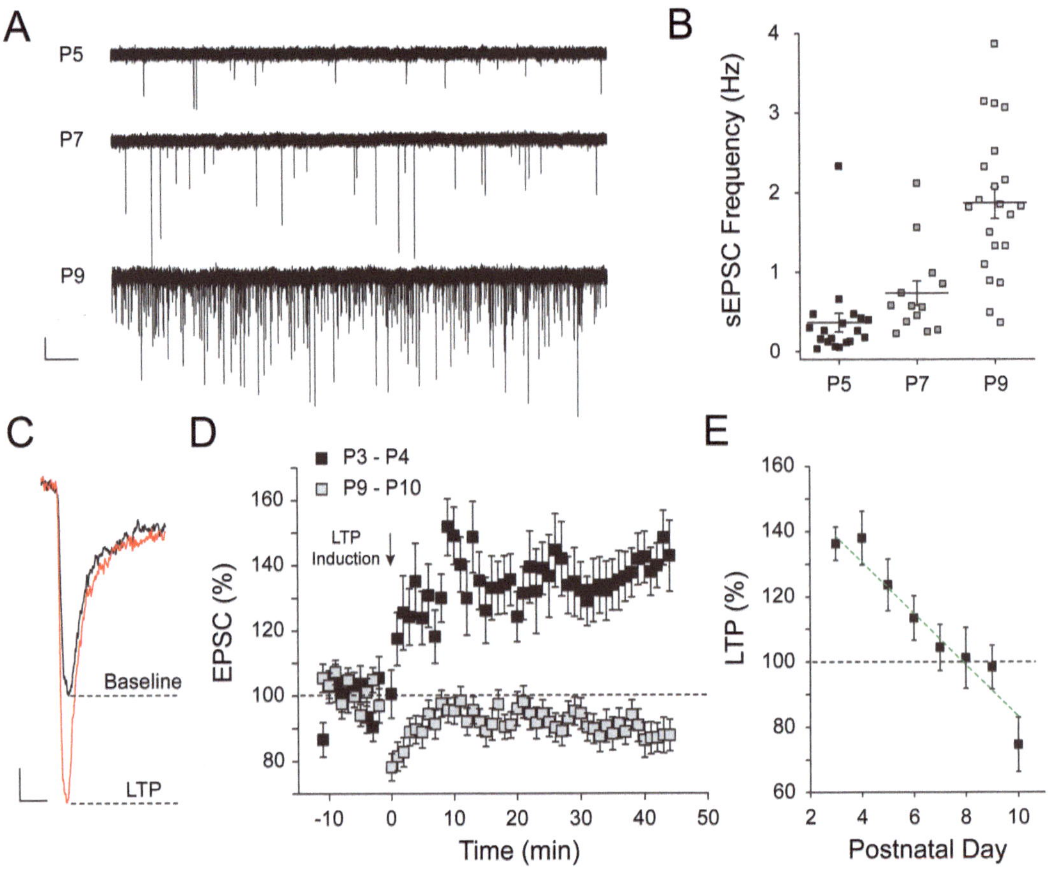

Fig. 7 Analysis of synaptic activity. (**a**) Example of sEPSC traces recorded from cortical interneurons in different developmental stages (P5, P7, and P9). Note the clear developmental increase in the number of synaptic events. Calibration: 2 s, 20 pA. (**b**) Collective data for sEPSC frequency during different developmental stages. Average and SEM are shown with horizontal and vertical bars, respectively, in each age. (**c**) Example of evoked EPSCs before (black) and after (red) LTP-inducing stimulation recorded from cortical excitatory neurons. Note the amplitude of EPSC increases in the LTP condition. Calibration: 2 ms, 20 pA. (**d**) Collective data for the time course of relative EPSC amplitude during LTP recordings at different developmental stages (P3–P4, 9 cells, and P9–P10, 7 cells). LTP-inducing stimuli were given at $t = 0$ after obtaining a stable 10-min baseline ($t = -10$ to 0). Note long-lasting (> 30 min) potentiation, i.e., LTP of EPSC amplitudes after LTP-inducing stimuli at an early developmental time point (P3–P4). LTP is not obvious, rather EPSC shows a small depression at a later time point (P9–P10), demonstrating that synaptic plasticity undergoes developmental alterations. (**e**) Collective data for the magnitude of LTP at different developmental time points. As indicated by the trendline (green dashed line), the LTP magnitude shows a developmental decrement. (**a**, **b**) modified from [13]. (**c**–**e**) modified from [39] with permission from Dr. Contractor

associated with the opening and closing of critical periods [19, 20]. The examples shown in Figs. 7c–e summarizes developmental changes in synaptic plasticity during the critical period in the murine somatosensory cortex.

4 Notes

1. The author prepares two types of 10× solutions for slicing solution (850 mM NaCl, 25 mM KCl, 12.5 mM NaH_2PO_4, 250 mM $NaHCO_3$) and for incubating and recording solutions (1.25 M NaCl, 24 mM KCl, 12 mM NaH_2PO_4, 250 mM $NaHCO_3$, 250 mM glucose). 10× solution for slicing can be kept at room temperature, but 10× solution for incubating and recording should be stored at 4 °C as it contains an organic material (glucose). Sucrose, glucose, DL-APV, and kynurenate are added daily for the slicing solution. DL-APV and kynurenate are added daily for incubating solutions. $CaCl_2$ and $MgCl_2$ are added daily for all solutions during which they should be equilibrated with 5% carbogen (95% O_2 and 5% CO_2) to avoid precipitation. When Ca/Mg precipitation develops, the color of the solution becomes cloudy. Applying stronger bubbling by carbogen may resolve the precipitation.
2. In slicing solution, the concentration of Na^+ should be reduced and replaced by other organic materials, such as sucrose, choline, or NMDG [21, 22]. High Na^+ concentration in slicing solution is known to cause cellular edema leading to cellular death. As the success rate of patch-clamp experiments heavily relies on how healthy the cells are, this cellular death or edema must be avoided as much as possible.
3. Normal ACSF same as the recording solution can also be used here. The author empirically uses this incubating solution containing low Ca^{2+} and blockers for glutamate receptors gradually perfused into the incubating chamber, which might ameliorate excitotoxicity.
4. The osmolarity of intracellular solutions makes a big difference in the success rate of forming a tight seal. It is typically slightly higher than that of extracellular solution. The author typically prepares solutions of ~300 mOsm/kg (extracellular: ~290 mOsm/kg). Osmolarity can be adjusted by adding some osmolytes such as sucrose [21].
5. Debris and precipitation in the pipette solution are the sources interrupting successful experiments. They obstruct the tip of the pipette and make forming a tight seal between the pipette and the cell membrane difficult. Filtering the pipette solution with a 0.22 μm centrifuge filter helps remove these in the solution.
6. How to dissect the brain varies depending on the purpose. Several manipulations should be considered when experimenters need to preserve or cut the synaptic connections of interest [23–25].

7. Vertical movement of the vibrating blade is known to be harmful to the slicing outcome [26].
8. You may also see small nonlinear transients, which originated from the capacitor component of the pipette (C_p). As the pipette electrode is made from an insulator (glass) which separates two conductive phases (pipette solution and extracellular solution), the pipette itself can be considered a capacitor. These nonlinear transients are relatively minor or even invisible in some cases as C_p is relatively small and can be neutralized by a small current injection (function of an amplifier, referred to as capacitance compensation). Therefore, I will skip this component hereafter for simplification. To the author's knowledge, most commercially available patch-clamp amplifiers equip a built-in circuit to cancel these components.
9. Patch-clamp electrode holder has a nodule where a tube can be connected for the pressure to come in. Choose a relatively hard tube. Using a too elastic tube results in failure in applying appropriate pressure through the pipette. Positive pressure sometimes blows out debris floating inside the pipette as well. In this case, you may see the R_p shows a slight change (a small decrease in many cases).
10. This explanation is simplified. An intact (not ruptured) cell membrane is NOT perfectly nonconductive. One easiest example is the fact that single-channel recordings are commonly conducted in a cell-attached configuration. For further reading, see [7, 8, 27, 28].
11. Empirically, cells showing an abrupt current reduction are unhealthy cells.
12. Empirically, if R_{leak} is not high enough, applying V_{hold} of ~ −70 mV may cause a disruption of the seal. If giga-seal can be obtained without applying V_{hold}, it can also be applied after giga-seal formation is confirmed (Fig. 2$d_{i,\ ii}$).
13. Large I_{hold} indicates that I_{leak} (current passing through R_{leak}) is large.
14. In an ideal situation (if the membrane could be perfectly ruptured and completely removed from the tip of the electrode), R_a would originate solely from the junction between the conductive pipette solution and conductive intracellular fluid, which would be extremely low. Practically, perfect rupture of the membrane is never possible.
15. This is solely based on the assumption that "$R_s << R_{leak}$." This means (1) giga-seal MUST be well maintained (very high R_{leak}: ~GΩ order) and (2) membrane MUST be ruptured successfully and MUST not interrupt the pipette solution being well connected to the intracellular environment (re-seal) (very low

R_s: less than tens of MΩ order) throughout the experiment. Both low R_{leak} and high R_s result in serious experimental errors and data acquired from these experiments can lead the experimenters to a critical misinterpretation.

16. As I_{leak} is an "artificial" current, passing through low R_{leak}, it is nonselective for any type of ions. It means that the reversal potential for I_{leak} is 0 mV theoretically. By setting V_{hold} at 0 mV, experimenters can distinguish I_{leak} from "real" currents passing through physiological membrane ion channels (if the patch electrode is electrically isolated, i.e., giga-seal is well maintained, a certain amount of currents is required to hold the cell at 0 mV).

17. The acceptable range of I_{hold} can vary depending on the purposes, and it cannot be strictly determined. As a nature of patch-clamp recordings, the very exact "physiological" resting membrane potential can never be measured. Practically, the author employs I_{hold} < a couple of hundreds of pA as criteria for recordings of synaptic plasticity as an example. The criteria might be less strict in these types of experiments in which a relative value of responses (baseline vs post-plasticity induction) matters. In some analyses in which more accurate voltage control is required such as measurements of voltage-gated Na^+ channel activity, however, the requirement may need to be more stringent. In that example, the author employs $I_{hold} < 100$ pA or so.

18. At the time a test pulse ($V_{command}$) is applied to the RC circuit, the voltage across C_m (V_{Cm}) is steady. Therefore, the current passing through the recording electrode (R_s) instantly after the $V_{command}$ is applied ($I_{instant}$) is $V_{command}/R_s$ based on Ohm's law. This instantaneous current ($I_{instant}$) shows an exponential decay, reflecting C_m being gradually charged. The decaying kinetics of $I_{instant}$ is subject to how fast the C_m is charged. The time constant of the C_m, tau (τ_m), is measured approximately as $\tau_m = C_m \times R_s$. When C_m is fully charged, V_{Cm} is steady again and the current recorded at this steady state ($I_{plateau}$) is given by $V_{command} / (R_s + R_m)$ from Ohm's equation. The sum of R_s and R_m is called "input resistance" (R_{in}), but as R_m is generally much larger than R_s, R_m and R_{in} are not strictly distinguished in some cases.

19. In many cases, electrophysiologists empirically set the data inclusion criteria of <15–20% change in R_s during recordings.

20. Applying hyperpolarizing pre-pulse might help more accurate measurements by releasing the voltage-dependent block of voltage-gated channels at the resting state in some cases.

21. In the majority classes of neurons, with the amount of injected current increases, the number of action potentials initially

increases (as shown in Fig. 3) and then starts to decrease, or even cells can abruptly stop firing action potentials. This phenomenon is due to spike adaptation (or accommodation) or depolarization block of voltage-dependent channels [29–32]. The neuron shown in Fig. 3 is a relatively rare type of neuron which lacks spike adaptation as its characteristics (fast-spiking interneuron; a subclass of cortical interneuron [33]).

22. Note that the sustained currents are not proportional to the magnitude of depolarization (≈ amount of injected current). This indicates that these currents are activated in a voltage-dependent manner, i.e., the responsible channels are not activated until the membrane voltage reaches a certain level (~ −10 mV in this particular cell). This is well contrasted to artificial currents (injected to neutralize capacitance components), which are roughly proportional to the magnitude of depolarization shown in Fig. 5c.
23. In that case, as discussed in Subheadings 3.3.11 and 3.3.12, poor seal and access (small R_{leak} and large R_s) result in a considerable experimental error, i.e., voltage is not clamped at all leading to poor isolation of excitatory or inhibitory transmissions. Therefore, the quality of the seal and access should be as high as possible, and intense monitoring of the parameters (small I_{hold} and large $I_{instant}$) is required.
24. There is a huge diversity in how synaptogenesis occurs depending on, e.g., brain region, cell type, or species. Figure 7a, b is just presenting one example.
25. As with synaptogenesis, there is also a huge variation in the critical period depending on a variety of factors.
26. Further reading: There are several video-based articles introducing materials and methods for patch-clamp available, which may be helpful for readers' better understanding [34–36]. Manuals for patch-clamp devices and software are also useful [37–40].

Acknowledgments

The author thanks Drs. John Marshall and Anis Contractor (Northwestern) for English editing and their helpful inputs to the manuscript. The author also thanks Dr. Contractor for his kind permission to present previously published data from his laboratory.

References

1. Neher E, Sakmann B (1976) Single-channel currents recorded from membrane of denervated frog muscle fibres. Nature 260(5554):799–802. https://doi.org/10.1038/260799a0
2. Hamill OP, Marty A, Neher E, Sakmann B, Sigworth FJ (1981) Improved patch-clamp techniques for high-resolution current recording from cells and cell-free membrane patches. Pflugers Arch 391(2):85–100. https://doi.org/10.1007/BF00656997
3. Sakmann B, Neher E (1984) Patch clamp techniques for studying ionic channels in excitable membranes. Annu Rev Physiol 46:455–472. https://doi.org/10.1146/annurev.ph.46.030184.002323
4. NobelPrize.org (1991). https://www.nobelprize.org/prizes/medicine/1991/summary/. Accessed 11 May 2023
5. Sakmann B, Edwards F, Konnerth A, Takahashi T (1989) Patch clamp techniques used for studying synaptic transmission in slices of mammalian brain. Q J Exp Physiol 74(7):1107–1118. https://doi.org/10.1113/expphysiol.1989.sp003336
6. Edwards FA, Konnerth A, Sakmann B, Takahashi T (1989) A thin slice preparation for patch clamp recordings from neurons of the mammalian central nervous system. Pflugers Arch 414(5):600–612. https://doi.org/10.1007/BF00580998
7. Ogden D (1994) Microelectrode techniques: the Plymouth workshop handbook, 2nd edn. The Company of Biologists, Cambridge
8. Molleman A (2003) Patch clamping: an introductory guide to patch clamp electrophysiology. Wiley
9. Williams SR, Mitchell SJ (2008) Direct measurement of somatic voltage clamp errors in central neurons. Nat Neurosci 11(7):790–798. https://doi.org/10.1038/nn.2137
10. Beaulieu-Laroche L, Harnett MT (2018) Dendritic spines prevent synaptic voltage clamp. Neuron 97(1):75–82 e73 https://doi.org/10.1016/j.neuron.2017.11.016
11. Hausser M, Clark BA (1997) Tonic synaptic inhibition modulates neuronal output pattern and spatiotemporal synaptic integration. Neuron 19(3):665–678. https://doi.org/10.1016/s0896-6273(00)80379-7
12. Williams SR, Christensen SR, Stuart GJ, Hausser M (2002) Membrane potential bistability is controlled by the hyperpolarization-activated current I(H) in rat cerebellar Purkinje neurons in vitro. J Physiol 539(Pt 2):469–483. https://doi.org/10.1113/jphysiol.2001.013136
13. Nomura T, Musial TF, Marshall JJ, Zhu Y, Remmers CL, Xu J, Nicholson DA, Contractor A (2017) Delayed maturation of fast-spiking interneurons is rectified by activation of the TrkB receptor in the mouse model of fragile X syndrome. J Neurosci 37(47):11298–11310. https://doi.org/10.1523/JNEUROSCI.2893-16.2017
14. Rakic P, Bourgeois JP, Goldman-Rakic PS (1994) Synaptic development of the cerebral cortex: implications for learning, memory, and mental illness. Prog Brain Res 102:227–243. https://doi.org/10.1016/S0079-6123(08)60543-9
15. Vitali I, Jabaudon D (2014) Synaptic biology of barrel cortex circuit assembly. Semin Cell Dev Biol 35:156–164. https://doi.org/10.1016/j.semcdb.2014.07.009
16. Vanderhaeghen P, Polleux F (2023) Developmental mechanisms underlying the evolution of human cortical circuits. Nat Rev Neurosci. https://doi.org/10.1038/s41583-023-00675-z
17. Wiesel TN, Hubel DH (1963) Single-cell responses in striate cortex of kittens deprived of vision in one eye. J Neurophysiol 26:1003–1017. https://doi.org/10.1152/jn.1963.26.6.1003
18. Hubel DH, Wiesel TN (1970) The period of susceptibility to the physiological effects of unilateral eye closure in kittens. J Physiol 206(2):419–436. https://doi.org/10.1113/jphysiol.1970.sp009022
19. Hensch TK (2005) Critical period plasticity in local cortical circuits. Nat Rev Neurosci 6(11):877–888. https://doi.org/10.1038/nrn1787
20. Crair MC, Malenka RC (1995) A critical period for long-term potentiation at thalamocortical synapses. Nature 375(6529):325–328. https://doi.org/10.1038/375325a0
21. Bischofberger J, Engel D, Li L, Geiger JR, Jonas P (2006) Patch-clamp recording from mossy fiber terminals in hippocampal slices. Nat Protoc 1(4):2075–2081. https://doi.org/10.1038/nprot.2006.312
22. Ting JT, Daigle TL, Chen Q, Feng G (2014) Acute brain slice methods for adult and aging animals: application of targeted patch clamp analysis and optogenetics. Methods Mol Biol 1183:221–242. https://doi.org/10.1007/978-1-4939-1096-0_14

23. Agmon A, Connors BW (1991) Thalamocortical responses of mouse somatosensory (barrel) cortex in vitro. Neuroscience 41(2–3):365–379. https://doi.org/10.1016/0306-4522(91)90333-j
24. Urban NN, Barrionuevo G (1998) Active summation of excitatory postsynaptic potentials in hippocampal CA3 pyramidal neurons. Proc Natl Acad Sci USA 95(19):11450–11455. https://doi.org/10.1073/pnas.95.19.11450
25. Kim J, Lee S, Tsuda S, Zhang X, Asrican B, Gloss B, Feng G, Augustine GJ (2014) Optogenetic mapping of cerebellar inhibitory circuitry reveals spatially biased coordination of interneurons via electrical synapses. Cell Rep 7(5):1601–1613. https://doi.org/10.1016/j.celrep.2014.04.047
26. Geiger JR, Bischofberger J, Vida I, Frobe U, Pfitzinger S, Weber HJ, Haverkampf K, Jonas P (2002) Patch-clamp recording in brain slices with improved slicer technology. Pflugers Arch 443(3):491–501. https://doi.org/10.1007/s00424-001-0735-3
27. Sakmann B, Neher E (1995) Single-channel recordlng. Springer, New York. https://doi.org/10.1007/978-1-4419-1229-9
28. Perkins KL (2006) Cell-attached voltage-clamp and current-clamp recording and stimulation techniques in brain slices. J Neurosci Methods 154(1–2):1–18. https://doi.org/10.1016/j.jneumeth.2006.02.010
29. Partridge LD, Stevens CF (1976) A mechanism for spike frequency adaptation. J Physiol 256(2):315–332. https://doi.org/10.1113/jphysiol.1976.sp011327
30. Fleidervish IA, Friedman A, Gutnick MJ (1996) Slow inactivation of Na+ current and slow cumulative spike adaptation in mouse and Guinea-pig neocortical neurons in slices. J Physiol 493(Pt 1):83–97. https://doi.org/10.1113/jphysiol.1996.sp021366
31. Torkkeli PH, Sekizawa S, French AS (2001) Inactivation of voltage-activated Na(+) currents contributes to different adaptation properties of paired mechanosensory neurons. J Neurophysiol 85(4):1595–1602. https://doi.org/10.1152/jn.2001.85.4.1595
32. Gollisch T, Herz AV (2004) Input-driven components of spike-frequency adaptation can be unmasked in vivo. J Neurosci 24(34):7435–7444. https://doi.org/10.1523/JNEUROSCI.0398-04.2004
33. Kawaguchi Y, Kubota Y (1997) GABAergic cell subtypes and their synaptic connections in rat frontal cortex. Cereb Cortex 7(6):476–486. https://doi.org/10.1093/cercor/7.6.476
34. Segev A, Garcia-Oscos F, Kourrich S (2016) Whole-cell patch-clamp recordings in brain slices. J Vis Exp 112. https://doi.org/10.3791/54024
35. Kida H, Sakimoto Y, Mitsushima D (2017) Slice patch clamp technique for analyzing learning-induced plasticity. J Vis Exp 129. https://doi.org/10.3791/55876
36. Weng W, Li D, Peng C, Behnisch T (2018) Recording synaptic plasticity in acute hippocampal slices maintained in a small-volume recycling-, perfusion-, and submersion-type chamber system. J Vis Exp 131. https://doi.org/10.3791/55936
37. Molecular Devices (2006) pCLAMP10: data acquisition and analysis for comprehensive electrophysiology; user guide. Molecular Devices
38. Molecular Devices (2012) The axon guide electrophysiology and biophysics laboratory techniques third edition. Molecular Devices, California
39. Harlow EG, Till SM, Russell TA, Wijetunge LS, Kind P, Contractor A (2010) Critical period plasticity is disrupted in the barrel cortex of FMR1 knockout mice. Neuron 65(3):385–398. https://doi.org/10.1016/j.neuron.2010.01.024
40. Scientifica Neurowire Blog. https://www.scientifica.uk.com/neurowire-blog. Accecced 11 May 2023

Chapter 20

In Vivo Whole-Cell Recording from the Mouse Brain

Masahiro Kawatani and Takayuki Yamashita

Abstract

Measuring the membrane potential dynamics of neurons offers a comprehensive understanding of the molecular and cellular mechanisms that form their spiking activity, thus playing a crucial role in unraveling the mechanistic processes governing brain function. Techniques for intracellular recordings of membrane potentials pioneered in the 1940s have witnessed significant advancements since their inception. Among these, whole-cell patch-clamp recording has emerged as a leading method for measuring neuronal membrane potentials due to its high stability and broad applicability ranging from cultured cells to brain slices and even behaving animals. This chapter provides a detailed protocol to acquire stable whole-cell recordings from neurons in the cerebral cortex of awake, head-restrained mice. Significant enhancements to our protocol include implanting a metal head-post using adhesive resin cement and preparing a recording pipette with a long shank for targeting deeper brain regions. This protocol, once implemented, enables whole-cell recordings up to 2.5 mM beneath the cortical surface.

Key words Mouse, Whole-cell recording, Patch-clamp, Intracellular recording, Membrane potential, Electrophysiology, In vivo, Behavior

1 Introduction

Neurons generate action potentials (APs) in an all-or-none fashion when the membrane potential (V_m) achieves a certain threshold. These APs traverse the axon, triggering depolarization of the nerve terminals and thus instigating neurotransmitter release for synaptic transmission. The resting V_m typically resides below this threshold, facilitating a myriad of subthreshold events, such as excitatory and inhibitory postsynaptic potentials (EPSPs and IPSPs) evoked at the dendrites and somata of the neurons. It is these subthreshold V_m events that neurons compute to elicit APs. In the cerebral cortex of the mammalian brain, neurons receive and dynamically integrate synaptic inputs from diverse sources, facilitating function within complex behavioral constructs. Hence, monitoring the subthreshold V_m of neurons unveils significant insights into the intricate computational processes they utilize to accomplish the intended

Koh-ichi Nagata (ed.), *Cerebral Cortex Development: Methods and Protocols*, Methods in Molecular Biology, vol. 2794,
https://doi.org/10.1007/978-1-0716-3810-1_20,

function. Furthermore, certain aspects of the subthreshold V_m of neurons in behaving animals demonstrate notable correlations with various behavioral and cognitive parameters. This highlights the pivotal importance of measuring neuronal V_m in awake, behaving animals, as it underscores the fundamental role that these subthreshold events play in cognition and behavior.

In 1946, Graham and Gerard pioneered a method now known as sharp-electrode intracellular recording to monitor the V_m of frogs' muscle fibers [1]. This technique involves penetrating the cellular membrane with an exceedingly fine glass pipette filled with potassium chloride solution, effectively serving as a microelectrode to measure the voltage differential between the interior and exterior of the muscle fiber. This technique was applied immediately to explore synaptic transmissions at frogs' neuromuscular junctions [2, 3] and to measure the V_m of motor neurons in the spinal cord of cats [4]. These applications provided fundamental insights into the mechanisms and functions of synaptic transmission in the peripheral and central nervous system. While the execution of sharp-electrode recording is relatively straightforward, the method has drawbacks. Penetration by the glass pipette may inflict occasional cell damage, necessitating negative current injection to hyperpolarize the cell until the damaged cellular membrane seals and the cell returns to a stable resting V_m. Despite this shortcoming, sharp-electrode intracellular recording remains in use, particularly for in vivo recordings from deep neurons within the brain [5–7]. Recent advancements have seen the development of sharp electrodes composed of quartz, offering greater strength and flexibility compared to traditional borosilicate glass. The advent of quartz intracellular pipettes has paved the way for stable recordings from primate cortical neurons [8].

In the early 1980s, Bert Sakmann, Erwin Neher, and their colleagues conceived an alternate method to monitor V_m known as whole-cell patch-clamp recordings [9]. Initially developed by Sakmann and Neher [10] to record min currents passing through single channels within a pipette tip, the patch-clamp recording technology employs a glass pipette (patch pipette) with a larger tip diameter than the sharp electrode. Upon attaching the patch pipette to the cellular membrane and applying negative pressure to the pipette's interior, the pipette tip approaches the membrane, creating a high-resistance seal (giga-ohm seal) between the membrane and the tip. This seal facilitates single-channel current recording. The technique then evolved to encompass a whole-cell configuration where further negative interior pressure, following the giga-ohm seal formation, ruptured the membrane patch at the tip. This whole-cell configuration allows V_m recording with relatively lower resistance between the electrode and the cell, providing superior voltage control, a higher signal-to-noise ratio, and facilitating the diffusion of drugs into the cell. This method's applicability extends beyond cell cultures to brain slices [11] and can be easily adapted to cells with a smaller diameter, where sharp electrode recording encounters difficulties due to damage from penetration. The whole-cell recording boasts a relatively stable recording feature, likely attributed to less cellular membrane damage and a larger surface area for giga-ohm sealing between the membrane and the pipette. This stability proves advantageous for *in vivo* recording in the intact brains of animals [12, 13]. Whole-cell patch-clamp recording has thus been used for V_m recording in awake, behaving animals, providing essential insights into subthreshold V_m dynamics of neurons and their correlation to spontaneous and learned behavior [14–21]. Furthermore, recent advances in optical technology have enabled targeted whole-cell recordings from fluorescently labeled neurons under two-photon excitation microscopy [22]. This configuration further enables targeting neurons with a specific molecular marker or an axonal projection target [19, 23, 24]. By introducing an expression of light-sensitive ion channels such as channelrhodopsin 2 (ChR2) in the targeted cell types, cell-type-specific recordings are also possible without directly observing the cells. In this setting, light stimulation can be done through an optical fiber inserted in the patch pipette and the ChR2 expression is confirmed by the V_m response to the stimulation [25].

In order to provide an insight into these two V_m measurement techniques, namely sharp-electrode intracellular recording and whole-cell patch-clamp recording, we present our recordings from putative pyramidal neurons at layer 2/3 of the mouse barrel cortex (Fig. 1). These neurons are typically characterized by a low spontaneous firing rate (approximately 0.1 Hz on average), a hyperpolarized membrane potential (approximately −70 mV on average), and slow V_m oscillation during periods of quiet wakefulness [14, 15]. During spontaneous whisker movement (whisking), the neurons depolarize, and the slow V_m oscillation decreases [14, 15]. Our experiment involved recording from these neurons using either sharp electrodes or patch pipettes while simultaneously filming the whisker movement of the mouse (Fig. 1a). We then analyzed spontaneous AP frequency, mean V_m, and the variance of V_m during quiet wakefulness (defined as a non-whisking state) and the changes induced by whisking. The recordings during quiet wakefulness revealed that the V_m variance obtained with sharp-electrode intracellular recording was significantly smaller than that with whole-cell patch-clamp recording (Fig. 1b). Presumably due to this, a markedly smaller change in the V_m variance upon whisking was observed in sharp electrode recording (Fig. 1c). Additionally, while the mean V_m during quiet wakefulness was not significantly different between the two conditions (Fig. 1b), V_m depolarization

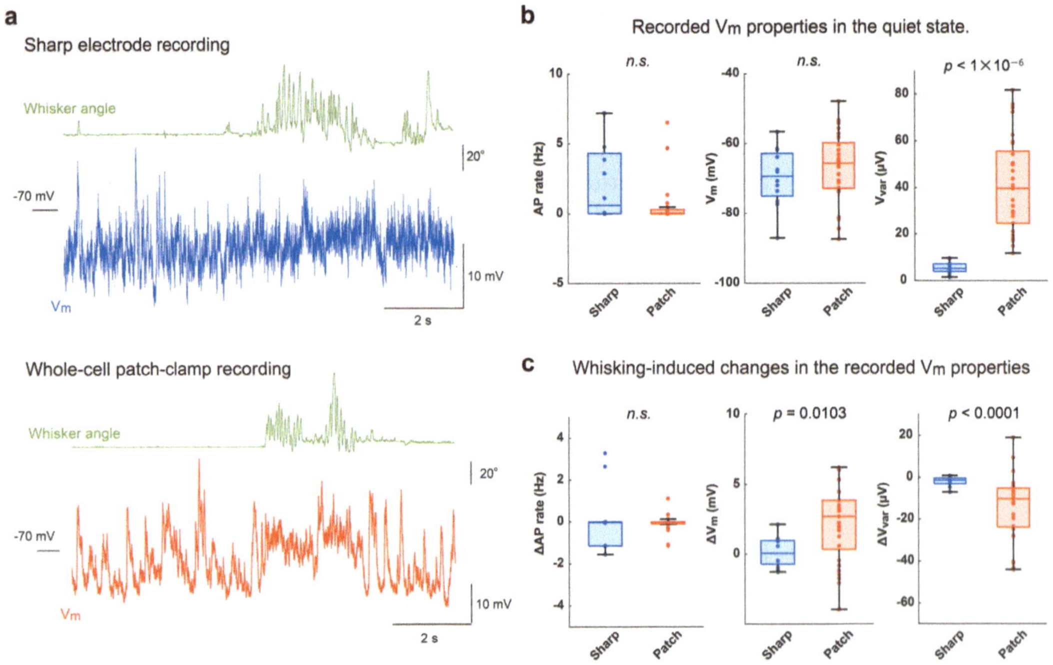

Fig. 1 Properties of V_m of L2/3 wS1 neurons in awake behaving mice, acquired using sharp quartz electrodes or patch pipettes. (**a**) Representative intracellular recording of V_m obtained with a sharp electrode (top, blue) and a patch pipette (bottom, orange). The green traces indicate the time course of the whisker angle. (**b**) Quantification of spontaneous AP rates (left), mean V_m (middle), and the variance of V_m (V_{var}, right) during quiet wakefulness (defined as non-whisking periods), recorded with sharp electrodes (Sharp) or patch pipettes (Patch). (**c**) Quantification of whisking-induced changes in the AP rates (left), mean V_m (middle), and V_{var} (right). The *p* value from the Wilcoxon rank sum test is indicated in each panel when a significant difference is observed. *n.s.* not significant

upon whisking was not detected in sharp-electrode intracellular recordings (Fig. 1c). These findings point to the limitations of sharp-electrode intracellular recording for use in the superficial layer of the mouse cortex and suggest a potential preference for whole-cell patch-clamp recording to capture the V_m dynamics during active states.

In this chapter, we will introduce the "blind" whole-cell patch-clamp method in awake, head-restrained mice [12, 26–28]. This method requires only a conventional electrophysiology setup and a head-fixation apparatus making it relatively accessible for most laboratories. Specifically, we provide an updated and sophisticated method that offers relatively unrestricted recording depth and potentially applies to many brain regions, including the thalamus or hippocampus.

2 Materials

2.1 Animals

Prepare mice (6-week-old or older) from animal distributors or inbred colonies.

2.2 Head-Post Implantation

- Anesthetic vaporizer and isoflurane.
- Stereomicroscope.
- Lighting equipment.
- Heating pad system.
- Surgical instruments: Fine scissors, forceps, and disposable scalpel blades with handle.
- L-shape small metal head-post (custom-made [28]).
- Ophthalmic gel.
- Antiseptic ointment (5% povidone-iodine cream).
- Antiseptic solution (diluted povidone-iodine solution).
- Cotton swabs.
- Iode deodorizer (10% hypoalcohol).
- Adhesive resin cement.
- Dental luting cement.
- Silicon elastomer.
- Analgetics: Carprofen.

2.3 Habituation for Head-Fixation

- Head-post holder (custom-made [28]).
- 30% sucrose water.

2.4 Craniotomy and Durotomy

- Anesthetic vaporizer and isoflurane.
- Mobile head-post holder (custom-made, Fig. 2).
- Stereomicroscope.
- Heating pad system.
- Silicon elastomer.
- Lightweight dental drill with a small diameter drill bit (tip diameter: 0.5 mM).
- Surgical instruments: Fine forceps and 30G needles.

2.5 Recording

- Glass capillaries: Borosilicate glass capillaries with filament.
- Pipette puller.
- Ringer's solution containing 5 mM HEPES, pH 7.3, 135 mM NaCl, 5 mM KCl, 1 mM $MgCl_2$, 1.8 mM $CaCl_2$; stored at 4 °C.
- Intracellular solution containing 10 mM HEPES, pH 7.3, 135 mM potassium gluconate, 4 mM KCl, 10 mM sodium

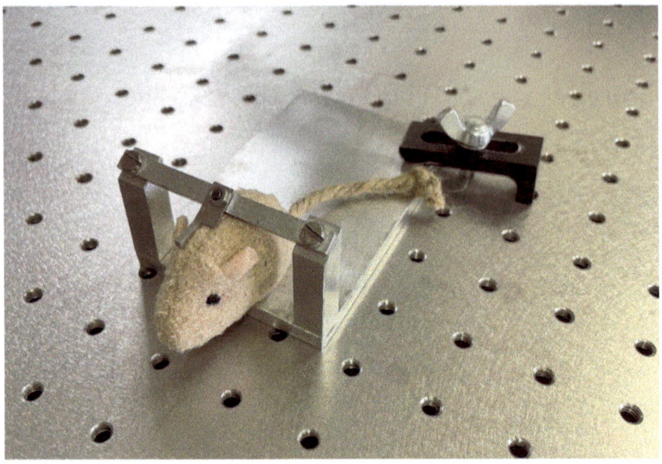

Fig. 2 The mobile head-post holder. A stuffed toy mouse, used for demonstration purposes, is head-restrained on the mobile head-post holder. The holder's frame is clamped to a surgical table

phosphocreatine, 4 mM Mg-ATP, and 0.3 mM Na-GTP. Aliquots of 1–1.5 mL each can be stored at −20 °C for a few months.

- Head-post holder (custom-made [28]).
- Stereomicroscope.
- Patch-clamp amplifier and head stage.
- A/D converter.
- Electrode holder.
- Ag/AgCl reference electrode.
- Four-axis micromanipulator.
- Oscilloscope.
- Pressure controller.

3 Methods

3.1 Surgery for Head-Post Implantation

1. Anesthetize the mouse with isoflurane (2–3% induction, 1–1.5% maintenance).
2. Position the mouse on a stereotaxic surgery apparatus with a heating pad to maintain body temperature during surgery (Fig. 3a).
3. Apply ophthalmic gel to the eyes of the mouse to prevent drying.
4. Administer povidone-iodine cream onto the fur and skin of the mouse head to sterilize and prevent fur from contaminating the skull (Fig. 3b).

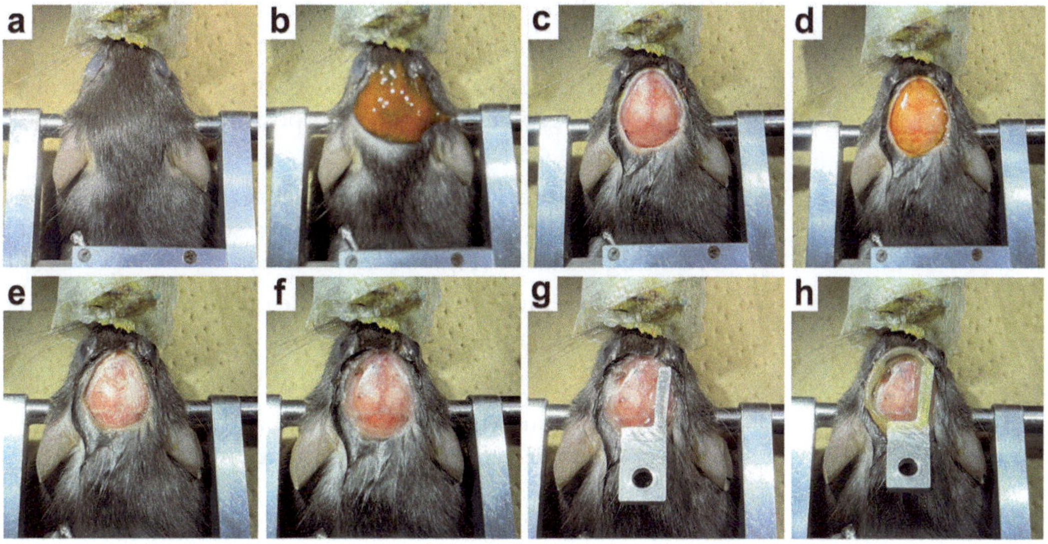

Fig. 3 Steps for head-post implantation. (**a**) The mouse is anesthetized and positioned on the stereotaxic setup. The eyes are covered with ophthalmic gel. (**b**) Antiseptic ointment (povidone-iodine cream) is applied to the fur and skin of the mouse head. (**c**) Skin from the head is removed, and soft tissue on the skull is excised. (**d**) The skull is treated with a reagent for cement adhesion. (**e**) The skull is scraped with a scalpel. (**f**) The entire skull surface is coated with a thin layer of adhesive resin cement. (**g**) The head-post is attached to the skull using adhesive resin cement. (**h**) A recording chamber is built using dental luting cement

5. Expose the skull by removing the skin covering from the cerebellum to the olfactory bulb.
6. Gently remove all connective tissues on the skull using cotton swabs or a scalpel blade (Fig. 3c).
7. Detach the neck muscle from the interparietal bone (*see* **Note 1**).
8. Stop bleeding from sutures, dissected skin, or other parts with cotton swabs.
9. Sterilize the surface of the skull with a povidone-iodine solution (*see* **Note 2**).
10. Clean the skull surface by wiping it with an iode deodorizer.
11. If the recording region is targeted stereotactically, mark the region by making a small incision using the scalpel blade or painting with a permanent marker.
12. Process the skull surface with a surface treatment reagent (Red activator, Superbond). After depositing a small volume of the reagent to cover the entire dorsal surface of the skull, wait for 1–2 min and then rinse with saline (Fig. 3d). If minor bleeding occurs, stop it with cotton swabs (*see* **Note 3**).
13. Scrape the skull surface to create shallow grooves with the scalpel blade. Avoid scratching the target region (Fig. 3e).

14. Cover the dissected skin edge and the skull with a thin layer of adhesive resin cement (Fig. 3f, *see* **Note 4**).
15. Put the L-shaped small metal head-post on the dental cement layer. Fill the gap between the head-post and the skull with the adhesive resin cement (Fig. 3g).
16. Prepare the luting cement to construct the recording chamber. Build a 3–4 mm wall so that Ringer's solution stays in the recording chamber (Fig. 3h, *see* **Note 5**).
17. Fill the recording chamber with silicon elastomer once the dental cement has hardened.
18. Administer carprofen (5 mg/kg) to the mouse subcutaneously.
19. Return the mouse to the home cage and allow 2–3 days to recover. Provide wet food to help the recovery.

3.2 Habituation

1. Bring the mouse to the experimental setup. Secure the head-post on the head-post holder with a nut [28]. Cover the body of the mouse with a small cylindrical cardboard tube (*see* **Note 6**).
2. Offer the mouse a small volume of sweetened water (30% sucrose solution) at the end of the habituation period.
3. Return the mouse to the home cage.
4. Gradually increase the time for each habituation session from 15 minutes to 1 h (*see* **Note 7**).

3.3 Craniotomy

1. Anesthetize the mouse with isoflurane (2–3% induction, 1–1.5% maintenance).
2. Position the mouse on the mobile head-post holder. Clamp the head-post holder on the surgery table. Remove the silicon elastomer and clean the chamber with Ringer's solution and cotton swabs.
3. Drill circularly around the target region to thin the bone. Frequently rinse the skull surface to prevent contamination with bone debris and overheating at the recording site from drilling (*see* **Notes 8** and **9**).
4. When the bone over the target area is sufficiently thinned, insert a sharp-tip 30G needle into the epidural space to open the cranium.
5. Clean the cranium carefully if bleeding occurs (*see* **Note 10**).
6. Remove the Ringer's solution from the recording chamber and fill the recording chamber with Kwik-Cast silicone sealant.
7. Return the mouse to the home cage for recovery.

3.4 Recording

1. Prepare recording pipettes (5–7 mega-ohm) by pulling the glass capillaries (*see* **Note 11**).
2. Unfreeze one aliquot of intracellular solution on ice. Optionally, 3 mg/mL biocytin can be dissolved into the solution with sonication. Keep the solution on ice to prevent degradation of ATP and GTP.
3. Place the mice on the recording setup and fix the head-post to the holder. Remove the silicon sealant and clean the interior of the chamber and the cranium under the stereomicroscope.
4. Place the Ag/AgCl reference electrode inside the recording chamber.
5. Fill the recording pipette with the internal solution and mount the pipette on the pipette holder.
6. Apply high positive pressure (200–300 mbar) to the pipette using the pressure controller (*see* **Note 12**).
7. Position the pipette tip into the cranium, very close to the surface of the dura, and fill the recording chamber with Ringer's solution.
8. Adjust the offset in the voltage-clamp mode. Continuously apply 10 mV square test pulses at 20 Hz and monitor the current deflection recorded by the recording pipette with an oscilloscope.
9. Advance the pipette using the micromanipulator and go through the dura and pia. When the pipette tip touches the surface of the dura, the current response decreases, reflecting the increase in resistance. Set the manipulator's readout to zero to measure the recording depth. After penetrating the dura/pia, the resistance typically returns to the same value as it was before penetration of the dura and pia. Again, adjust the offset (*see* **Note 13**).
10. Reduce the pipette pressure to 50–70 mbar and move the pipette close to the target region (50–100 μm above the target depth).
11. Reduce the pipette pressure to 20–30 mbar and advance the pipette by 2 μm steps until a sudden increase in pipette resistance is seen at the target depth. Contact with neurons is typically associated with a transient oscillation of current responses.
12. Release the positive pressure of the pipette and at the same time apply a slight negative pressure to the pipette to facilitate a giga-ohm seal formation. Successful giga-ohm seal formation is indicated by an immediate increase in the pipette resistance to a giga-ohm order after releasing the pipette pressure.

13. Lower the holding potential to −70 mV to prevent depolarization after transitioning to a whole-cell configuration. This step also facilitates a tighter seal formation.
14. Retract the pipette 2–4 μm after the giga-ohm seal formation. The electrode capacitance should be compensated at this stage.
15. Apply transient negative pressure (suction) to the pipette to open the membrane. After break-in, we recommend applying small positive pressure (+3–5 mbar) continuously to the pipette in order to prevent resealing.
16. Switch the amplifier to the current-clamp mode and record V_m. If low-amplitude, broad action potentials are observed, it indicates a high access resistance. Applying pulses of positive or negative pressure to the pipette or moving the pipette one step back may reduce the access resistance and improve the recording quality (*see* **Note 14**).
17. Once stable and high-quality recording conditions have been achieved, proceed with the experimenter's own recording protocol.
18. After finishing the recording, slowly retract the recording pipette. If post hoc anatomical inspection of the recorded neuron is desired, carefully monitor the current response to the 10 mV square step pulses while retracting the pipette. An increase in the resistance to more than hundreds of mega-ohms after withdrawing the pipette indicates a formation of an outside-out patch-clamp configuration [9], suggesting a successful pipette retraction without cellular damage.
19. After the end of the recording, the mouse can be transcardially perfused with 4% paraformaldehyde solution for post hoc anatomical investigation.

4 Notes

1. The *temporalis* muscle may be separated from the bone using a scalpel when the target is in lateral brain areas (e.g., secondary somatosensory cortex and auditory cortex). Use cotton swabs to move the muscle laterally in such a case.
2. This sterilization step prevents microbial infection and subsequent inflammation and maintains cleanliness within the recording chamber.
3. Intrinsic optical imaging [29] with an intact skull can be done simply by applying saline to the skull and waiting until the bone becomes transparent.
4. Covering with clear dental cement improves the stability of the recording chamber and allows transcranial intrinsic imaging

without drilling [30]. Preparing a smooth surface to cover the skull improves the visualization of the skull surface.

5. Ensure that the recording pipette can reach the target area without touching the wall of the chamber.
6. In the initial habituation trial, mice can be aggressive. Mildly anesthetizing when clamping them to the head-post holder can be helpful.
7. Adjust the habituation paradigm according to the experimental purpose: longer habituation can cause lower spontaneous activity of the mouse on the setup.
8. A smaller craniotomy leads to more stable recordings. To further improve recording stability, we recommend not making a durotomy.
9. For optogenetic experiments, thinning the skull next to the cranium would be needed to place the tip of an optical fiber. To ensure recording stability, we do not recommend making a large craniotomy so that the optic fiber can touch the brain surface.
10. Do not scrub the craniotomy site with cotton swabs to avoid damage to the superficial cortex.
11. For recording from deeper brain regions, we recommend using a pipette with a relatively long shank with an average diameter of 75–125 μm (Fig. 4), which seems to stabilize the superficial areas. Implanting a small chamber covering the cranium may also be advantageous for stable recording in behaving mice [31].
12. A pressure controller such as a manual-seal-sucker from Sigmann Elektronik helps control the pressure inside the pipette.

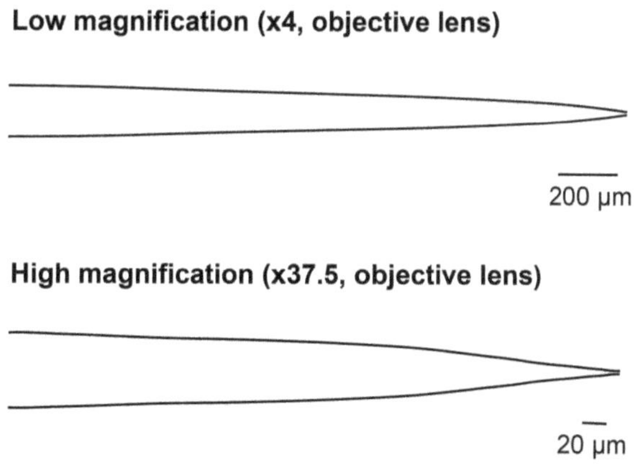

Fig. 4 Schematic drawings of the recording patch pipette. The pipette with a relatively long shank is suitable for whole-cell recording from deeper brain areas

13. After penetration of the dura and pia, the pipette resistance may not return to the same value. When the pipette tip is broken, the pipette resistance essentially decreases. When the pipette is clogged by tissue, the resistance increases. The pipette should be replaced with a new one in such cases. When targeting a deep brain area, inserting a dummy pipette with a relatively thicker shank (~200 μm diameter) above the target region and retracting it before recording trials for once would be a good option to create easier access for the recording pipette.
14. The recording pipette occasionally pushes the recorded neuron during approaching. Retraction of the recording pipette by a few 2 μm steps usually reduces series resistances, but sometimes it causes detachment of the pipette from the recorded cell.

References

1. Graham J, Gerard RW (1946) Membrane potentials and excitation of impaled single muscle fibers. J Cell Comp Physiol 28:99–117. https://doi.org/10.1002/jcp.1030280106
2. Fatt P, Katz B (1952) Spontaneous subthreshold activity at motor nerve endings. J Physiol 117:109–128. https://doi.org/10.1113/jphysiol.1952.sp004735
3. Del Castillo J, Katz B (1954) Quantal components of the end-plate potential. J Physiol 124: 560–573. https://doi.org/10.1113/jphysiol.1954.sp005129
4. Brock LG, Coombs JS, Eccles JC (1952) The recording of potentials from motoneurones with an intracellular electrode. J Physiol 117: 431–460. https://doi.org/10.1113/jphysiol.1952.sp004759
5. Reynolds JNJ, Avvisati R, Dodson PD et al (2022) Coincidence of cholinergic pauses, dopaminergic activation and depolarisation of spiny projection neurons drives synaptic plasticity in the striatum. Nat Commun 13:1296. https://doi.org/10.1038/s41467-022-28950-0
6. Bellavance MA, Takatoh J, Lu J et al (2017) Parallel inhibitory and excitatory trigemino-facial feedback circuitry for reflexive vibrissa movement. Neuron 95:673–682. https://doi.org/10.1016/j.neuron.2017.06.045
7. Sturdy CB, Wild JM, Mooney R (2003) Respiratory and telencephalic modulation of vocal motor neurons in the zebra finch. J Neurosci 23:1072–1086. https://doi.org/10.1523/jneurosci.23-03-01072.2003
8. Gao L, Wang X (2020) Intracellular neuronal recording in awake nonhuman primates. Nat Protoc 15:3615–3631. https://doi.org/10.1038/s41596-020-0388-3
9. Hamill OP, Marty A, Neher E et al (1981) Improved patch-clamp techniques for high-resolution current recording from cells and cell-free membrane patches. Pflügers Arch 391:85–100. https://doi.org/10.1007/BF00656997
10. Neher E, Sakmann B (1976) Single-channel currents recorded from membrane of denervated frog muscle fibres. Nature 260:799–802. https://doi.org/10.1038/260799a0
11. Edwards FA, Konnerth A, Sakmann B et al (1989) A thin slice preparation for patch clamp recordings from neurones of the mammalian central nervous system. Pflügers Arch 414:600–612. https://doi.org/10.1007/bf00580998
12. Margrie TW, Brecht M, Sakmann B (2002) In vivo, low-resistance, whole-cell recordings from neurons in the anaesthetized and awake mammalian brain. Pflügers Arch 444:491–498. https://doi.org/10.1007/s00424-002-0831-z
13. Pei X, Volgushev M, Vidyasagar TR et al (1991) Whole cell recording and conductance measurements in cat visual cortex in-vivo. Neuroreport 2:485–488. https://doi.org/10.1097/00001756-199108000-00019
14. Crochet S, Petersen CCH (2006) Correlating whisker behavior with membrane potential in barrel cortex of awake mice. Nat Neurosci 9: 608–610. https://doi.org/10.1038/nn1690
15. Poulet JFA, Petersen CCH (2008) Internal brain state regulates membrane potential synchrony in barrel cortex of behaving mice. Nature 454:881–885. https://doi.org/10.1038/nature07150

16. Harvey CD, Collman F, Dombeck DA et al (2009) Intracellular dynamics of hippocampal place cells during virtual navigation. Nature 461:941–946. https://doi.org/10.1038/nature08499
17. Kim E, Bari BA, Cohen JY (2021) Subthreshold basis for reward-predictive persistent activity in mouse prefrontal cortex. Cell Rep 35:109082. https://doi.org/10.1016/j.celrep.2021.109082
18. Inagaki HK, Fontolan L, Romani S et al (2019) Discrete attractor dynamics underlies persistent activity in the frontal cortex. Nature 566:212–217. https://doi.org/10.1038/s41586-019-0919-7
19. Yamashita T, Petersen CCH (2016) Target-specific membrane potential dynamics of neocortical projection neurons during goal-directed behavior. elife 5:e15798. https://doi.org/10.7554/eLife.15798
20. Sachidhanandam S, Sreenivasan V, Kyriakatos A et al (2013) Membrane potential correlates of sensory perception in mouse barrel cortex. Nat Neurosci 16:1671–1677. https://doi.org/10.1038/nn.3532
21. Sreenivasan V, Esmaeili V, Kiritani T et al (2016) Movement initiation signals in mouse whisker motor cortex. Neuron 92:1368–1382. https://doi.org/10.1016/j.neuron.2016.12.001
22. Margrie TW, Meyer AH, Caputi A et al (2003) Targeted whole-cell recordings in the mammalian brain in vivo. Neuron 39:911–918. https://doi.org/10.1016/j.neuron.2003.08.012
23. Gentet LJ, Avermann M, Matyas F et al (2010) Membrane potential dynamics of GABAergic neurons in the barrel cortex of behaving mice. Neuron 65:422–435. https://doi.org/10.1016/j.neuron.2010.01.006
24. Yamashita T, Pala A, Pedrido L et al (2013) Membrane potential dynamics of neocortical projection neurons driving target-specific signals. Neuron 80:1477–1490. https://doi.org/10.1016/j.neuron.2013.10.059
25. Muñoz W, Tremblay R, Rudy B (2014) Channelrhodopsin-assisted patching: in vivo recording of genetically and morphologically identified neurons throughout the brain. Cell Rep 9:2304–2316. https://doi.org/10.1016/j.celrep.2014.11.042
26. Petersen CCH, Hahn TT, Mehta M et al (2003) Interaction of sensory responses with spontaneous depolarization in layer 2/3 barrel cortex. Proc Natl Acad Sci USA 100:13638–13643. https://doi.org/10.1073/pnas.2235811100
27. Wang HY, Eguchi K, Yamashita T et al (2020) Frequency-dependent block of excitatory neurotransmission by isoflurane via dual presynaptic mechanisms. J Neurosci 40:4103–4115. https://doi.org/10.1523/jneurosci.2946-19.2020
28. Crochet S (2012) Intracellular whole-cell patch-clamp recordings of cortical neurons in awake head-restrained mice. In: Fellin T, Halassa M (eds) Neuronal network analysis: concepts and experimental approaches. Humana Press, Totowa
29. Ferezou I, Bolea S, Petersen CCH (2006) Visualizing the cortical representation of whisker touch: voltage-sensitive dye imaging in freely moving mice. Neuron 50:617–629. https://doi.org/10.1016/j.neuron.2006.03.043
30. Takeuchi RF, Ito KN, Ueda R et al (2022) Posteromedial cortical networks encode visuomotor prediction errors. bioRxiv. https://doi.org/10.1101/2022.08.16.504075
31. Dacre J, Sánchez Rivera M, Schiemann J et al (2023) A cranial implant for stabilizing whole-cell patch-clamp recordings in behaving rodents. J Neurosci Methods 390:109827. https://doi.org/10.1016/j.jneumeth.2023.109827

Chapter 21

In Vivo Brain Microdialysis

Makoto Kondo

Abstract

Many biological molecules in the brain interstitial fluid are involved in neuronal functions. Therefore, measuring the levels of these molecules in the extracellular fluid would provide deep insights into the physiological/pathological mechanisms underlying brain functions/disorders. In vivo microdialysis is a powerful technique used to examine the extracellular levels of various molecules in the brains of living animals. In neuroscience research, this technique has been widely used to investigate relatively small molecules including neurotransmitters and amino acids. However, recent advances in technology have made it possible to assess large molecules in the brain interstitial fluid, such as signaling peptides and proteins, using microdialysis probes with high-molecular-weight cutoff membranes. This chapter describes an in vivo microdialysis method to collect and measure the levels of large biological molecules in the extracellular fluid of the brains of freely moving mice.

Key words In vivo microdialysis, Mouse brain, Large biological molecules, Brain interstitial fluid, Freely moving animal

1 Introduction

Many biological molecules in the brain interstitial fluid are involved in brain functions. These molecules are involved in various biological processes and disease mechanisms. Therefore, it is important to monitor these biological molecules in live animal brains to improve the understanding of physiological mechanisms of brain functions and pathological mechanisms of brain disorders.

In vivo microdialysis is a powerful technique used to collect biological molecules in the brain interstitial space from awake, freely moving animals [1, 2]. In this method, a microdialysis probe with a semipermeable membrane is inserted into the target brain region, which is perfused with a perfusion buffer. During perfusion with continuous sampling, biological molecules that are sufficiently small to pass through the microdialysis membrane passively diffuse from the brain extracellular fluid into the perfusion buffer according to the concentration gradient and can be collected

Koh-ichi Nagata (ed.), *Cerebral Cortex Development: Methods and Protocols*, Methods in Molecular Biology, vol. 2794, https://doi.org/10.1007/978-1-0716-3810-1_21,

as a dialysate for further analysis. The total movement of molecules of interest across the membrane is dependent on a number of factors including the molecular weight, aqueous solubility, and protein binding of the molecule [1, 2].

In vivo microdialysis has several advantages over other sampling methods [3, 4]. It is possible to continuously sample dynamic changes in a specific molecule from the extracellular compartment in a target brain region without tissue extraction and mechanical homogenization after sacrificing the animal. Moreover, while sampled cerebrospinal fluid likely reflects the level of a specific molecule derived from the entire central nervous system, microdialysis can detect its focal release/secretion at a target brain region where the microdialysis probe is inserted. In addition, because the microdialysis technique enables successive sample collection from a single living animal for several hours, each animal can serve as its own intrinsic control. Thus, the intrinsic (i.e., baseline) level of a target molecule can be measured and compared with the levels obtained before and after any treatment for each animal. This advantage minimizes inter-animal variability and can reduce the number of animals used for experiments [3, 4].

Originally, microdialysis was widely used to measure relatively small molecules including neurotransmitters, amino acids, and low-molecular-weight compounds in neuroscience studies for many years [1, 2]. Meanwhile, although it is necessary to sample large biological molecules in the brain interstitial fluid to understand a variety of physiological/pathological processes in animal brains, the ability to recover large molecules has been limited by the availability of suitable microdialysis probes and perfusion techniques. However, the recent development of microdialysis probes with higher-molecular weight cutoff membranes and a push–pull perfusion system has made it possible to collect various large biological molecules, such as signaling peptides and proteins, from the brains of awake, freely moving animals [5, 6]. Moreover, improved microdialysis probes with vent holes to equalize the pressure accumulated inside the probe and a push–pull perfusion technique are employed to minimize fluid leakage from the probe membrane into the surrounding brain tissue [3, 4]. Using these advanced in vivo microdialysis materials and techniques, many studies have performed stable sampling and monitoring of the dynamic changes in levels of important large molecules, such as neurotrophic factors, inflammatory cytokines, and beta-amyloid peptides, in the brains of freely moving mice following biological stimulation [3, 4]. My recent studies have involved collecting extracellular fluid from a particular mouse brain region using this microdialysis method and examining insulin-like growth factor 1 (IGF-1, molecular weight 7.7 kDa) levels [7, 8]. IGF-1 is a neurotrophic factor and is known to play important roles in various brain functions, including cognition and emotion [7]. These studies have shown that systemic pharmacological treatment increased IGF-1 levels in extracellular fluid from the mouse hippocampus [7] and medial prefrontal cortex [8].

This chapter describes an in vivo microdialysis method used to collect large biological molecules from the brain interstitial fluid of live mice. I use the medial prefrontal cortex as a target brain region and describe the protocol to collect IGF-1 from brain interstitial fluid. First, I describe the stereotaxic surgery necessary for guide cannula implantation. Then, I describe how to prepare the in vivo microdialysis system and how to sample large biological molecules from the brain interstitial fluid of freely moving mice using microdialysis probes.

2 Materials

2.1 Stereotaxic Surgery

1. Mice used for stereotaxic surgery.
2. Stereotaxic instruments for mice, including the stereotaxic frame, manipulator, adaptor, holder with clamp, ear bar, nose clamp, and incisor bar.
3. Isoflurane for inhalation anesthesia.
4. Stereotaxic adaptor (shaft length 4 mM).
5. Guide cannula (shaft length 4 mM).
6. Dummy cannula (shaft length 4 mM).
7. Cap nut.
8. Dental drill.
9. Dental cement.
10. Spatula for dental cement application.
11. Bone screws.
12. Screwdriver for bone screws.
13. Surgical clipper.
14. Scalpel or razor blade.
15. Sterile needles.
16. Cotton swabs.

2.2 In Vivo Microdialysis

1. Ringer's solution: 147 mM NaCl, 4 mM KCl, and 2.3 mM $CaCl_2$. Add 8.6 g of NaCl, 0.3 g of KCl, and 0.25 g of $CaCl_2$ to 1 L of distilled water and stir the solution until completely dissolved. Filter the solution through a 0.22 μm syringe filter (*see* **Note 1**).
2. Perfusion buffer: 0.15% bovine serum albumin (BSA) in Ringer's solution. Filter the solution through a 0.2 μm syringe filter unit (*see* **Note 2**).

3. 70% ethanol.
4. Microdialysis probes with high-molecular-weight cutoff membranes (shaft length 4 mM, membrane length 2 mM) (*see* **Note 3**).
5. Syringe pump.
6. Glass gastight 2.5 mL syringe.
7. Disposable 1 mL syringe.
8. Cage kit for a freely moving mouse, including a mouse cage, liquid swivel, and balancer (*see* **Note 4**).
9. Peristaltic pump (roller pump).
10. Peristaltic tubes for the peristaltic pump (0.25 mM i.d.).
11. Connection needle.
12. Fraction collector with a cooling unit.
13. Teflon tubing (0.1 mM i.d.).
14. Fluorinated ethylene propylene (FEP) tubing (0.25 mM i.d.).
15. Joint tubing (0.3 mM i.d.).
16. 1% liquid bleach.
17. Methanol.

3 Methods

3.1 Stereotaxic Surgery for Guide Cannula Implantation

1. Anesthetize the mouse with isoflurane.
2. Shave the hair on the head of the mouse with surgical clippers.
3. Mount the mouse on the stereotaxic frame by placing the ear bars into the ear canal and clamping the head with the ear bars. Then, fix the mouse using nose clamps and the incisor bar. Make sure that the mouse head is stable and cannot be moved from side to side.
4. Make a midline incision in the skin of the head with a scalpel or razor blade.
5. Wipe the connective tissue and blood from the skull surface using cotton swabs (*see* **Note 5**).
6. Set a guide cannula on the arm of the stereotaxic adaptor and fix it with a cap nut. Set the stereotaxic adaptor on the stereotaxic holder with a clamp and mount it onto the manipulator of the stereotaxic frame.
7. Identify the bregma on the skull and position the tip of the guide cannula at the bregma. Determine the anterior/posterior, lateral, and vertical coordinates (*see* **Note 6**).
8. Determine the coordinates needed for placement of the probe in the target brain region using a mouse brain atlas [9]. In this

example, the stereotaxic coordinates (anterior/posterior: 1.8 mM, lateral: 0.3 mM, vertical: 1.4 mM) are used to place the microdialysis probe with 2 mM of membrane length in the medial prefrontal cortex [8] (*see* **Note 7**).

9. Move the guide cannula to the target point by calculating the coordinates by addition or subtraction from the bregma coordinates. Make a mark with ink at this point on the skull.
10. Drill a hole at the marked point through the width of the skull using a dental drill. Puncture the dura with a sterile needle (*see* **Note 8**).
11. Move the guide cannula and lower it slowly until the proper ventral coordinate in the brain is reached.
12. Make another hole on the skull for an anchor screw using a dental drill and implant a bone screw with the screwdriver (*see* **Note 9**).
13. Apply dental cement to the skull surface using a spatula to cover the base of the guide cannula, the bone screw, and the remaining exposed skull (*see* **Note 10**).
14. Allow the dental cement to dry completely. Remove the stereotaxic holder from the stereotaxic adaptor. Remove the cap nut and pull the stereotaxic adaptor out of the guide cannula. Then, place the dummy cannula into the guide cannula and fasten the cap nut (*see* **Note 11**).
15. Release the mouse from the stereotaxic instrument and place it in an individual cage until the day of in vivo microdialysis (*see* **Note 12**).

3.2 In Vivo Microdialysis in Freely Moving Mice

3.2.1 Construction of the Microdialysis Circuit

1. Connect the tubing as indicated in Figs. 1 and 2. At this point, use the connection needle instead of a microdialysis probe to connect the two FEP tubes.
2. Fill a gastight glass 2.5 mL syringe with perfusion buffer (0.15% BSA in Ringer's solution, as described in **step 2** in Subheading 2.2) ensuring that no air bubbles are present.
3. Mount the syringe filled with perfusion buffer onto the syringe pump and connect it to the microdialysis circuit.
4. Start the syringe pump and fill the entire microdialysis circuit with the perfusion buffer. When the circuit is filled with the perfusion buffer, stop the syringe pump (*see* **Note 13**).
5. Mount the peristaltic tube on the tubing guide of the peristaltic pump (*see* **Note 14**).

3.2.2 Preparation of the Microdialysis Probe and Connection to the Circuit

1. Perform a quality check of the microdialysis probe before inserting it into the mouse brain as follows. Fill a disposable 1 mL syringe with distilled water and connect it to the outlet of the probe using a joint tube. Then, cover the vent holes in the

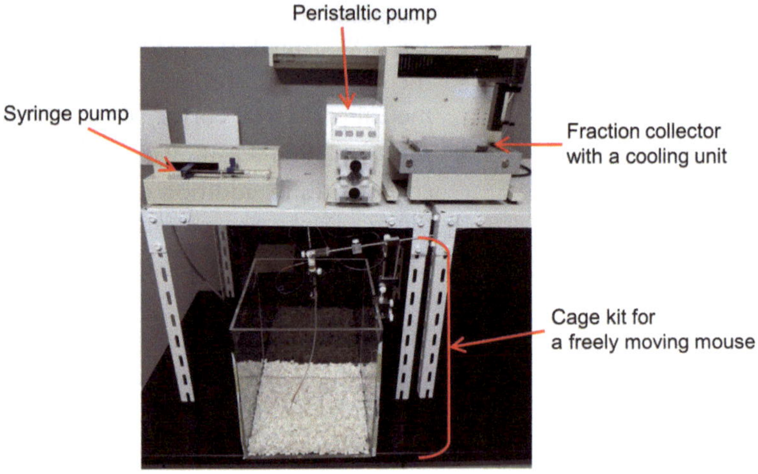

Fig. 1 Example of the in vivo microdialysis system setup. A syringe pump, cage kit for a freely moving mouse, peristaltic pump, and fraction collector with a cooling unit are shown

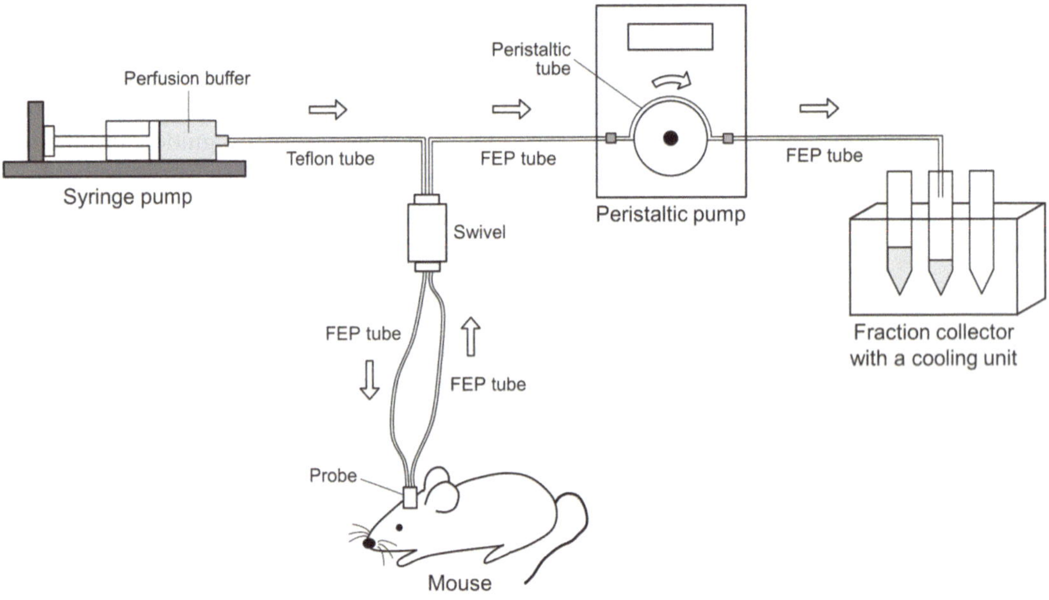

Fig. 2 Construction of the microdialysis circuit with the push–pull perfusion technique for in vivo studies in freely moving mice

probe using your fingers and gently infuse distilled water into the probe. Confirm that water passes from the inlet and that there are no leaks from the tip of the microdialysis membrane or the connection of the membrane and shaft.

2. Activate the probe by submerging the microdialysis membranes in 70% ethanol for 1 second, and then infuse water again into the probe with a syringe (*see* **Note 15**).

3. Remove the connection needle (described in **step 1** in Subheading 3.2.1) and place the activated microdialysis probe with the cap nut by connecting the free ends of the FEP tubing to the inlet and outlet ports of the microdialysis probe.
4. Start the syringe pump at 10 μL/min, and then start the peristaltic pump at 10 μL/min (*see* **Note 16**).

3.2.3 Insertion of the Probe into the Mouse Brain and Initiation of Sampling

1. Anesthetize a mouse with an implanted guide cannula (as described in Subheading 3.1) and move it into a cage for in vivo microdialysis (*see* **Note 17**).
2. Loosen the cap nut, remove the dummy cannula, and then insert the microdialysis probe into the brain slowly via the guide cannula. Tighten the cap nut (Fig. 3).
3. Maintain a 10 μL/min flow rate for 3 h by continuously running both the syringe pump and the peristaltic pump (*see* **Notes 18** and **19**).

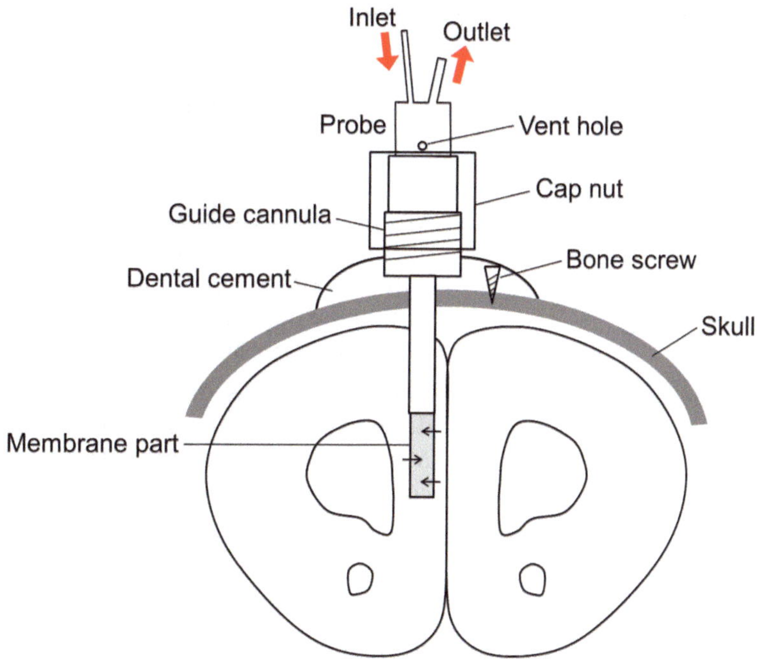

Fig. 3 Schematic representation of microdialysis probe implantation into the mouse medial prefrontal cortex. The microdialysis probe is inserted into the mouse brain through a guide cannula. The guide cannula and bone screw were implanted, and the base of the guide cannula and bone screw were wrapped with dental cement in advance during the stereotaxic surgery. The construct depicted in the schema consists of a guide cannula, microdialysis probe, cap nut, bone screw, and consolidated dental cement on the skull. The gray bar indicates the membrane part of the microdialysis probe, across which biological molecules in the brain interstitial space are harvested

4. Stop the peristaltic pump and then stop the syringe pump.
5. Set the syringe pump and peristaltic pump to 1.0 μL/min (*see* **Note 20**). Start the syringe pump and then the peristaltic pump (*see* **Note 21**).
6. Start sample collection using a fraction collector with a cooling unit (*see* **Notes 22** and **23**).
7. After completion of sampling, loosen the cap nut and remove the probe from the mouse brain.
8. Replace the probe in the microdialysis circuit with a connection needle. Remove the peristaltic tubing from the pump while ensuring it is not compressed. Fill the entire tubing with 1% liquid bleach and let it stand for at least 20 min. Then, flush the circuit thoroughly with distilled water. Run methanol and then air through the tubing and let it dry (*see* **Note 24**).
9. IGF-1 concentrations in dialysate samples can be analyzed by enzyme-linked immunosorbent assays (ELISAs) (*see* **Note 25**).

4 Notes

1. An alternative is artificial cerebrospinal fluid (aCSF): 122 mM NaCl, 25 mM $NaHCO_3$, 3 mM KCl, 1.3 mM $CaCl_2$, 1.2 mM $MgSO_4$, and 0.4 mM KH_2PO_4, pH 7.35 [6].
2. The perfusion buffer should be prepared on the day of microdialysis and used within 24 h. BSA is added to the perfusion buffer to prevent perfusate loss into tissue via the large pore membrane because of its high osmolarity and to avoid nonspecific adsorption of target large molecules to the tubing [5].
3. The shaft lengths of the microdialysis probe, guide cannula, dummy cannula, and stereotaxic adaptor should be the same. Microdialysis probes with 1000 kDa cutoff membranes are commercially available [3, 4].
4. A cage kit for freely moving mice is commercially available. A general cage kit includes a liquid swivel and a balancer to keep tubing and fluid lines from twisting.
5. It is important to keep the exposed skull surface dry for subsequent procedures.
6. The bregma is the anatomical point on the skull where the coronal suture perpendicularly intersects the sagittal suture. The bregma is often used as a reference point to calculate the coordinates of a target mouse brain region in stereotaxic surgery [5].
7. The coordinates and probe length are dependent on the target brain region. In this case, a microdialysis probe with 2 mM of

membrane length is used. Regarding the vertical coordinate, the guide cannula tip should be placed 2 mM above the microdialysis probe. After in vivo microdialysis, histological analysis should be performed to verify the probe position [7, 8].

8. Drill a hole carefully and check to determine whether the guide cannula will clear the hole without touching the sides. Keep drilling and checking until the cannula can clear in a straight path [10].
9. Another hole for a bone screw should usually be made on the contralateral side of the cranial bone.
10. The skull surface should be dried. Make sure that the base of the guide cannula and the bone screw are sufficiently covered to secure them.
11. A dummy cannula is placed for temporary occupation of the guide cannula until insertion of the microdialysis probe (as described in **step 2** in Subheading 3.2.3).
12. A warmed cage should be used to prevent hypothermia in mice. Observe the mouse until it recovers from anesthesia. After surgery, the mouse should be housed individually to prevent the implant from being damaged by other mice. The mouse can undergo in vivo microdialysis 2 days after the surgery.
13. Air bubbles should be removed from the circuit before connecting the microdialysis probe.
14. Peristaltic tubing should be free of twists or turns to maintain stable flow.
15. Do not keep the probes in 70% ethanol for longer than 5 s because this can damage the probe by weakening the connecting portion.
16. To avoid air entry into the probe outlet line, when starting/stopping the perfusion system, start/stop the peristaltic pump first and then do the same for the syringe pump. If air bubbles are generated, remove them before insertion of the microdialysis probe into the mouse brain.
17. When the mouse is calm, procedures in **steps 1** and **2** in Subheading 3.2.3 can be performed without anesthesia.
18. Because various substances induced by tissue damage are collected immediately after insertion of the probe, a relatively fast flow rate can effectively prevent tubing blockage caused by protein aggregation. Furthermore, perfusion for 3 h before sampling is necessary to reduce undesirable effects caused by acute local tissue damage because of the insertion of the microdialysis probe [6]. This step can contribute to stable baseline sampling.

19. Verify whether the push–pull perfusion is evenly balanced. If there is perfusate leakage from the vent hole, increase the peristaltic pump speed to produce a stronger pull. Conversely, in the case of air entry into the probe outlet line, decrease the peristaltic pump speed to produce a weaker pull.
20. If necessary, refill the syringe with perfusion buffer at this time.
21. Confirm that the push–pull perfusion is evenly balanced again. If necessary, adjust the peristaltic pump as described in **Note 19**.
22. The fraction collector and sample collection tubes should be cooled in advance using the cooling unit of the fraction collector.
23. The first few successive samples should be obtained as baseline samples. Then, perform the experimental treatment, such as drug administration, behavioral tests, or other biological stimulation. Thus, samples collected following any experimental procedures from each mouse can be compared with the baseline samples.
24. Microdialysis circuits can be used repeatedly if maintenance is appropriately performed.
25. Concentrations of targeted molecules in dialysate samples can be measured using appropriate methods, such as ELISAs and high-performance liquid chromatography (HPLC). For example, an ELISA kit for measuring IGF-1 content is commercially available [7, 8].

Acknowledgments

This work was supported by JSPS KAKENHI Grant Numbers JP19K11440 and JP22K11498 and by AMED grant numbers JP21wm0525026 and JP20lm0203007.

References

1. Chefer VI, Thompson AC, Zapata A et al (2009) Overview of brain microdialysis. Curr Protoc Neurosci 47:7.1.1–7.1.28
2. Anderzhanova E, Wotjak CT (2013) Brain microdialysis and its applications in experimental neurochemistry. Cell Tissue Res 354:27–39
3. Takeda S, Sato N, Ikimura K et al (2011) Novel microdialysis method to assess neuropeptides and large molecules in free-moving mouse. Neuroscience 186:110–119
4. Yamada K (2017) In vivo microdialysis of brain interstitial fluid for the determination of extracellular tau levels. In: Smet-Nocca C (ed) Tau protein. Methods in molecular biology, vol 1523. Humana Press, New York, pp 285–296
5. Kushikata T, Hirota K (2011) Neuropeptide microdialysis in free-moving animals. In: Merighi A (ed) Neuropeptides. Methods in molecular biology, vol 789. Humana Press, Totowa, pp 261–269
6. Yamada K (2018) In vivo microdialysis method to collect large extracellular proteins from brain interstitial fluid with high-molecular weight cut-off probes. J Vis Exp 139:e57869
7. Kondo M, Koyama Y, Nakamura Y et al (2018) A novel 5HT3 receptor-IGF1 mechanism

distinct from SSRI-induced antidepressant effects. Mol Psychiatry 23:833–842

8. Deyama S, Kondo M, Shimada S et al (2022) IGF-1 release in the medial prefrontal cortex mediates the rapid and sustained antidepressant-like actions of ketamine. Transl Psychiatry 12:178

9. Franklin KBJ, Paxinos G (2007) The mouse brain in stereotaxic coodinates, 3rd edn. Elsevier, Burlington

10. Geiger BM, Frank LE, Caldera-Siu AD et al (2008) Survivable stereotaxic surgery in rodents. J Vis Exp 20:e880

Chapter 22

CT Scan Data Analysis in Malformations of Cortical Development

Marcello Scala and Mariasavina Severino

Abstract

Malformations of cortical development (MCDs) are a diverse group of disorders that result from abnormal neuronal migration, proliferation, and differentiation during brain development. Head computed tomography (CT) has limited use in the diagnosis of MCDs and should be reserved for selected cases with specific indications or when magnetic resonance imaging is not available or contraindicated. CT can detect brain calcifications associated with MCDs, thus helping in the differential diagnosis between acquired and genetic MCDs or in the identification of different genetic patterns. Moreover, CT can provide high-resolution images of the skull and bones, thus identifying associated malformations, such as craniosynostosis, inner and middle ear malformations, and vertebral anomalies. In this chapter, we review the CT scan technique, data analysis, and indications in the investigation of MCDs.

Key words Cortical development, Brain malformations, Malformations of cortical development, Computed tomography, CT scan, Radiations

1 Introduction

A computed tomography (CT) scan, formerly called computed axial tomography (CAT) scan, is an X-ray imaging technique developed in the 1970s by Allan MacLeod Cormack and Godfrey Hounsfield, who were awarded the Nobel Prize in Physiology or Medicine in 1979 [1–4]. The first CT scanners were installed in 1974, and continuous advancements in technology and mathematics have enabled single images to be computed into two-dimensional (2D) or three-dimensional (3D) informative medical images. Since its development, CT scanning has provided excellent clinicopathological correlations and increased the diagnostic rate of different medical disorders, especially in the neurology field [5, 6]. Indeed, noncontrast head CT has rapidly substituted cerebral X-ray, angiography, and pneumography for detecting brain infarcts, hemorrhages, trauma, and tumors, thus

Koh-ichi Nagata (ed.), *Cerebral Cortex Development: Methods and Protocols*, Methods in Molecular Biology, vol. 2794, https://doi.org/10.1007/978-1-0716-3810-1_22,

becoming a versatile imaging technique, especially useful in the emergency setting [6–8]. A contrast-enhanced CT scan can be performed to detect vascular abnormalities, such as aneurysms and arteriovenous malformations (AVMs). Contrast agents such as iodinated contrast media may be administered intravenously and rapidly distributed throughout the blood vessels of the brain [9, 10]. This results in enhanced visibility of the vasculature, thus obtaining CT angiography, or it might be used to perform CT perfusion, thus allowing better detection of abnormalities.

Malformations of cortical development (MCDs) are a heterogeneous group of brain malformations that result from abnormal neuronal migration, proliferation, and organization during embryonic development (*see* Table 1) [11–13]. MCDs can manifest as a range of neurological disorders, including epilepsy, developmental delay, intellectual disability, and movement disorders [14]. MCD diagnosis relies on neuropathological findings, which are seldom accessible except in cases of epilepsy surgery or autopsy. Consequently, the initial diagnosis is mostly based on neuroimaging characteristics, complemented by associated clinical phenotyping and genetic discoveries to aid in subclassification [15].

Currently, brain magnetic resonance imaging (MRI) is the gold standard for identifying these malformations, due to the high spatial resolution, the superior soft tissue contrast, and the lack of ionizing radiation. A head CT scan exposes a child to approximately

Table 1
Current classification scheme of malformations of cortical development

Malformations secondary to abnormal neuronal and glial proliferation and apoptosis
1. Severe congenital microcephaly
2. Megalencephaly
3. Cortical dysgenesis with abnormal cell proliferation but without neoplasia
4. Cortical dysgenesis with abnormal cell proliferation and neoplasia
Malformations due to abnormal cortical migration
1. MCDs with neuroependymal abnormalities: Periventricular heterotopia
2. MCDs due to generalized abnormal transmantle migration
3. MCDs presumably due to localized abnormal late radial or tangential transmantle migration
4. MCDs due to abnormal terminal migration and defects in pial limiting membrane
Malformations due to abnormal postmigrational development
1. MCDs with polymicrogyria or cortical malformations resembling polymicrogyria
2. Cortical dysgenesis secondary to inborn error of metabolism
3. Focal cortical dysplasia due to late developmental disturbances
4. Postmigrational developmental microcephaly

Adapted from Barkovich et al. [13]

2 mSv (millisieverts) of radiation, which is equivalent to more than 1 year of natural background radiation. Head CT has many limitations in the investigation of MCDs, such as the limited ability to differentiate gray and white matter, the low spatial resolution, and the inability to assess myelination. Additionally, CT implies exposure to ionizing radiation, which is a concern especially in pediatric patients. Therefore, despite its large availability, low cost, and speed, head CT scan has a limited use in the diagnosis of MCDs and should be reserved for selected cases with specific indications or when magnetic resonance imaging is not available or contraindicated.

2 Materials

2.1 CT Scan Machine

The CT scanner is a sophisticated medical imaging device that uses X-rays and provides detailed cross-sectional images of the body [16–18]. CT scan machines consist of a large, cylindrical gantry that houses an X-ray tube and a detector array [19, 20]. In a spiral or helical scan, the table moves continuously as the X-ray source and detectors rotate, thus producing 3D imaging in shorter acquisition times, with important consequences in the emergency setting. Advancements in CT technology, such as multi-detector CT and dual-energy CT, have improved the speed and accuracy of brain imaging. These techniques allow for faster scans with higher resolution and reduced radiation exposure for the patient. Contrast agents such as iodinated contrast media are not required to study MCDs, and CT perfusion or CT angiography are not performed in this clinical scenario [15].

3 Methods

3.1 Scanning of the Patient

The patient lies on a table that slides in and out of the gantry's central opening, which is shaped like a doughnut. As the table moves through the gantry, the X-ray tube and detector rotates around the patient, taking multiple images or "slices" of the brain from different angles. The X-ray tube emits a narrow beam of X-rays, which pass through the patient's head and are detected by the detector array [21]. The detector array then measures the intensity of the X-rays that pass through the patient, generating a digital signal that is sent to a computer (Fig. 1). The computer uses this information to construct a 2D image of a thin slice of the brain, with a thickness ranging from 1 to 10 mM, which is displayed on a monitor [21]. The CT machine then rotates slightly, and the process is repeated to obtain images from different angles. The computer combines these 2D images to create detailed, 3D images of the brain. This process is known as "tomographic reconstruction"

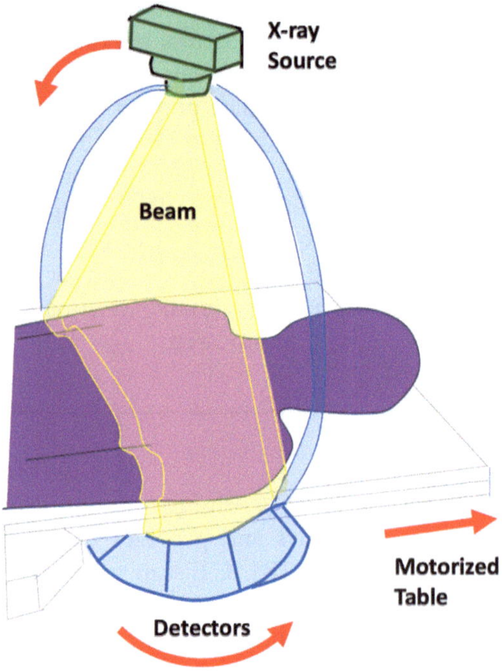

Fig. 1 During a CT scan, a thin X-ray beam rotates around an area of the body while the patient moves on a motorized platform, thus generating a 3D image of the internal structures

and allows physicians to view the brain from different angles and perspectives [19, 20]. Low-dose, age-specific protocols for pediatric brain CT scans should be applied without jeopardizing image quality. CT scans are acquired by a radiologic technician and interpreted and reported by a trained radiologist.

3.2 CT Scan Images

The images created are made up of pixels with radiosensitivity, which are displayed using Hounsfield units (HU). The Hounsfield Units (HU) represent the density of various tissues, with higher values for denser material and vice versa. For instance, the HU of water is 0, air is negative 1000, and bone is positive 400–2000. To display/view CT scans, windowing (also known as gray-level mapping), contrast stretching, histogram modification, or contrast enhancement is used to transform HU numbers into grayscale values [20, 21]. The window width and level can be set in order to obtain different tissue contrasts, such as the brain parenchyma, soft tissues, or the bone (Fig. 2). This process highlights key anatomy and pathology features.

3.3 CT Data Analysis

The CT data can be postprocessed in order to have multiplanar or 3D images. Multiplanar reformation or reconstruction (MPR) involves the process of converting data from an imaging modality acquired in a certain plane, usually axial, into another plane, such as coronal, sagittal, or oblique (Fig. 3a–c) [22]. Three-dimensional

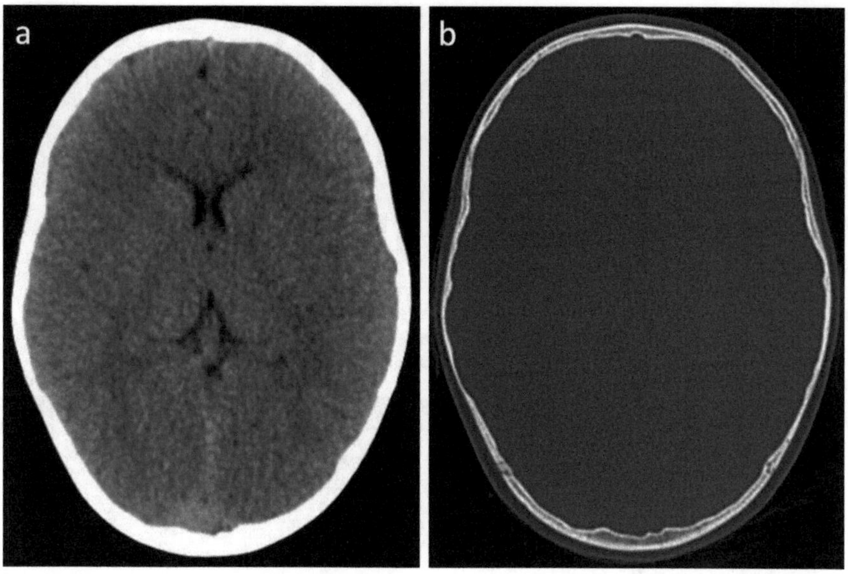

Fig. 2 Windowing for head CT. (**a**) Brain window image at the level of the basal ganglia shows differences among different types of soft tissues, such as brain, blood, vasculature, air-filled structures, and fluid-containing spaces. (**b**) Bone window image at the same level visualizes details of bone structures and identifies subtle skull lesions

(3D) computed tomography (CT) reformations can also be obtained, and are particularly useful to study the skull and vertebral bones (Fig. 3d–f). It is important to note that current CT machines display the image opposite to the patient's side as the image is produced as viewed from the patient's foot, with the right side of the image being the patient's left side.

4 Notes

1. CT uses ionizing radiation, which can be harmful to the developing brain and increase the risk of cancer. Therefore, CT should be used judiciously and only when necessary, such as in emergency situations, when MRI is not available or contraindicated, and for specific indications [23].
2. CT has limited sensitivity and specificity in the detection of MCDs, especially when they are small and focal, such as in the case of focal cortical dysplasia, dysgyria, and small heterotopias (Fig. 3) [15, 24].
3. Major cortical abnormalities, such as severe lissencephaly, severe microcephaly or megalencephaly, hemimegalencephaly, extensive areas of polymicrogyria, large periventricular nodular heterotopias, and schizencephaly, can be identified on a CT scan (Fig. 4) [23].

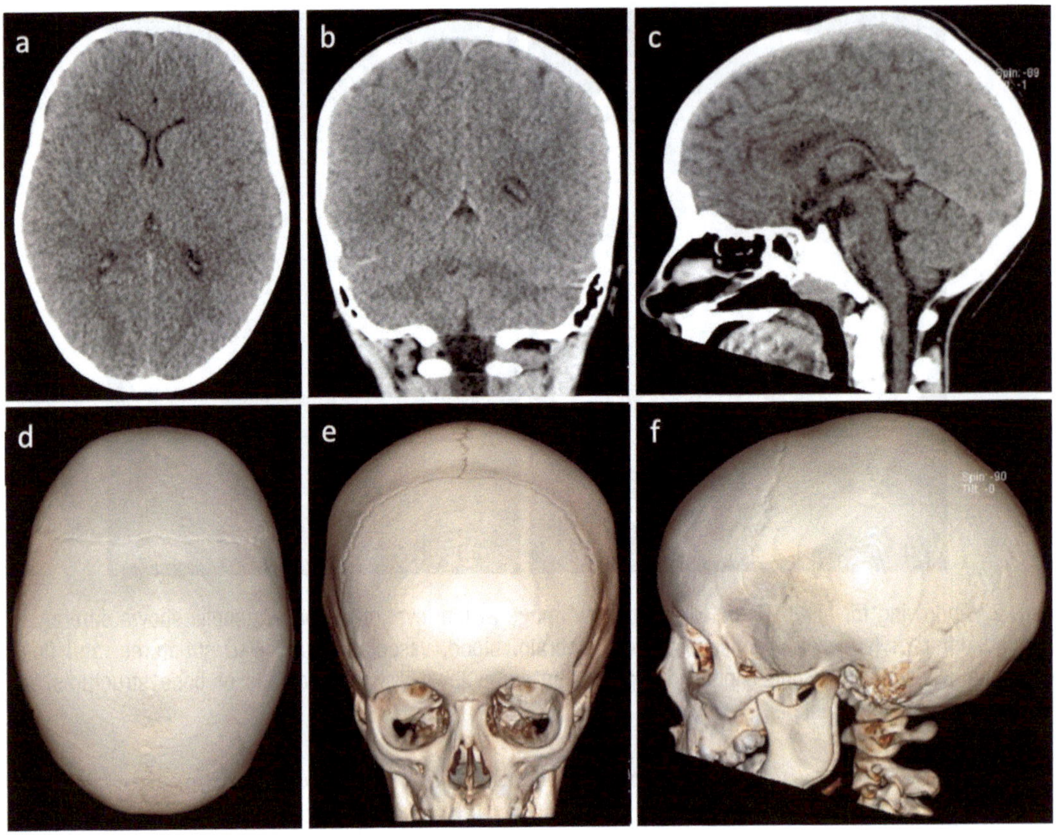

Fig. 3 Multiplanar and 3D visualization of head CT images. Axial (**a**), coronal (**b**), and sagittal (**c**) MPR-CT brain window images show the anatomical and pathological findings on the three planes. (**d–f**) Different views of the 3D reformatted images depict normal skull bones and sutures

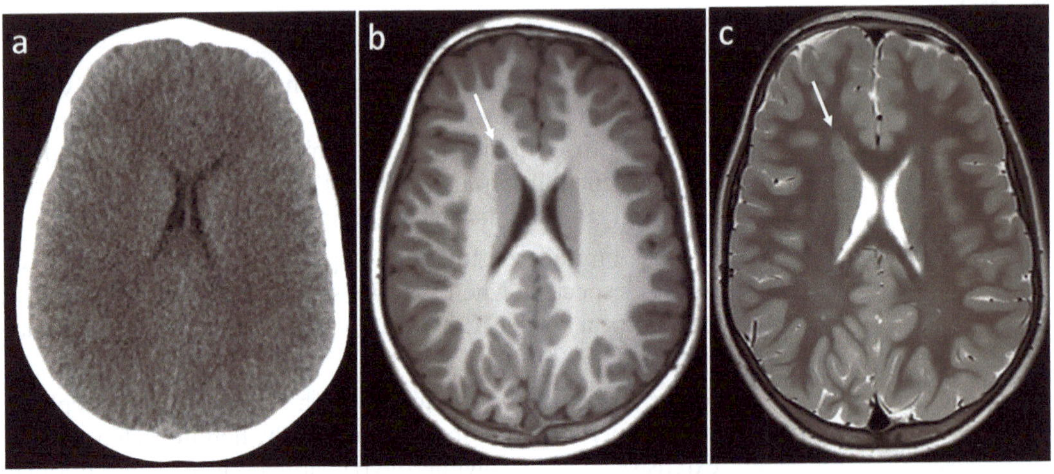

Fig. 4 Head CT (**a**) and brain MRI (**b**, **c**) in periventricular nodular heterotopia. (**a**) Axial CT does not depict this small abnormality, while corresponding axial T1-weighted (**b**) and T2-weighted (**c**) images well demonstrate the small nodule of gray matter heterotopia at the level of the right frontal periventricular region (arrows)

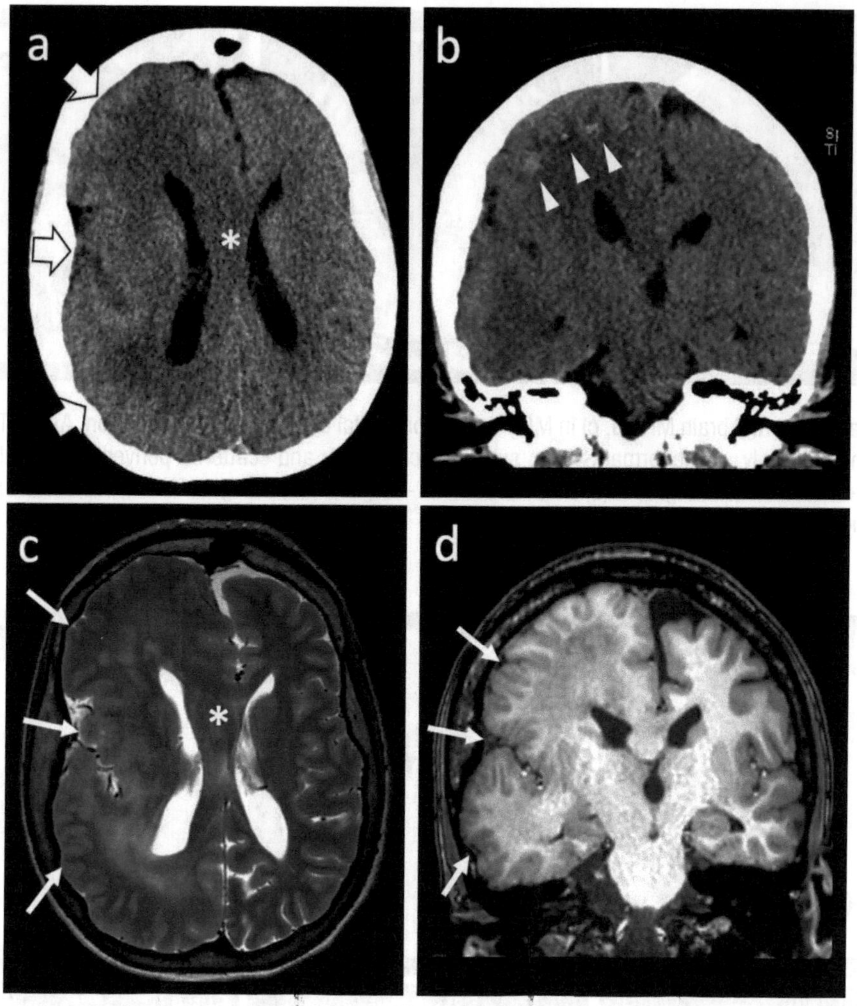

Fig. 5 Head CT (**a**, **b**) and brain MRI (**c**, **d**) in hemimegalencephaly. In this case, both axial CT (**a**) and T2-weighted (**c**) images reveal the large malformation of cortical development characterized by an enlarged right hemisphere (thick arrows) with thick septum pellucidum (asterisks) and abnormal white matter. Axial CT fails in demonstrating the associated fronto-temporo-parietal polymicrogyria, well described on axial T2-weighted (**c**) and coronal T1-weighted (**d**) images (small arrows), due to its low gray–white matter contrast and spatial resolution. However, coronal CT (**b**) better depicts small subcortical calcifications in the right frontal lobe (arrowheads)

4. Some associated brain malformations may be identified on CT scans, including hydrocephalus, corpus callosum agenesis, and severe ponto-cerebellar hypoplasia (Fig. 5) [23, 25].
5. CT can detect brain calcifications associated with MCDs, thus helping in the differential diagnosis between acquired and genetic MCDs or in the identification of different genetic patterns (Fig. 6) [20].
6. CT can provide high-resolution images of the skull and bones, thus identifying associated malformations, such as craniosynostosis, inner and middle ear malformations, and vertebral anomalies (Fig. 7) [23].

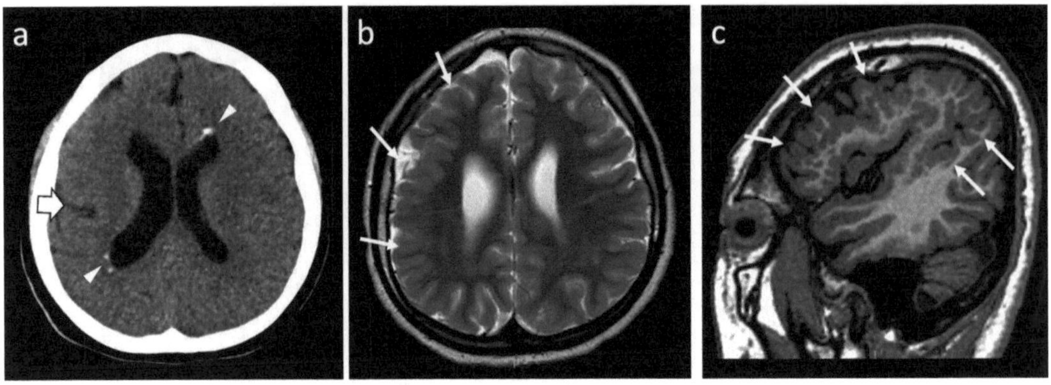

Fig. 6 Head CT (**a**) and brain MRI (**b**, **c**) in MCD due to congenital cytomegalovirus infection. Axial CT image (**a**) reveals microcephaly, an abnormal Sylvian sulcus (thick arrow), and scattered periventricular calcifications (arrowheads). Axial T2-weighted (**b**) and sagittal T1-weighted (**c**) images better depict the perisylvian polymicrogyria (small arrows)

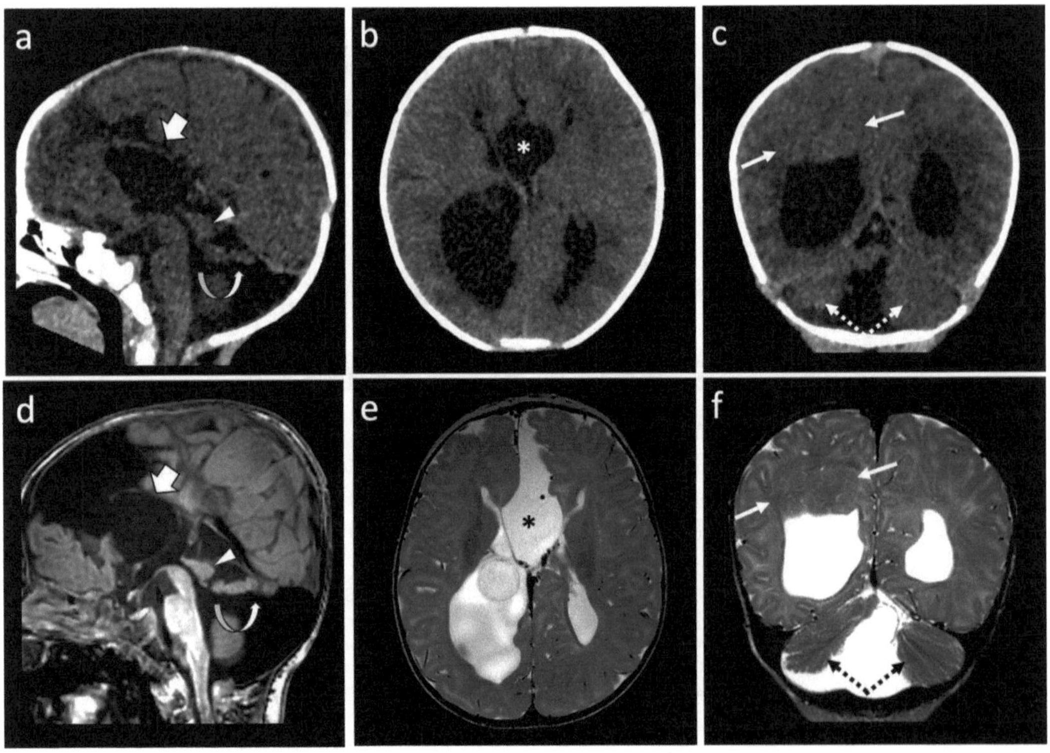

Fig. 7 Head CT (**a**–**c**) and brain MRI (**d**–**f**) in a patient with oral–facial–digital syndrome. Sagittal- (**a**), axial- (**b**), and coronal-reformatted (**c**) CT images and corresponding sagittal T1-weighted (**d**), axial T2-weigthed (**e**) and coronal T2-weighted (**f**) images show complete callosal agenesis (thick arrows), thick midbrain tectum (arrowheads), and hypodysplastic rotated cerebellar vermis (curved arrows). There is also an interhemispheric cyst (asterisks) associated with hypoplasia of cerebellar hemispheres (dashed arrows). Note that the large area of heterotopic gray matter in the right parietal lobe is better visualized on coronal T2-weighted images compared to the CT scan (thin arrows)

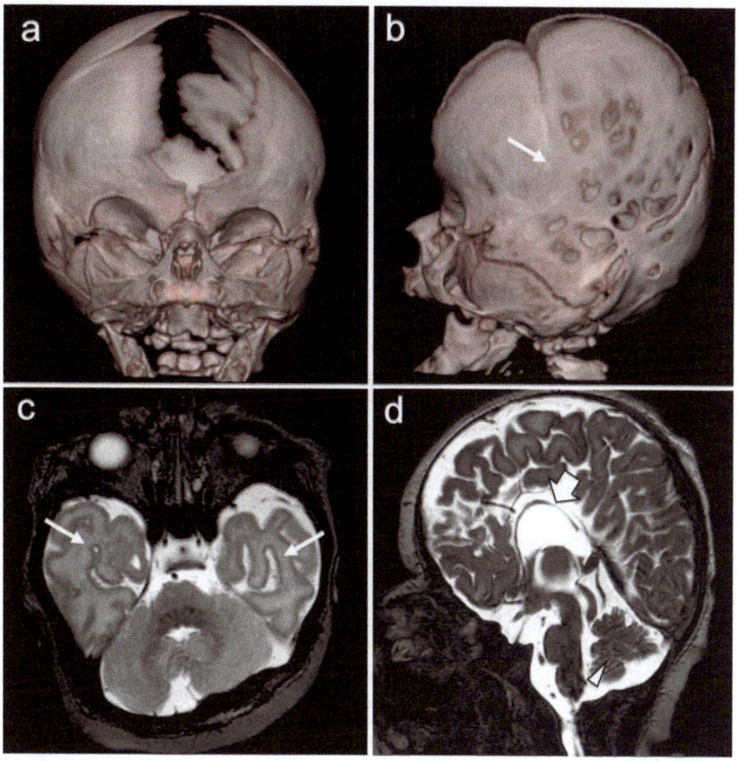

Fig. 8 Head CT (**a**, **b**) and brain MRI (**c**, **d**) in a patient with Apert syndrome. CT scan with 3D views (**a**, **b**) demonstrate premature fusion of coronal sutures (thin arrow) resulting in skull and facial deformities. Axial T2-weighted (**c**) and sagittal DRIVE T2-weighted (**d**) images show bilateral dysgyria at the level of the temporal lobes (thin arrows), callosal hypoplasia (thick arrow), and small inferior cerebellar vermis (arrowhead)

7. CT cannot provide information on the functional integrity of the brain, which is important in the diagnosis and management of neurological disorders [25] (Fig. 8).

4.1 Concluding Remarks

In conclusion, CT has several limitations, including its use of ionizing radiation and limited sensitivity and specificity in the detection of MCDs. Therefore, CT should be used judiciously and after MR imaging to provide a comprehensive evaluation of MCDs.

Acknowledgments

We thank the Neuroradiology Unit of the IRCCS Istituto Giannina Gaslini for the technical support.

References

1. Petrik V, Apok V, Britton JA, Bell BA, Papadopoulos MC (2006) Godfrey Hounsfield and the dawn of computed tomography. Neurosurgery 58:780–787
2. Bhattacharyya KB (2016) Godfrey Newbold Hounsfield (1919-2004): the man who revolutionized neuroimaging. Ann Indian Acad Neurol 19:448–450
3. Tan SY, Poole PS (2020) Allan MacLeod Cormack (1924-1998): discoverer of computerised axial tomography. Singapore Med J 61:4–5
4. Raju TN (1999) The Nobel chronicles. 1979: Allan MacLeod Cormack (b 1924); and Sir Godfrey Newbold Hounsfield (b 1919). Lancet 354:1653
5. Kistler JP, Hochberg FH, Brooks BR, Richardson EP, New PF, Schnur J (1975) Computerized axial tomography: clinicopathologic correlation. Neurology 25:201–209
6. Ambrose J (1973) Computerized transverse axial scanning (tomography). 2. Clinical application. Br J Radiol 46:1023–1047
7. Merino-deVillasante J, Taveras JM (1976) Computerized tomography (CT) in acute head trauma. AJR Am J Roentgenol 126: 765–778
8. Ambrose J, Gooding MR, Richardson AE (1975) An assessment of the accuracy of computerized transverse axial scanning (EMI scanner) in the diagnosis of intracranial tumour. A review of 366 patients. Brain 98:569–582
9. Siegel MJ, Schmidt B, Bradley D, Suess C, Hildebolt C (2004) Radiation dose and image quality in pediatric CT: effect of technical factors and phantom size and shape. Radiology 233(2):515–522
10. Rydberg J, Buckwalter KA, Caldemeyer KS, Phillips MD, Conces DJ Jr, Aisen AM et al (2000) Multisection CT: scanning techniques and clinical applications. Radiographics 20: 1787–1806
11. Barkovich AJ, Kuzniecky RI, Jackson GD, Guerrini R, Dobyns WB (2005) A developmental and genetic classification for malformations of cortical development. Neurology 65: 1873–1887
12. Guerrini R, Dobyns WB (2014) Malformations of cortical development: clinical features and genetic causes. Lancet Neurol 13:710–726
13. Barkovich AJ, Guerrini R, Kuzniecky RI, Jackson GD, Dobyns WB (2012) A developmental and genetic classification for malformations of cortical development: update 2012. Brain 135 (5):1348–1369. https://doi.org/10.1093/brain/aws019
14. Desikan RS, Barkovich AJ (2016) Malformations of cortical development. Ann Neurol 80: 797–810
15. Severino M, Geraldo AF, Utz N, Tortora D, Pogledic I, Klonowski W et al (2020) Definitions and classification of malformations of cortical development: practical guidelines. Brain 143:2874–2894
16. Baker HL, Campbell JK, Houser OW, Reese DF, Sheedy PF, Holman CB (1974) Computer assisted tomography of the head. An early evaluation. Mayo Clin Proc 49:17–27
17. Paxton R, Ambrose J (1974) The EMI scanner. A brief review of the first 650 patients. Br J Radiol 47:530–565
18. New PF, Scott WR, Schnur JA, Davis KR, Taveras JM (1974) Computerized axial tomography with the EMI scanner. Radiology 110: 109–123
19. Liguori C, Frauenfelder G, Massaroni C, Saccomandi P, Giurazza F, Pitocco F et al (2015) Emerging clinical applications of computed tomography. Med Devices 8:265–278
20. Wesolowski JR, Lev MH (2005) CT: history, technology, and clinical aspects. Semin Ultrasound CT MR 26:376–379
21. McCollough CH, Leng S, Yu L, Fletcher JG (2015) Dual- and multi-energy CT: principles, technical approaches, and clinical applications. Radiology 276:637–653
22. Dalrymple NC, Prasad SR, Freckleton MW, Chintapalli KN (2005) Informatics in radiology (infoRAD): introduction to the language of three-dimensional imaging with multidetector CT. Radiographics 25:1409–1428
23. Rossi A, Argyropoulou M, Zlatareva D, Boulouis G, Pizzini FB, van den Hauwe L et al (2023) European recommendations on practices in pediatric neuroradiology: consensus document from the European Society of Neuroradiology (ESNR) European Society of Paediatric Radiology (ESPR) and European Union of Medical Specialists Division of Neuroradiology (UEMS). Abstr Pediatr Radiol 53(1):159–168. https://doi.org/10.1007/s00247-022-05479-4
24. Santos AC, Escorsi-Rosset S, Simao GN, Terra VC, Velasco T, Neder L et al (2014) Hemispheric dysplasia and hemimegalencephaly: imaging definitions. Childs Nerv Syst 30: 1813–1821
25. Sisodiya SM (2004) Malformations of cortical development: burdens and insights from important causes of human epilepsy. Lancet Neurol 3:29–38

Chapter 23

MRI Data Analysis in Malformations of Cortical Development

Mariasavina Severino, Domenico Tortora, and Marcello Scala

Abstract

Brain magnetic resonance imaging (MRI) is a noninvasive imaging modality that utilizes powerful magnets and radio waves to generate detailed images of the brain, making it a valuable tool for investigating malformations of cortical development (MCD). Various MRI techniques, including 3D T1-weighted, multiplanar thin-sliced T2-weighted, and 3D fluid-attenuated inversion recovery (FLAIR) sequences, can provide high-resolution images with excellent spatial and contrast resolution, allowing for a detailed visualization of cortical anatomy and abnormalities. Almost all MCD can be detected and characterized using MRI. Advanced techniques, such as arterial spin labeling MR perfusion, diffusion tensor imaging (DTI), and functional MRI (fMRI), may be used to improve the detection rate of these malformations and to plan surgery in case of drug-resistant epilepsy. However, there are also limitations related to high cost, relatively low availability, need for sedation or anesthesia, and limited sensitivity for detecting subtle focal cortical malformations. Despite these limitations, brain MRI plays a crucial role in the investigation of MCD, providing valuable information for diagnosis, treatment planning, and patient management.

Key words Brain MRI, T1-weighted sequences, T2-weighted sequences, Arterial spin labeling, Diffusion-weighted imaging, Malformations of cortical development

1 Introduction

Brain magnetic resonance imaging (MRI) is a noninvasive imaging modality that utilizes powerful magnets and radio waves to generate detailed images of the brain [1]. It provides high-resolution images with excellent contrast resolution, allowing for detailed visualization of brain anatomy and abnormalities [1]. Various MRI techniques, such as T1-weighted, T2-weighted, fluid-attenuated inversion recovery (FLAIR), and post-contrast imaging, can be employed to capture different aspects of brain tissue properties and pathology [1]. Diffusion-weighted imaging (DWI) has greatly improved the detection of cytotoxic edema, thus enhancing the diagnosis of cerebrovascular diseases [1]. Susceptibility-weighted imaging (SWI) is a high-resolution MRI technique that

Koh-ichi Nagata (ed.), *Cerebral Cortex Development: Methods and Protocols*, Methods in Molecular Biology, vol. 2794,
https://doi.org/10.1007/978-1-0716-3810-1_23,

can be used to detect subtle local changes in the susceptibility of the brain parenchyma, thus detecting iron deposition, hemorrhages, and calcifications [2]. Advanced techniques such as diffusion tensor imaging (DTI) and functional MRI (fMRI) can provide additional information about the microstructural and functional aspects of the brain [3, 4]. For example, perfusion MRI techniques using gadolinium-based contrast media, such as dynamic susceptibility contrast (DSC) MR perfusion, or noninvasive techniques, such as arterial spin labeling (ASL) MR perfusion, are helpful in assessing cerebral perfusion in several conditions, including stroke, tumors, and epilepsy [5]. Diffusion tensor imaging (DTI) can be used to reconstruct in vivo white matter tracts and assess their integrity [4]. Functional MRI (fMRI) can be used to detect changes in blood flow in the brain that are associated with neural activity, allowing researchers to study brain function in real time [3].

Since its development in the 1970s and 1980s, brain MRI has played a crucial role in the investigation of various brain disorders, including malformations of cortical development, providing valuable information for diagnosis, treatment planning, and patient management [1] (*See* **Note 1**). Malformations of cortical development (MCD) are a diverse group of disorders that result from abnormal neuronal migration, proliferation, and differentiation during brain development (*see* Table 1) [6–8]. These neurodevelopmental disorders are characterized by abnormal brain cortical architecture, which can result in various cognitive, motor, and sensory deficits [9, 10]. A classic example is provided by the neurodevelopmental disorders associated with pathogenic genetic variations in the RAC genes, encoding for crucial regulators of actin

Table 1
Current classification of MCDs

Malformations secondary to abnormal neuronal and glial proliferation and apoptosis
1. Severe congenital microcephaly
2. Megalencephaly
3. Cortical dysgenesis with abnormal cell proliferation but without neoplasia
4. Cortical dysgenesis with abnormal cell proliferation and neoplasia
Malformations due to abnormal cortical migration
1. MCDs with neuroependymal abnormalities: periventricular heterotopia
2. MCDs due to generalized abnormal transmantle migration
3. MCDs presumably due to localized abnormal late radial or tangential transmantle migration
4. MCDs due to abnormal terminal migration and defects in pial limiting membrane
Malformations due to abnormal postmigrational development
1. MCDs with polymicrogyria or cortical malformations resembling polymicrogyria
2. Cortical dysgenesis secondary to inborn error of metabolism
3. Focal cortical dysplasia due to late developmental disturbances
4. Postmigrational developmental microcephaly

Adapted from Barkovich et al. [6]

cytoskeleton dynamics with a pivotal role during corticogenesis [9]. The abnormal function of RAC GTPases results in a disruption of fundamental processes during brain development, leading to a spectrum of malformations of cortical development, featuring nodular heterotopias, gyration defects, and white matter abnormalities [9]. The use of brain MRI and the introduction of new imaging techniques have further increased the diagnostic accuracy of MCD [11].

2 Materials

2.1 Magnetic Resonance Scanner

MRI is a noninvasive diagnostic imaging technique that uses a powerful magnetic field, radio waves, and a computer to produce detailed images of the body's internal structures [12, 13]. An MRI scanner consists of a large cylindrical magnet that surrounds the patient and a bed that slides in and out of the magnet [14] (*See* **Note 2**). The scanner is controlled by a computer, which processes the data obtained during the scan to create images of the body. For brain imaging, the most commonly used machines are 1.5 and 3 Tesla (T) MRI scanners. These scanners produce a magnetic field that is at least 30,000 times stronger than the Earth's magnetic field [14]. In addition to 1.5 and 3 T MRI scanners, there are also higher field strength MRI scanners, such 7 T scanners, which are mostly used for research purposes (*see* **Note 5**) [22].

3 Methods

3.1 Imaging Acquisition

During an MRI brain scan, the patient lies down on a bed that slides into the scanner (Fig. 1). The scanner produces a strong magnetic field, causing the hydrogen atoms in the brain tissues to align with the field [12, 13] (*see* **Note 3**). Radio waves are then directed at the patient's head, causing the aligned hydrogen atoms to emit signals that are detected by the scanner (Fig. 2). The signals received by the scanner are processed by a computer to create detailed images of the brain structures, including the cerebral cortex, white matter, and subcortical structures [15]. By varying the parameters of the pulse sequence, different contrasts may be generated between tissues based on the relaxation properties of the hydrogen atoms. The images can be viewed in different planes, such as axial, coronal, and sagittal, allowing physicians to examine the brain from various angles [12]. Different MRI techniques can be used to investigate specific aspects of brain function or structure.

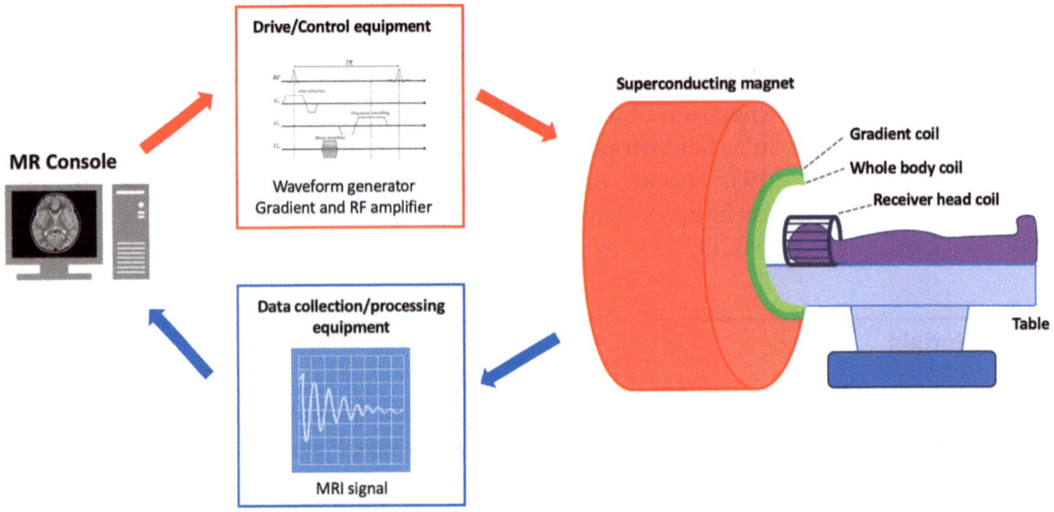

Fig. 1 During a brain MRI examination, the patient is placed inside a large magnet and should remain still during the long imaging acquisition process. A superconductive magnet, radio waves, and a computer create detailed images of the brain. The magnet generates the magnetic field by coiling electrical wire and running a current through the wire. Gradient coils induce precise magnetic field changes, thus allowing image slices of the brain to be created. A head coil functions as an antenna that emits the "exciting" radiofrequency pulse and receives the MRI signals from the "relaxing" protons

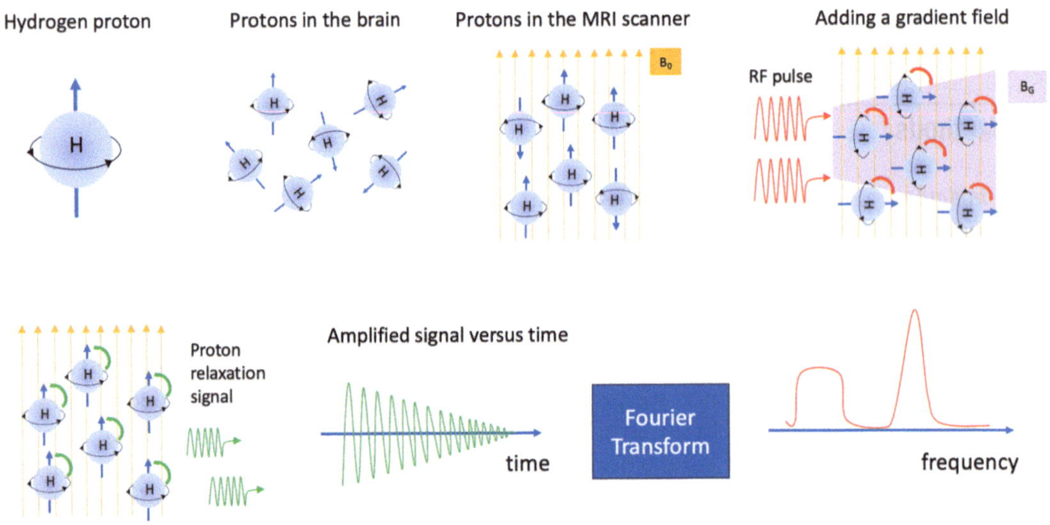

Fig. 2 Basic MRI physics concepts. For imaging purposes, the hydrogen nucleus (a single proton) is used because of its abundance in soft tissues. Under normal circumstances, these hydrogen protons spin in the brain with their axes randomly aligned. When the brain is placed inside the magnetic field (B_0) of the scanner, the magnetic moment of the protons lines up to be either parallel or antiparallel to the direction of the field. Most protons align parallel to B_0 as this is a lower energy state. Pulses of radio waves excite the nuclear spin energy transition, and magnetic field gradients localize the polarization in space. The receiving coil detects weak radio frequency radiation from the protons as they "relax" from this magnetic interaction. This signal is detected as a function of time and then is converted to signal strength as a function of frequency by means of a Fourier transformation

3.2 Brain MRI Protocol for Studying MCD

The MRI protocols employed for the investigation of MCD in clinical practice should include a series of optimized structural sequences that may be acquired both with 1.5 T and 3 T scanners [16, 17]. This "basic protocol" mainly corresponds to the Harmonized Neuroimaging of Epilepsy Structural Sequences (HARNESS-MRI) protocol developed by the International League Against Epilepsy (ILAE) Neuroimaging Task Force in 2019 [18]. To this core protocol, several advanced sequences may be obtained to identify subtle MCD like focal cortical dysplasia, to plan the intervention when surgery for epilepsy is considered, and for research purposes (*see* **Note 4**).

Sequences that should be included in the "core protocol" are (Fig. 3):

- T1-weighted imaging: This sequence provides detailed images of the brain's anatomy showing high signal intensity in areas with high fat content, such as white matter, and low signal intensity in areas with low fat content, such as cerebrospinal fluid. The best techniques to study MCD are the magnetization-prepared rapid gradient-echo (MP-RAGE) sequence as well as the equivalent 3D spoiled gradient echo and 3D turbo field echo protocols. These sequences should be acquired with isotropic voxels (i.e., identical dimensions across

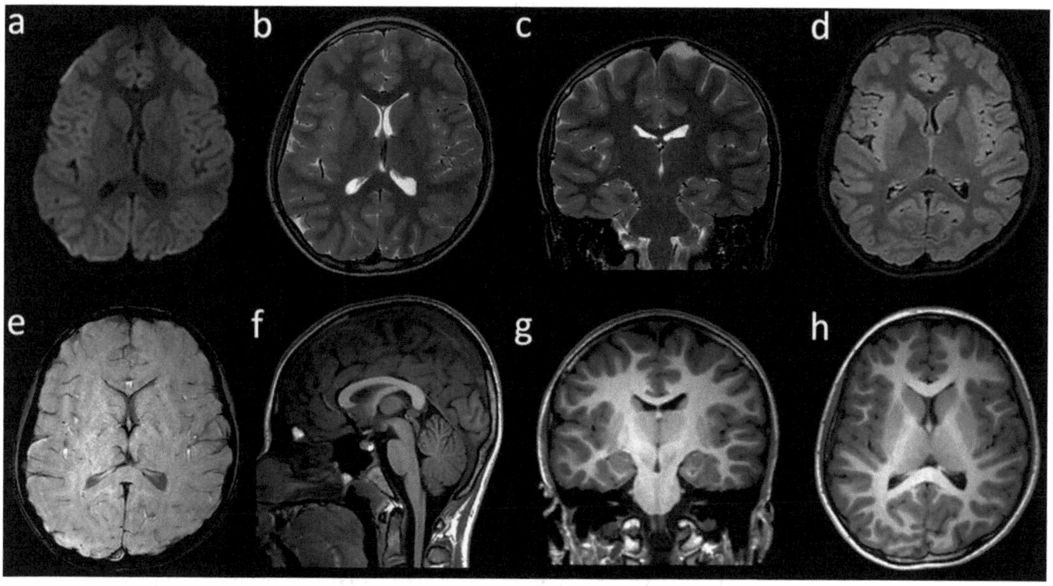

Fig. 3 Core brain MRI protocol for studying MCDs. This protocol should include (**a**) diffusion-weighted imaging; (**b**) axial and (**c**) coronal T2-weighted images; (**d**) axial-reformatted 3D FLAIR image; (**e**) susceptibility-weighted imaging; and (**f**) sagittal-, (**g**) coronal-, and (**h**) axial-reformatted 3D T1-weighted images

planes), allowing high-contrast reformatting of the images in different orientations.

- T2-weighted imaging: This sequence allows one to assess both brain anatomical details and associated signal alterations of the gray and white matter. T2-weighted images have high signal intensity in areas with high water content, such as cerebrospinal fluid and edema, and low signal intensity in areas with low water content, such as gray matter. The best technique for MCD is the 2D turbo spin echo sequence with a high in-plane resolution (2- or 3 mM thick slices) acquired in the axial and coronal planes (*see* **Note 6**).
- 3D fluid-attenuated inversion recovery (FLAIR) imaging: This sequence has high sensitivity for signal changes in the brain, including hyperintensities related to gliosis and increased extracellular space. This sequence (named CUBE, VISTA, or SPACE, depending on the MR vendor) may be acquired as 2D (with 3 mM thick slices) in different planes or as 3D (with slices <1 mM thick). The 3D FLAIR is preferred when studying MCD, since it allows one to reformat the images in three planes with a higher detection rate of focal cortical dysplasia (*see* **Notes** 7 and **8**).
- Diffusion-weighted imaging (DWI): This sequence is used to assess the movement of water molecules in the brain parenchyma. DWI is sensitive to changes in tissue microstructure and can distinguish between cytotoxic or intramyelinic edema (characterized by restricted diffusion) and vasogenic edema (characterized by facilitated diffusion). In its simplest acquisition scheme (i.e., with three diffusion-encoding gradients), this sequence has become part of every basic brain MR protocol in clinical practice. In subjects with MCD, it allows one to detect seizure-related brain changes or ischemic complications.
- Susceptibility-weighted imaging (SWI) or T2*-weighted imaging: These sequences have high sensitivity for the presence of metal deposition (such as iron or manganese), hemosiderin, and calcification in the brain parenchyma [2]. They have also become part of basic neuroimaging protocols in clinical practice. In MCD, they are useful for depicting brain calcifications or previous hemorrhages that may help in the differential diagnosis.

Additional advanced sequences that might be used to study MCD include:

- Arterial spin labeling (ASL) MR perfusion: This technique employs magnetically labeled blood water to noninvasively estimate cerebral blood flow providing a brain perfusion map that may be analyzed in patients with MCD to detect regions of hyper- or hypoperfusion related to seizures (Fig. 4).

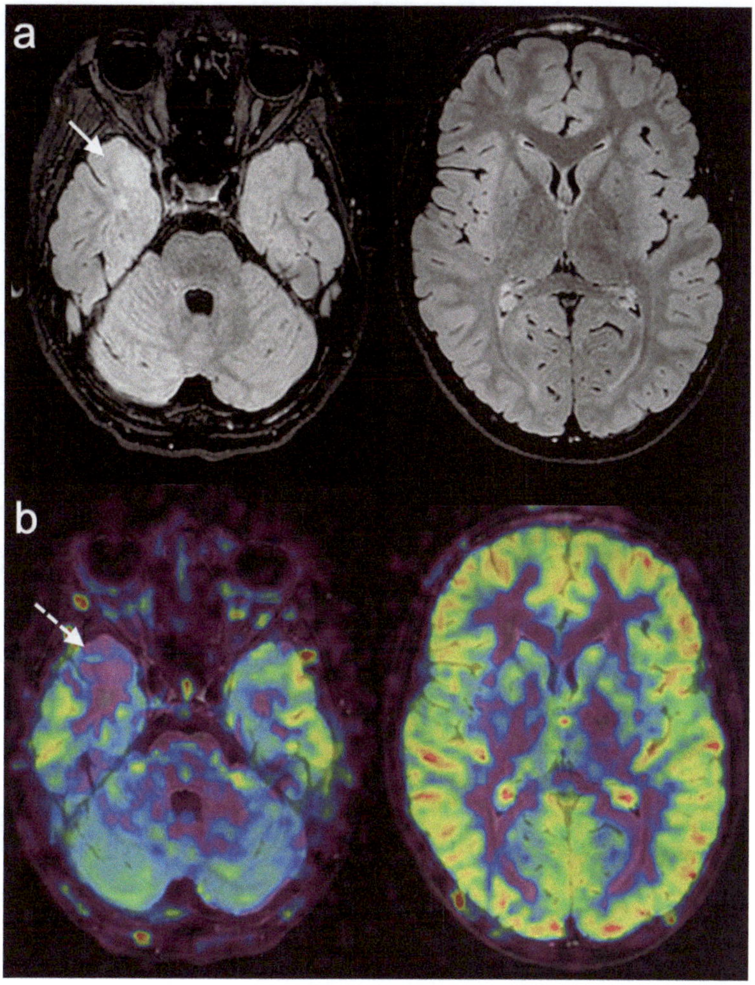

Fig. 4 Arterial spin labeling (ASL) findings in focal cortical dysplasia. (**a**) Axial-reformatted 3D FLAIR images depicts an area of gray and white matter blurring with faint white matter hyperintensity in the right temporal lobe (arrow) consistent with a focal cortical dysplasia. (**b**) Axial ASL perfusion MR images superimposed to the FLAIR sequence reveal a postictal area of reduced perfusion at the level of the cortical malformation (dashed arrow)

- Diffusion tensor imaging (DTI): Since the diffusion of water is limited by the presence of myelin and axonal membranes, the directionality of white matter tracts can be inferred from studies of water diffusion patterns. DTI uses diffusion vectors to measure the directionality of white matter bundles, thus enabling to reconstruct the main white matter tracts in vivo. This may be useful in epilepsy surgery planning to avoid damage to the corticospinal tracts, optic radiations, arcuate fasciculus, and other important white matter tracts during the intervention (Fig. 5) [4].

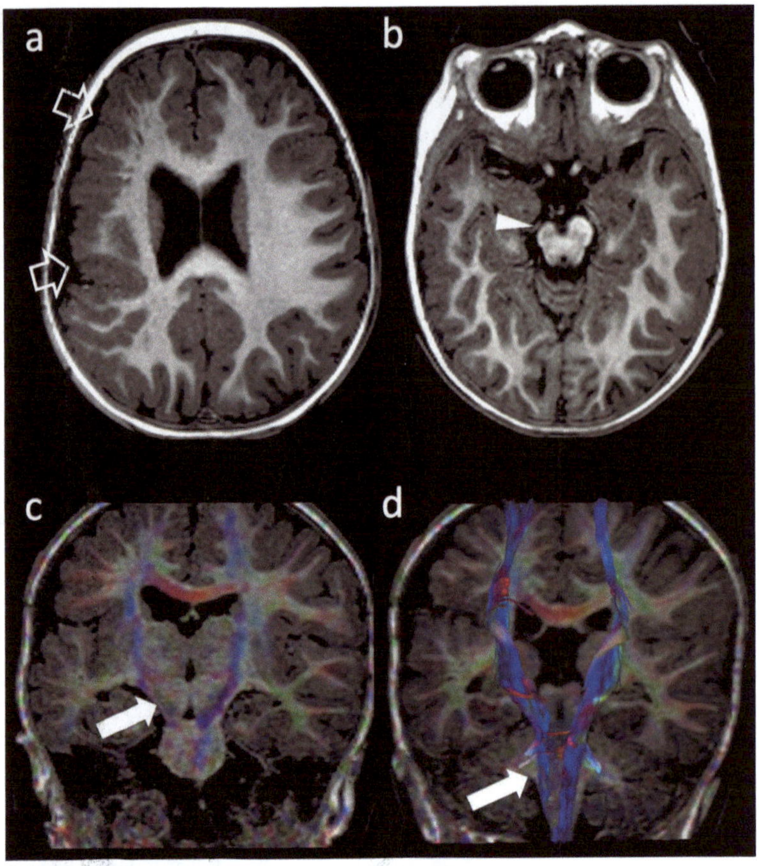

Fig. 5 Diffusion tensor imaging (DTI) findings in polymicrogyria. (**a**, **b**) Axial-reformatted 3D T1-weighted images reveal right frontoparietal polymicrogyria (empty arrows) associated with a small right cerebral peduncle (arrowhead) suggesting ipsilateral corticospinal tract hypoplasia. (**c**) Coronal color-coded fractional anisotropy maps superimposed to the 3D T1-weighted images and (**d**) 3D reconstruction of the corticospinal tracts confirm that the right corticospinal tract is hypoplastic (thick arrows)

- Functional magnetic resonance imaging (fMRI): this technique measures changes in blood oxygenation levels in response to neural activity and is used to study brain function [3]. fMRI is commonly used to map brain activity in response to various stimuli and is also used to localize brain function prior to surgery [3]. As for DTI, this sequence may reduce the damage to eloquent cortical areas during intervention.

3.3 Data Analysis and Clinical Implications

The comprehensive analysis of brain MRI data is essential for accurate diagnosis and classification of MCD and for understanding the underlying pathophysiology of these complex neurodevelopmental disorders. Analysis of brain MRI data typically involves careful examination of the brain cortical surface, gyrification

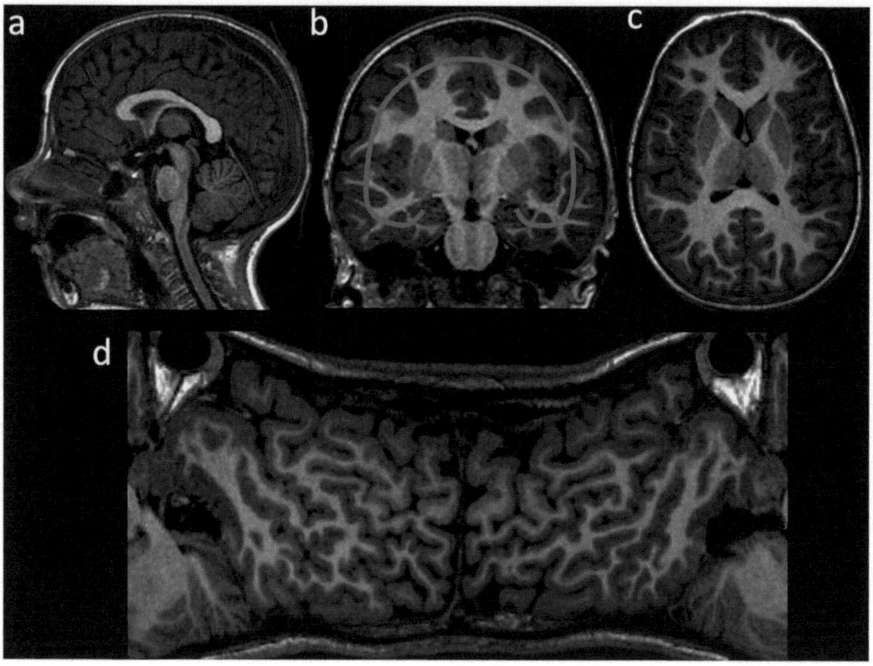

Fig. 6 Images reformatting for studying MCD. 3D T1-weighted images can be reformatted in the sagittal (**a**), coronal (**b**), and axial (**c**) planes. In addition, curvilinear reconstructions (d) may be performed, showing the entire cortex in one slice

patterns, cortical thickness, and sulcal morphology to identify any abnormalities or irregularities [7, 8] (*see* **Note 9**). In addition, it is important to scrutinize the white matter from the periventricular regions to the cortex to identify gray matter heterotopias or signal alterations consistent with focal cortical dysplasia [1].

Multiplanar, curvilinear, and 3D reformatting enable comprehensive evaluation of cortical malformations from different perspectives (Fig. 6) [1]. Advanced computational tools, image processing techniques, and automated algorithms, such as surface-based machine-learning algorithms, can also be employed to improve the detection rate of subtle MCD, such as focal cortical dysplasia [19] (*see* **Note 10**). Brain MR images should be finally assessed to identify other cerebral or midbrain–hindbrain malformations and/or associated acquired lesion.

4 Notes

1. Brain MRI is a noninvasive imaging modality that does not involve ionizing radiation, making it safe for repeated imaging studies, particularly in pediatric patients [1, 15].
2. Brain MRI can be expensive and may not be readily available in all healthcare settings, limiting access for some patients and populations [20, 21].

3. Young children (usually <6 years of age) or patients with movement/cognitive disorders may require sedation to undergo MRI, which can pose risks and challenges [21].
4. Gadolinium-based contrast agents' administration is usually not needed in the diagnosis of MCD [1, 7].
5. If available, 3 Tesla scanners are preferred over 1.5 T scanners when studying MCD [16, 17].
6. MR sequences must be adapted to the brains of newborns and infants to obtain good contrast between the gray and white matter [21].
7. Due to incomplete myelination, the FLAIR sequence is usually not performed in children below the age of 2 years (Fig. 7) [21].
8. Manipulation of the pixel values displayed in an image to gain optimal contrast (i.e., windowing) may be useful to detect subtle MCD, such as focal cordical dysplasia (FCD) [1, 16, 17].
9. MCD in infants may be associated with distinct MR appearances, different from the pattern seen in a matured brain [6, 7].
10. MRI may have limitations in detecting subtle cortical abnormalities or very small lesions, particularly in cases of FCD [19].

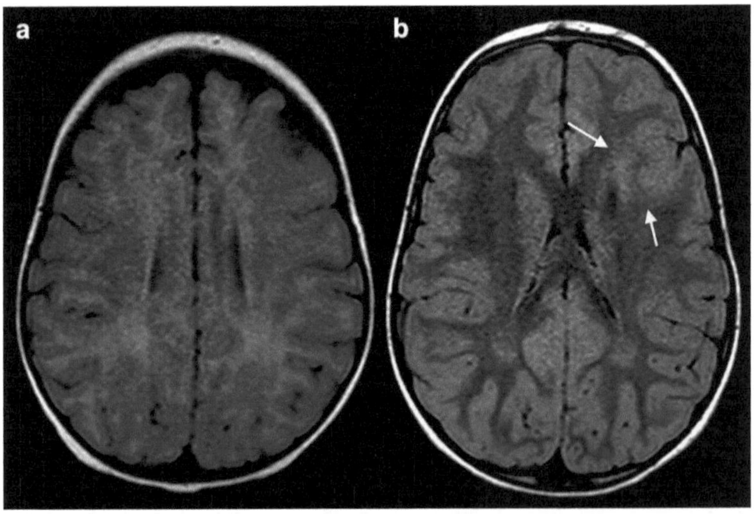

Fig. 7 FLAIR imaging in focal cortical dysplasia. Axial FLAIR images performed at 8 months (**a**) and 2.5 years of age (**b**) show that the right frontal cortical dysplasia (arrows) is not visible at 8 months of age due to incomplete myelination with poor contrast between gray and white matter

5 Conclusions

In conclusion, MRI can provide high-resolution images of the brain with excellent contrast resolution, allowing for detailed visualization of cortical anatomy and malformations. This technique represents the diagnostic gold standard for MCD and is the first radiological examination that should performed in children with developmental delay and/or epilepsy.

Acknowledgments

We thank the Unit of Neuroradiology of the Institute Giannina Gaslini for the technical support.

References

1. Barkovich AJ, Kuzniecky RI (2005) Neuroimaging of malformations of cortical development. Neurol Clin 23:569–581
2. Haller S, Haacke EM, Thurnher MM, Barkhof F (2021) Susceptibility-weighted imaging: technical essentials and clinical neurologic applications. Radiology 299:3–26
3. Logothetis NK (2008) What we can do and what we cannot do with fMRI. Nature 453: 869–878
4. Soares JM, Marques P, Alves V, Sousa N (2013) A hitchhiker's guide to diffusion tensor imaging. Front Neurosci 7:31
5. Tortora D, Cataldi M, Severino M, Consales A, Pacetti M, Parodi C, Sertorio F, Ramaglia A, Cognolato E, Nobile G, Mancardi M, Prato G, Siri L, Giacomini T, Striano P, Arnaldi D, Piatelli G, Rossi A, Nobili L (2022) Comparison of qualitative and quantitative analyses of MR-arterial spin labeling perfusion data for theassessment of pediatric patients with focal epilepsies. Diagnostics 12(4):811. https://doi.org/10.3390/diagnostics12040811. PMID: 35453858; PMCID: PMC9032819
6. Barkovich AJ, Guerrini R, Kuzniecky RI, Jackson GD, Dobyns WB (2012 May) A developmental and geneticclassification for malformations of cortical development: update 2012. Brain 135(Pt 5):1348–1369. https://doi.org/10.1093/brain/aws019. Epub 2012 Mar 16. PMID: 22427329; PMCID: PMC3338922
7. Severino M, Geraldo AF, Utz N, Tortora D, Pogledic I, Klonowski W et al (2020) Definitions and classification of malformations of cortical development: practical guidelines. Brain 143:2874–2894
8. Guerrini R, Dobyns WB (2014) Malformations of cortical development: clinical features and genetic causes. Lancet Neurol 13:710–726
9. Scala M, Nishikawa M, Ito H, Tabata H, Khan T, Accogli A et al (2022) Variant-specific changes in RAC3 function disrupt corticogenesis in neurodevelopmental phenotypes. Brain 145:3308–3327
10. Nishikawa M, Scala M, Umair M, Ito H, Waqas A, Striano P et al (2023) Gain-of-function p.F28S variant in RAC3 disrupts neuronal differentiation, migration and axonogenesis during cortical development, leading to neurodevelopmental disorder. J Med Genet 60:223–232
11. Scala M, Nishikawa M, Nagata KI, Striano P (2021) Pathophysiological mechanisms in neurodevelopmental disorders caused by Rac GTPases dysregulation: what's behind neuro-RACopathies. Cells 10:3395
12. McGowan JC (2008) Basic principles of magnetic resonance imaging. Neuroimag Clin N Am 18(4):623–636. https://doi.org/10.1016/j.nic.2008.06.004
13. Shah A, Aran S (2023) A review of magnetic resonance (MR) safety: the essentials to patient safety. Cureus 15(10):e47345. https://doi.org/10.7759/cureus.47345. PMID: 38021512; PMCID: PMC10657250
14. Jezzard P, Balaban RS (1995) Correction for geometric distortion in echo planar images from B0 field variations. Magn Reson Med 34:65–73

15. McGowan JC (2008) Basic principles of magnetic resonance imaging. Neuroimaging Clin N Am 18:623–636
16. Bernasconi A, Cendes F, Theodore WH, Gill RS, Koepp MJ, Hogan RE et al (2019) Recommendations for the use of structural magnetic resonance imaging in the care of patients with epilepsy: a consensus report from the international league against epilepsy neuroimaging task force. Epilepsia 60:1054–1068
17. Wang I, Bernasconi A, Bernhardt B, Blumenfeld H, Cendes F, Chinvarun Y, Jackson G, Morgan V, RamppS VAE, Federico P (2020) MRI essentials in epileptology: a review from the ILAE imaging taskforce. Epileptic Disord 22(4):421–437. https://doi.org/10.1684/epd.2020.1174
18. Larivière S, Federico P, Chinvarun Y, Jackson G, Morgan V, Rampp S, Vaudano AE, Wang I, Cendes F, Boelman CG, Bernasconi A, Bernasconi N, Bernhardt BC, Schrader DV (2021) ILAE neuroimaging task forcehighlight: harnessing optimized imaging protocols for drug-resistant childhood epilepsy. Epileptic Disord 23(5):675–681. https://doi.org/10.1684/epd.2021.1312
19. Spitzer H, Ripart M, Whitaker K, D'Arco F, Mankad K, Chen AA et al (2022) Interpretable surface-based detection of focal cortical dysplasias: a multi-centre epilepsy lesion detection study. Brain 145:3859–3871
20. Sammet S (2016) Magnetic resonance safety. Abdom Radiol (NY) 41:444–451
21. Rossi A, Argyropoulou M, Zlatareva D, Boulouis G, Pizzini FB, van den Hauwe L, Raissaki M, Pruvo JP, Rosendahl K, Hoffmann C, Sundgren PC, ESNR Pediatric Neuroradiology Subspecialty Committee (2023) ESPR Neuroradiology taskforce. European recommendations on practices in pediatric neuroradiology:consensus document from the European Society of Neuroradiology (ESNR), European Society of PaediatricRadiology (ESPR) and European Union of Medical Specialists Division of Neuroradiology (UEMS). Pediatr Radiol 53(1):159–168. https://doi.org/10.1007/s00247-022-05479-4. Epub 2022 Sep 5. PMID: 36063184;PMCID: PMC9816178
22. Vachha B, Huang SY (2021) MRI with ultrahigh field strength and high-performance gradients: challenges and opportunities for clinical neuroimaging at 7 T and beyond. Eur Radiol Exp 5:35

Chapter 24

Genomic Copy Number Analysis Using Droplet Digital PCR: A Simple Method with EvaGreen Single-Color Fluorescent Design

Takeaki Tamura, Taichi Imaizumi, Keiko Shimojima Yamamoto, and Toshiyuki Yamamoto

Abstract

Droplet digital PCR (ddPCR) is an emerging method for the absolute quantification of PCR products, and it can detect DNA copy numbers accurately. It analyzes the end-point absolute fluorescence signals of the PCR-positive droplets and calculates the target concentration. EvaGreen is a nonspecific double-stranded DNA-binding fluorescent dye, and the ddPCR system also supports assays using this cost-effective hydrolysis probe. Here, we describe a simple method of quantification for DNA copy numbers using the EvaGreen single-color fluorescent design.

Key words Droplet digital PCR, ddPCR, EvaGreen, DNA copy numbers, Single color, QuantaSoft software

1 Introduction

Genomic copy number variations (CNVs) are one of the major causes of pediatric neurodevelopmental disorders [1, 2]. Chromosomal microarray testing is recommended for the first-tier genetic testing of pediatric patients with neurodevelopmental disorders because it can comprehensively screen CNVs in high resolution [3]. Once possible pathogenic CNVs were detected through chromosomal microarray testing, it is required to analyze parental samples to confirm whether those were of de novo origin or not. Parental samples can be analyzed by chromosomal microarray testing, the same as patients' samples. However, it is not cost-effective to analyze only one candidate locus using chromosomal microarray testing. For this reason, several attempts have been made to quantify polymerase chain reaction (PCR) products. Quantitative PCR (qPCR) using a real-time PCR machine has been often used to

Koh-ichi Nagata (ed.), *Cerebral Cortex Development: Methods and Protocols*, Methods in Molecular Biology, vol. 2794, https://doi.org/10.1007/978-1-0716-3810-1_24,

overcome this problem. However, qPCR requires a calibration curve and at least analyzes thrice for more precise results because qPCR is a semiquantitative method [4, 5].

In comparison, droplet digital PCR (ddPCR) is a high-throughput quantification tool for PCR products [4], which analyzes the end-point absolute fluorescence signals of the PCR-positive droplets and calculates the target concentration. Therefore, ddPCR does not require calibration curves to determine the target concentrations and enables us to detect DNA copy number more sensitively and precisely than qPCR [6–8].

As with qPCR, a TaqMan hydrolysis probe is applicable to ddPCR. Thus, the copy number analysis through ddPCR is usually achieved with a two-color fluorescent TaqMan design, whereas, with the usage of EvaGreen, a nonspecific double-stranded DNA--binding fluorescent dye, the analysis could be performed through a single-color fluorescent design [9]. This approach is cost-effective, flexible, and easier to design because all we have to prepare for the day-to-day analyses in laboratories is only a pair of the target PCR primers. In our assay, the concentration of the targeted and reference genes for each single sample is analyzed in different wells and compared to determine the genomic copy numbers of the region of interest. Genomic copy numbers can be calculated using the reference genes as follows: genomic copy numbers = (A/B) × C, where A is the concentration of the target region, B is the concentration of the reference gene, and C is the copy number of the reference gene [10, 11].

ddPCR requires only a small amount of DNA for a single reaction. To make this powerful technology more applicable to daily operations in laboratories, we describe its intelligible operation manual. ddPCR assay starts with the PCR primer design and sample preparation. Through the droplet generation step, the PCR sample is divided into numerous independent partitions, which are named droplets, and these act as individual PCR microreactors. After PCR amplification, only in genomic DNA-contained droplets, the target sequence would be amplified. Those amplification-positive droplets are analyzed through the droplet reading step, and the concentration of the target sequence is determined using Poisson's statistics [4].

2 Materials

Prepare all solutions using ultrapure distilled water and analytical grade reagents. Prepare and store all reagents at room temperature except for the enzyme.

1. Template DNA: 10 ng/μL (*see* **Note 1**).
2. Region of interest target-specific PCR primer pair with forward and reverse: 50 μM (*see* **Note 2**).
3. Reference target PCR primer pair with forward and reverse: 50 μM (Table 1).
4. Tris–EDTA buffer (TE Buffer): 10 mM Tris–HCl, pH 8.0, 1 mM EDTA-2Na, pH 8.0.
5. QX200™ ddPCR™ EvaGreen Supermix (Bio-Rad Laboratories, California, USA).
6. QX200™ Buffer Control for EvaGreen (Bio-Rad Laboratories).
7. QX200™ Droplet Generation Oil for EvaGreen (Bio-Rad Laboratories) (*see* **Note 3**).
8. DG8™ cartridges for QX200™/QX100™ Droplet Generator (Bio-Rad Laboratories).
9. DG8 cartridge holder (Bio-Rad Laboratories).
10. DG8™ gaskets for QX200™/QX100™ Droplet Generator (Bio-Rad Laboratories).
11. Pierceable foil plate seals.
12. 96-well PCR plates (*see* **Note 4**).
13. 8-channel multi-pipette (*see* **Note 5**).
14. Filtered pipette tips.
15. PX1 PCR Plate Sealer (Bio-Rad Laboratories).
16. QX200™ Droplet Generator (Bio-Rad Laboratories).
17. QX200 Droplet Reader (Bio-Rad Laboratories).
18. Thermal cycler (*see* **Note 6**).
19. QuantaSoft software.

3 Methods

3.1 PCR Primer Design

1. Get the genomic sequence of the targeted genes. The human genome reference sequence is available on the UCSC Genome Browser (https://genome.ucsc.edu).
2. Choose the appropriate exon of the targeted genes, which is located in the target CNV region. Long exons (>200 nucleotides) would be better.
3. Design exon-specific primers; each primer should be designed inside the targeted exons. The appropriate PCR product size, primer size, primer melting temperature (Tm), and primer GC content (GC%) are as follows: PCR product size: 70–150 bp, primer size: 18–24 bp (optimum: 20 bp),

Table 1
Summary of the primer sequences provided for reference genes

The reference genes	Sense primer sequence	Anti sense primer sequence	Product length (bp)
RPP30	5′-GATTTGGACCTGCGAGCG-3′	5′-GCGGCTGTCTCCACAAGT-3′	62
RPPH1	5′-GTCAGACTGGGCAGGAGATG-3′	5′-TGGCCGTGAGTCTGTTCC-3′	75
AR	5′-CCAGCAGAAATGATTGCACTA-3′	5′-CATTTCCGAAGACGACAAGA-3′	70
XIST	5′-TGAGACCTGAGGACTGCAAA-3′	5′-AGCTTGGCCAGATTCTCAAA-3′	77

Table 2
Appropriate primer settings

Product size range (bp)	70 ~ 150
Primer size (bp)	18 ~ 24 (optimum: 20)
Primer Tm (°C)	58 ~ 60 (optimum: 60)
Primer GC% (%)	40 ~ 60 (optimum: 50)

primer Tm: 58–60 °C (optimum: 60 °C), and primer GC%: 40–60% (optimum: 50%), summarized in the Table 2 (*see* **Note 7**).

4. Check the specificity of a primer pair using the UCSC Genome Browser BLAT (https://genome.ucsc.edu/cgi-bin/hgBlat).

3.2 Sample Preparation (Fig. 1, Table 3)

Carry out all procedures at room temperature. To prevent contamination, use filter tips and perform procedures inside a clean bench:

1. Prepare the ddPCR primer mix: Mix 1 μL of forward primer (50 μM), 1 μL of reverse primer (50 μM), and 48 μL of TE Buffer (*see* **Note 8**).
2. Prepare the pre-mix: If the number of the target is "n," prepare "n + 1" volumes. For one tube; mix 10 μL of EvaGreen Supermix, 7 μL of nuclease-free water, and 1 μL of DNA (10 ng/μL).
3. Divide exactly same amount of pre-mix with 18 μL into separate 1.5 mL microtubes.
4. Add 2 μL of the ddPCR primer mix (the targets and the reference) into each tube.
5. Mix up the reaction mix using a vortex mixer (*see* **Note 9**).

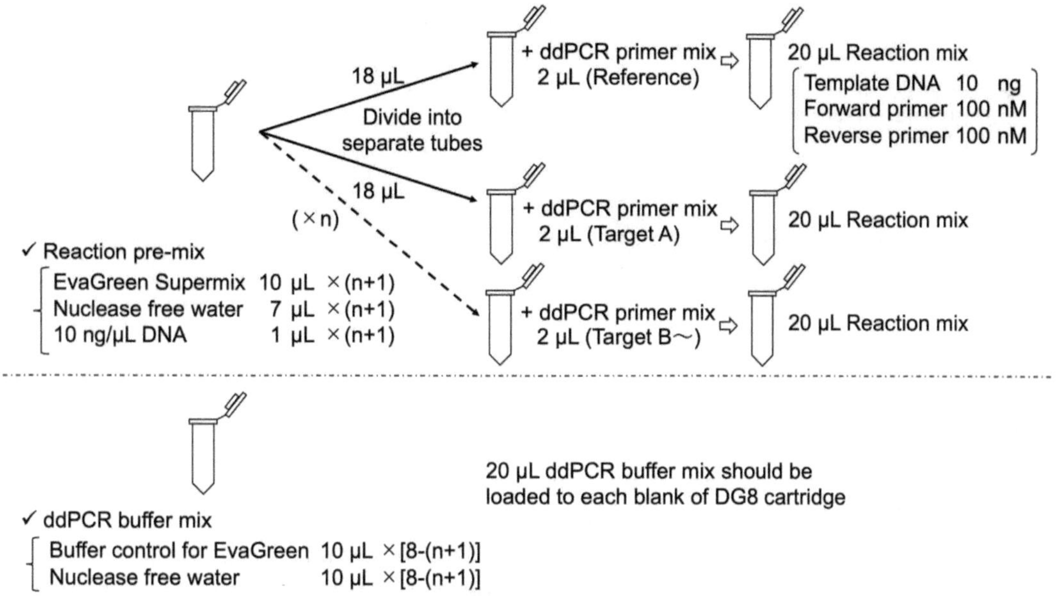

Fig. 1 Schema for sample preparation: how to prepare the Reaction mix and the ddPCR buffer mix. The number of targeted genes is indicated as "n." The concentration of target genes should be compared with a reference gene; thus, the concentration of at least one reference gene should be analyzed simultaneously in the same reaction. A maximum of eight samples could be applied to a DG8 cartridge. The unused wells in the DG8 cartridge should be filled with 20 μL ddPCR buffer mix. Thus, the number of blanks to be filled with 20 μL ddPCR buffer mix is indicated as "8-(n + 1)"

Table 3
Content of final reaction mix

EvaGreen supermix	10 μL
Nuclease free water	7 μL
10 ng/μL DNA	1 μL
ddPCR primer mix	2 μL
Total	20 μL

6. If less than 8 wells of DG8 cartridge are planned to be used, prepare "8-(n + 1)" volumes of ddPCR buffer mix for blank cartridges. For each blank, ddPCR buffer mix requires 10 μL of Buffer Control for EvaGreen and 10 μL of nuclease-free water (*see* **Note 10**).
7. Mix up the buffer mix using a vortex mixer.

3.3 Droplet Generation (Fig. 2, See Note 11)

1. Set a DG8™ cartridge for QX200™/QX100™ Droplet Generator in the cartridge holder.
2. Load 20 μL of the reaction mixture into the middle lane of the DG8™ cartridge, starting from the left well.

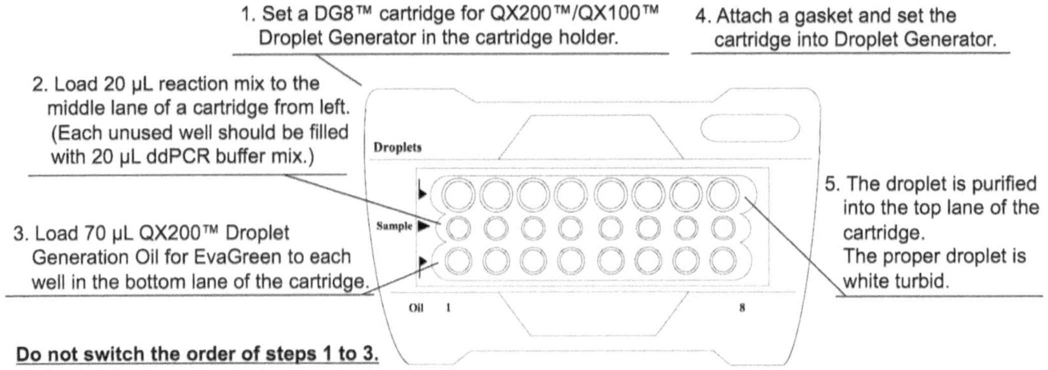

Fig. 2 Order of running the droplet generation

3. All eight wells in the same lane should be filled with the samples. Otherwise, the unused blank wells should be filled with 20 µL of the ddPCR buffer mix.
4. Load 70 µL of QX200™ Droplet Generation Oil for EvaGreen into the lower lanes of the cartridge (*see* **Note 12**).
5. Attach a gasket on the top of the DG8™ cartridge to cover the wells.
6. Place the DG8™ cartridge into the QX200™ Droplet Generator. After closing the slide cover of the generator, the droplet generation starts automatically. The right LED indicator flickers and changes to lit when the droplet generation is complete.
7. Remove the gasket, and the generated droplet will be found in the wells of the top lane of the cartridge. The proper droplet is white turbid.

3.4 PCR (See Note 13)

1. The droplets should be carefully transferred to a 96-well PCR plate by pipetting with 8-channel multi-pipette (*see* **Note 14**).
2. Insert the tips of the multi-pipette vertically to the bottom of the DG8 cartridge wells, tilt 30–45°, and aspirate droplets in about 3–5 s at a uniform speed.
3. Load 40 µL of droplets into a 96-well PCR plate. Discharge the droplets to the 96-well PCR plate in about 3–5 s at a uniform speed (*see* **Note 15**).
4. Fill the top two wells of the leftmost row of the three rows of the divisible 96-well PCR plate with 40 µL of QX200™ Droplet Generation Oil for EvaGreen (Fig. 3).
5. The PCR plate should be heat sealed with the usage of PX1™ PCR Plate Sealer; cover the PCR plate with PCR Plate Heat Seal with the red line in a seal on the top (*see* **Note 16**).
6. Place the PCR plate in a thermal cycler and run with the following recommended thermal cycling protocol: 5 min at

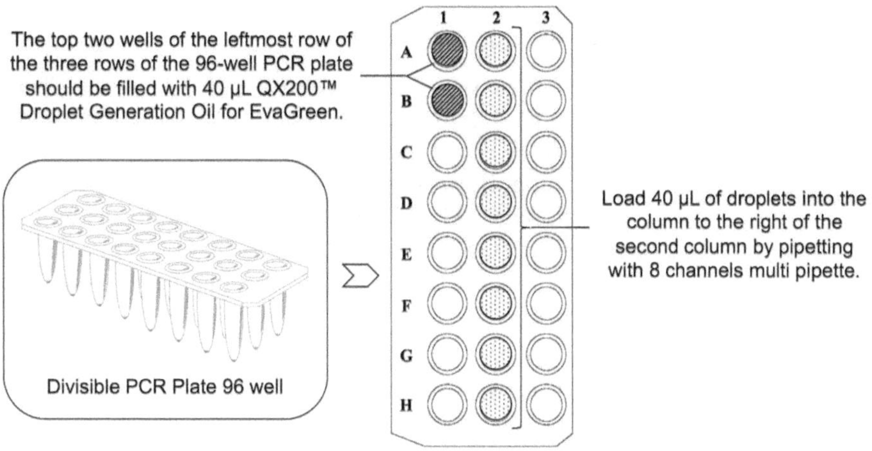

Fig. 3 appropriate wells of the 96-well PCR plate where the generated droplets should be loaded

Table 4
Recommended thermal cycling protocol for EvaGreen assays

Temparature, °C	Time	Ramp rate	Number of cycles
95	5 min	2 °C/s	1
95	30 s		40
60	1 min		
4	5 min		1
90	5 min		1
4	Infinite		1

95 °C, 40 cycles [30 s at 95 °C, 1 min at 60 °C], 5 min at 4 °C, 5 min at 90 °C, and an infinite hold at 4 °C. Use a 2 °C/sec ramp rate (Table 4) (*see* **Note 17**).

3.5 Setting Up QuantaSoft™ Software

1. Start up the QX200 Droplet Reader and switch on the attached computer. Launch QuantaSoft software.
2. The well settings: To use the well editor and define the settings for the wells, double-click on the well (*see* **Note 18**). Enter the well settings according to the arrangement of the samples on the 96-well PCR plate.
3. Fill the Sample panel: Enter the sample name, select the Experiment from the stepdown menu, and choose "ABS" (Absolute Quantification). Select the Supermix from the stepdown menu, and choose "QX200 ddPCR EvaGreen Supermix."
4. Fill the Target 1 panel: Enter the sample name, select the Type from the stepdown menu, and choose "Ch1 Unknown." When

both sample and target 1 panel are completed, click apply > OK (*see* **Note 19**).

5. When the experiment is running with EvaGreen® assays, the top two wells of the leftmost rows of the three rows of the 96-well PCR plate are filled with the Droplet Generation Oil. Regarding these two wells, in the Sample panel, select the Experiment "ABS" (Absolute Quantification) and select the Supermix "QX200 ddPCR EvaGreen Supermix." In the Target 1 panel, select the Type from the stepdown menu and choose "Blank."
6. When all the well settings are completed, click "Save as" and save the template. Before the experiment runs, the template must be saved at least once.

3.6 Droplet Reading (See Note 20)

1. Check the status indicator lights on the QX200 Droplet reader and ascertain whether the bottle-shaped indicator is solid green or flashing green. When the indicator is flashing green, the run less than 96 wells can be started.
2. Press the silver button in the middle of the green lid of the QX200 Droplet reader to open.
3. Place the 96-well PCR plate into the plate holder inside the QX200 Droplet reader. The plate holder is covered by the top. Lift up the release tabs located on both sides of the top of the plate holder to remove. Place the 96-well PCR plate into the base of the plate holder, in the position the well A1 of the plate must be on the top left. Cover with the top of plate holder and lay down both side tabs firmly (*see* **Note 21**).
4. Press the silver button in the middle of the green lid of the QX200 Droplet reader to close.
5. The system should be cleaned once in a day before any experiments are run. Press the "Flush system" button inside the "Instrument Routines" panel, located on the top right of the screen.
6. After the clean-up of the system is complete, press the "Run" button.

3.7 Data Analysis (See Note 22)

1. After the experiment has been completed, the data analysis window opens automatically.
2. Click the "Events" tab to check the number of droplets measured. Select the wells which you want to display from the well selector located at top right of screen (*see* **Note 23**). Check the "total" button displayed at the right side of the window, and the total droplet count will be displayed as a bar at the right side of the positive droplet count. When the total

number of the droplet count is less than 10,000/well, the droplets will not be auto-analyzed in the software.

3. Click the "1D Amplitude" tab to visualize the data of the fluorescence intensity of the droplets; the fluorescence intensity and the numbers of the positive and negative droplets can be displayed in the graph. The threshold line can be adjusted manually in this tab, and the threshold could be assigned either for single well or multiple wells.
4. In the "Concentration" tab, the result of the determined concentrations of the target sequences can be identified.

4 Notes

1. Template DNA concentration is critical. The concentration of genomic DNA should be quantified with the usage of appropriate DNA spectrophotometers. Nuclease-free water is recommended to adjust the template DNA concentration to 10 ng/μ L. In this assay, the final detected concentrations of the reference and target genes are compared, and consequently the genomic copy number of the region of interest will be determined. Thus, the concentrations of genomic DNA contained in each 20 μL reaction mix must be equal. In the QX200 ddPCR system, the recommended amount of genomic DNA in each reaction is 1–120,000 copies/20 μL reaction. The digested DNA contained in 20 μL reaction mix should be less than 1 μg.
2. Not only the targeted genes but also at least one reference gene should be analyzed in the same reaction. The genes which will not be influenced by any CNV are recommended as the reference gene, for example, the ribonuclease P/MRP 30 kDa subunit gene (*RPP30*; MIM* 606115) or the ribonuclease P RNA component H1 gene (*RPPH1*; MIM*608513) for the reference for the autosomal chromosomes, and the X inactivation–specific transcript gene (*XIST*; MIM* 314670) or the androgen receptor gene (*AR*; MIM* 313700) for the reference for the X-chromosome [10]. The primer sequences for the reference genes are summarized in Table 1.
3. When the oil appears opaque, it is not recommended to be used.
4. Divisible 96-well PCR plates are also available for this assay and are easier to handle.
5. Pipet-Lite Multi Pipette L8-50XLS+ (Mettler Toledo) and Pipette Tips RT LTS 200 μL F 960A/10 (Mettler Toledo) are recommended for use. To prevent the fracture of droplets,

the maximum volume of the multi-pipette is recommended to be 50 μL or less.

6. In this assay, the thermal cycling protocol requires appropriate ramp rate. Prepare PCR thermal cyclers which can adjust the ramp rate.
7. Use the web interface (e.g., Primer3Plus: https://www.primer3plus.com) to design the primers. The pair of primers that exhibit complementarity between 3′ ends might result in primer dimers, and it could prevent PCR amplification. Avoid primer dimers as well as any secondary structures such as hairpins, self-dimers, and cross-dimers, when possible. It also could be checked using the web interface (Primer3Plus: https://www.primer3plus.com). The fluorescent amplitude of EvaGreen is reported to be varied based on the amplicon length [9]; thus, it is recommended that the amplicon lengths of the target and reference be not much different.
8. The recommended final concentration of each primer in the reaction mix (20 μL) is 100–200 nM. If 2 μL ddPCR primer mix is added into the reaction mix, the final concentration of the primer in the reaction mix will be 100 nM.
9. The reaction mix (EvaGreen Supermix) foams easily; do not tap the tubes to mix.
10. Please calculate the appropriate total amount of ddPCR buffer mix; each unused well in a DG8 cartridge needs to be filled with 20 μL ddPCR buffer mix. Buffer Control for EvaGreen also foams easily; do not tap the tubes to mix.
11. In this step, the PCR sample is divided and partitioned into 20,000 independent droplets. Each droplet contains either a few or no genomic DNAs [4].
12. Be careful not to introduce air bubbles when pouring the Reaction Mix or Buffer Mix into the cartridge wells. To prevent bubbles from being mixed into the bottom of the cartridge wells, samples should be loaded along the walls of the wells. After loading samples, it would be better to remove visible bubbles with the point of microtips. At least, the visible bubbles should not remain at the bottom of the wells.
13. In this step, the target sequence will be amplified in the DNA-contained droplets. Each droplet acts as an independent PCR microreactor [4].
14. For this purpose, Pipet-Lite Multi Pipette L8-50XLS+ (Mettler Toledo) is strongly recommended for use.
15. After transferred to the 96-well PCR plate, the proper droplets are confirmed to be separated into two layers. Each liquid level of the wells might be slightly different.

16. The appropriate temperature and time in order to heat seal the PCR plate might be different from materials to be used. For example, when the Eppendorf twin.tec® PCR Plate 96 (divisible; Eppendorf, Hamburg, Germany) is planned to be heat sealed with PCR Plate Heat Seal (foil, pierceable; Bio-Rad Laboratories), 180 degrees for 5 seconds is appropriate.
17. The PCR plate can be stored at 4 °C for 3–4 days after a PCR run.
18. In case of the usage of the same settings as in the previous run, load and copy the saved settings from another experiment. Click Template > Load. Copy the settings of the optional wells using Ctrl + C, then open the new project setting, and paste using Ctrl + V.
19. When the experiment is running with EvaGreen® or TaqMan (FAM) assay, fill the Target 1 panel. The Target 2 panel is used when the fluorescence of hexachlorfluorescein (HEX) or V fluorescent dye (VIC, LifeTechnologies) is measured. In case of repeated usage of the settings of the optional wells, copy that and paste to the other well, click and select the well, copy using Ctrl + C, and paste using Ctrl + V. You can select several contiguous wells using shift + click. Non-contiguous wells can be selected using Ctrl + click.
20. In the droplet reader, each individual droplet will be analyzed. Fluorescence from positive droplets, which contain at least one copy of the target sequence, is measured to quantify the concentration of the target.
21. In the case of using a separated 96-well PCR plate, the plate must be placed in the middle of the plate holder. In the case of the separated plate on the far left, place another separated 96-well PCR plate on the opposite side of the plate holder as a balancer.
22. With the usage of QuantaSoft software, the fluorescence intensity and the number of each positive and negative droplet can be viewed as a 1D plot. The concentration of the target sequence is determined using Poisson's statistics.
23. Multiple wells can be used simultaneously. In such cases, select all wells in the row/column by clicking the well at the top of the row/column.

References

1. Harel T, Lupski JR (2018) Genomic disorders 20 years on-mechanisms for clinical manifestations. Clin Genet 93(3):439–449
2. Parenti I, Rabaneda LG, Schoen H et al (2020) Neurodevelopmental disorders: from genetics to functional pathways. Trends Neurosci 43(8):608–621
3. Miller DT, Adam MP, Aradhya S et al (2010) Consensus statement: chromosomal microarray is a first-tier clinical diagnostic test for

individuals with developmental disabilities or congenital anomalies. Am J Hum Genet 86(5):749–764

4. Quan PL, Sauzade M, Brouzes E (2018) dPCR: a technology review. Sensors (Basel) 18(4):1271
5. Bustin SA, Benes V, Garson JA et al (2009) The MIQE guidelines: minimum information for publication of quantitative real-time PCR experiments. Clin Chem 55(4):611–622
6. Butchbach ME (2016) Applicability of digital PCR to the investigation of pediatric-onset genetic disorders. Biomol Detect Quantif 10: 9–14
7. Hindson BJ, Ness KD, Masquelier DA et al (2011) High-throughput droplet digital PCR system for absolute quantitation of DNA copy number. Anal Chem 83(22):8604–8610
8. Hindson CM, Chevillet JR, Briggs HA et al (2013) Absolute quantification by droplet digital PCR versus analog real-time PCR. Nat Methods 10(10):1003–1005
9. Miotke L, Lau BT, Rumma RT et al (2014) High sensitivity detection and quantitation of DNA copy number and single nucleotide variants with single color droplet digital PCR. Anal Chem 86(5):2618–2624
10. Imaizumi T, Yamamoto-Shimojima K, Yamamoto H et al (2020) Establishment of a simple and rapid method to detect *MECP2* duplications using digital polymerase chain reaction. Congenit Anom (Kyoto) 60(1):10–14
11. Imaizumi T, Yamamoto-Shimojima K, Yamamoto T (2019) Advantages of ddPCR in detection of *PLP1* duplications. Intractable Rare Dis Res 8(3):198–202

Chapter 25

Locomotor Assays in *Drosophila* Larvae and Adult Flies

Leo Tsuda, Huynh Man Anh, and Hideki Yoshida

Abstract

Brain defects often lead to motor dysfunctions in humans. *Drosophila melanogaster* has been one of the most useful organisms in the study of neuronal biology due to its similarities with humans and has contributed to a more detailed understanding of the effects of genetic dysfunctions in the brain on behavior. We herein present modified protocols for the crawling assay with larvae and the climbing assay with adult flies that are simple to perform as well as a series of commands for ImageJ to automatically analyze data for the crawling assay.

Key words *Drosophila* larval and adult locomotor ability, Crawling assay, Climbing assay, Automatic analysis

1 Introduction

1.1 Crawling Assay Using Larvae

Larval locomotor is executed by commands from the brain that control muscles via peripheral nerves. To examine the effects of gene dysfunctions in the central nervous system, drug treatments, toxic substances, and changes in the rearing environment on behavior in *Drosophila*, the larval locomotor assay is a simple and useful method that does not require expensive equipment [1–4]. The assay reported by Connolly et al. to measure larval locomotor ability in 1975 [5] is employed with some modifications [1–4]. A recent significant improvement is the modification of the ImageJ/FIJI plugin wrMTrck, originally designed to analyze *Caenorhabditis elegans* behaviors, which tracks *Drosophila* larvae and analyses the length of the traveling path and crawling speed from the data obtained [6–11]. We herein provide the macro to automatically analyze larval crawling video data using the ImageJ plugin wrMTrck.

1.2 Climbing Assay Using Adult Flies

The climbing assay using *Drosophila* adult flies has contributed to the study of neurodegenerative diseases, such as Alzheimer's disease, Parkinson's disease, and amyotrophic lateral sclerosis [12–

Koh-ichi Nagata (ed.), *Cerebral Cortex Development: Methods and Protocols*, Methods in Molecular Biology, vol. 2794, https://doi.org/10.1007/978-1-0716-3810-1_25,

14]. This assay utilizes the property "negative geotaxis," namely the movement of flies in the opposite direction to gravity, and has long been used as a behavioral test of flies [15–19]. When adult flies are tapped to the bottom of a plastic vial, they will climb back up toward the top. A camera records flies climbing up the side of the vial, and the number of flies that reach a certain distance from the bottom within a specified time limit is recorded and scored. Since the regulation of working speed is supported by the functions of the complex neural network of the central nervous system, abnormalities in the central nervous system may be easily detected using the climbing assay [20]. We herein provide the basic method to study climbing activity.

2 Materials

2.1 Crawling Assay Using Larvae

1. Video camera.
2. Tripod.
3. Trace stand (recommended).
4. Deionized water.
5. 2% agar plate made on a 15 cm plastic Petri dish (*see* **Note 1**).
6. Soft paintbrush.
7. Wandering third instar larvae.
8. Plastic containers.
9. Stereo microscope.
10. ImageJ/FIJI 1.53 t.
11. Plugin wrMTrck.
12. The macro, which is a program to automate a series of ImageJ commands, is provided in this chapter (*see* **Note 2**).
13. Video converter to convert video files. We recommend the free MOV to AVI converter from Pazera software.

2.2 Climbing Assay Using Adult Flies

1. Cylindrical plastic tube (diameter of 2 cm and height of 10 cm) with a line drawn 5 cm from the bottom of the tube with a marker.
2. Alternatively, a white background with a line drawn 5 cm from the bottom.
3. Sponge lids.
4. Stopwatch.
5. Video camera.

3 Methods

3.1 Crawling Assay Using Larvae

1. Rear larvae under adequate conditions, such as sufficient space to prevent overcrowding (*see* **Note 3**).
2. Set up the video camera, 2% agar plate, and ruler as shown in Fig. 1 (*see* **Note 4**).
3. Transfer wandering larvae from a culture vial with a wet soft paintbrush to the container, wash them with deionized water to quickly remove food, and then transfer the larvae to another container (*see* **Note 5**).
4. Select male larvae under a stereo microscope (*see* **Note 6**).
5. Transfer less than four larvae with a wet soft paintbrush to the center of the 2% agar plate after removing excess deionized water with tissue paper (*see* **Note 7**).
6. Take a video with a video camera for 1 min after all larvae start to move (Fig. 2).

Fig. 1 The equipment to take a video of crawling larvae

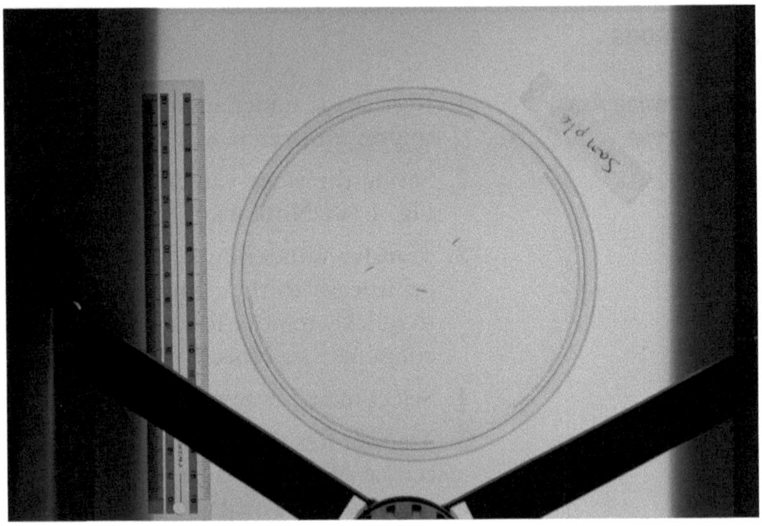

Fig. 2 A scene from a video recording of crawling larvae

7. Analyze movie files using the FIJI/ImageJ plugin wrMTrck to perform the procedure described below.
8. Convert the movie files (MOV) to AVI files and keep the files in one folder.
9. Open FIJI/ImageJ.
10. Reopen the macro file (IJM file) we provide, press "Run," and select the folder containing the AVI files.
11. Assign the length of 5 cm of the ruler on the image and press "OK."
12. The results of the analysis as an Excel file and image files of the travel path of larvae as a TIFF file are saved in the same folder containing the AVI files.
13. Perform appropriate statistical analyses with the data obtained (*see* **Note 8**).

3.2 Climbing Assay Using Adult Flies

1. Groups of 20 flies are placed in an empty vial.
2. Leave them to stand for 10 min.
3. Set up the video camera and vials as shown in Fig. 3.
4. Start video recording.
5. Flies are gently tapped to the bottom of the plastic vial and allowed to stand.
6. Immediately start to count the time.
7. After 10 s, video recordings are stopped. Allow flies to stand for 3 min.
8. Restart video recordings.

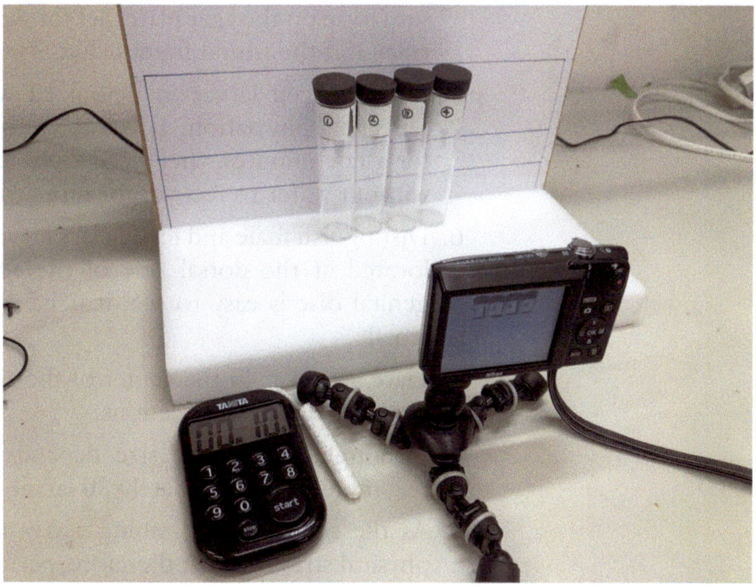

Fig. 3 The equipment to take a video of climbing adult flies

9. The number of flies crossing the 5 cm mark is recorded.
10. Experiments are performed with 4 vials, using a total of 80 animals, and the average value of 2 experiments is calculated (*see* **Note 9**).

4 Notes

1. One gram of agar is melted in 50 mL of deionized water with heating in a microwave oven. During heating in the microwave, the solution needs to be stirred sometimes to avoid sudden boiling. When the agar is completely melted, pour the solution into a 15 cm plastic Petri dish.
2. The macro (IJM file) is present in the following URL: https://drive.google.com/file/d/1xfE7hPs7jKNhqpqY9FPqKr4JOJiedeZj/view?usp=drive_link.
3. To minimize genetic background effects, flies need to be outcrossed with an appropriate fly line at least five times before performing the crawling assay.
4. The resolution for the video camera depends on the diameter of the agar plate (minimum of approximately 3 px/mm). We recommend a resolution of 480 px for a 9 cm plastic Petri dish plate and 720 px for a 15 cm plastic Petri dish plate. The brightness of the trace stand needs to be adjusted to clearly

see larvae on the agar plate. Set the agar plate not to overlap the ruler and the tripod frames (Fig. 2).

5. Do not keep larvae in deionized water for a long time to prevent asphyxiation. If larvae are placed in deionized water for a few minutes, they need to be kept in a plastic container without water to recover for more than 5 min.
6. Distinguish male and female larvae with genital discs, which are located at the dorsal side of the abdomen. While the male genital disc is easy to see, it is hard to recognize that of the female.
7. Place all larvae in the center of the agar plate with their heads facing in different directions.
8. Although the sample size depends on the experiment, we recommend approximately 30 larvae for one strain.
9. As the number of climbing assays increases, toxicity due to physical stress occurs; therefore, perform as few assays as possible. Furthermore, since the climbing efficiency of flies varies depending on the condition of the plastic container (e.g., old vs. new), these conditions need to be unified.

References

1. Shaver SA, Riedl CAL, Parkes TL, Sokolowski MB, Hilliker AJ (2000) Isolation of larval behavioral mutants in *Drosophila melanogaster*. J Neurogenet 4:193–205
2. Nichols CD, Becnel J, Pandey UB (2012) Methods to assay *Drosophila* behavior. J Vis Exp 61:3795
3. Dhar G, Mukherjee S, Nayak N, Sahu S, Bag J, Rout R, Mishra M (2020) Various behavioural assays to detect the neuronal abnormality in flies. In: Fundamental approaches to screen abnormalities in *Drosophila*. Springer, New York, pp 223–251
4. Even A, Morelli G, Turchetto S, Shilian M, Bail RL, Laguesse S, Krusy N, Brisker A, Brandis A, Inbar S, Chariot A, Saudou F, Dietrich P, Dragatsis I, Brone B, Broix L, Rigo JM, Weil M, Nguyen L (2021) ATP-citrate lyase promotes axonal transport across species. Nat Commun 12:5878
5. Sewell D, Burnet B, Connolly K (1975) Genetic analysis of larval feeding behaviour in *Drosophila melanogaster*. Genet Res Camb 24: 163–173
6. Rasband (1997–2018) WS ImageJ, U. S. National Institutes of Health, Bethesda. https://imagej.nih.gov/ij/
7. Abramoff MD, Magalhaes PJ, Ram SJ (2004) Image processing with ImageJ. Biophoton Int 11:36–42
8. Schneider CA, Rasband WS, Eliceiri KW (2012) NIH image to ImageJ: 25 years of image analysis. Nat Methods 9:671–675
9. Schindelin J, Arganda-Carreras I, Frise E, Kaynig V, Longair M, Pietzsch T, Preibisch S, Rueden C, Saalfeld S, Schmid B, Tinevez JY, White DJ, Hartenstein V, Eliceiri K, Tomancak P, Cardona A (2012) Fiji: an open-source platform for biological-image analysis. Nat Methods 9:676–682
10. Husson SJ, Costa WS, Schmitt C, Gottschalk (2013) A keeping track of worm trackers. WormBook:1–17
11. Brooks DS, Vishal K, Kawakami J, Bouyain S, Geisbrecht ER (2016) Optimization of wrMTrck to monitor *Drosophila* larval locomotor activity. J Insect Physiol 93-94:11–17
12. Feany MB, Bender WW (2000) A *Drosophila* model of Parkinson's disease. Nature 404:394–398
13. Jahn TR, Kohlhoff KJ, Scott M, Tartaglia GG, Lomas DA, Dobson CM, Vendruscolo M, Crowther DC (2011) Detection of early locomotor abnormalities in a *Drosophila* model of Alzheimer's disease. J Neurosci Methods 197: 186–189

14. Borg R, Cauchi RJ (2013) The Gemin associates of survival motor neuron are required for motor function in *Drosophila*. PLoS One 8: e83878
15. Benzer S (1967) Behavioral mutants of *Drosophila* isolated by countercurrent distribution. Proc Natl Acad Sci USA 58:1112–1111
16. Ganetzky B, Flanagan JR (1978) On the relationship between senescence and age-related changes in two wild-type strains of *Drosophila melanogaster*. Exp Gerontol 13:189–196
17. Gargano JW, Martin I, Bhandari P, Grotewiel MS (2005) Rapid iterative negative geotaxis (RING): a new method for assessing age-related locomotor decline in *Drosophila*. Exp Gerontol 40:386–395
18. Inagaki HK, Kamikouchi A, Ito K (2010) Methods for quantifying simple gravity sensing in *Drosophila melanogaster*. Nat Protoc 5:20–25
19. Madabattula ST, Strautman JC, Bysice AM, O'Sullivan JA, Androschuk A, Rosenfelt C, Doucet K, Rouleau G, Bolduc F (2015) Quantitative analysis of climbing defects in a *Drosophila* model of neurodegenerative disorders. J Vis Exp 100:e52741
20. Bidaye SS, Laturney M, Chang AK, Bidaye SS, Laturney M, Chang AK, Liu Y, Bockemühl T, Büschges A, Scott K (2020) Two brain pathways initiate distinct forward walking programs in *Drosophila*. Neuron 108:469–485

Chapter 26

Analyses of Genetic Regulation of the Nervous System in the Nematode *Caenorhabditis elegans*

Tzu-Ting Huang and Ikue Mori

Abstract

This chapter aims to provide a comprehensive overview of the methodologies available to dissect genetic regulation of the nervous systems in the nematode *Caenorhabditis elegans*. These techniques encompass genetic screens and genetic tools to unravel the spatial–temporal contribution of genes on neural structure and function. Unbiased genetic screens on random mutations induced by ethyl methanesulfonate (EMS) or target gene silencing by genome-wide RNA interference (RNAi) help progress our understanding of the genetic control of neural development and functions. Complement to unbiased genetic approaches, gene- and protein-targeted manipulation by Cre/LoxP recombination system and auxin-inducible degron (AID) protein degradation system, respectively, helps identify tissues/cells and the time window critical for gene and protein function during the proper execution of a particular behavior. Considering the remarkable conservation of genetic pathways between *C. elegans* and mammalian systems, elucidating the genetic underpinnings of neural functions and learning behaviors in *C. elegans* may furnish invaluable insights into analogous processes in more complex organisms. As shown in the following chapter, leveraging these diverse methodologies enable researchers to elucidate the intricate network governing neural function and structure, laying the foundation for innovating strategies to ameliorate cognitive alterations.

Key words *C. elegans*, Genetic screens, CRISPR/Cas9, Cre/LoxP, Auxin-inducible degron (AID) system

1 Introduction

In the 1950s, seminal discoveries in molecular biology of resolving DNA structures and its genetic codes transforming DNA sequences into protein sequences ignited a shift of focus toward understanding genomic encoding of complex phenotypes like neural development and behavioral plasticity. Foreseeing this shift before the postgenomic era inspired Sydney Brenner to select and establish the nematode *Caenorhabditis elegans* as an ideal organism for these investigations in laboratories [1]. The subsequent research dedicated to the whole-genome sequence of *C. elegans* has also

Koh-ichi Nagata (ed.), *Cerebral Cortex Development: Methods and Protocols*, Methods in Molecular Biology, vol. 2794,
https://doi.org/10.1007/978-1-0716-3810-1_26,

facilitated a solid research foundation for genetic regulation on neural development and behavioral plasticity [2–5].

To identify genes integral to neural functions and structures, unbiased forward genetic screens based on visible phenotypes are conducted to infer genetic encoding of neural morphology, locomotory, or learning behaviors [6–10]. Large-scale forward genetic screens typically entail the introduction of random mutations using mutagens such as ethyl methanesulfonate (EMS) or trimethylpsoralen (TMP). EMS and TMP induce mutations with rates of approximately 2.5×10^{-3} and 1×10^{-3}, respectively [11]. RNA interference (RNAi) is another effective technique to dissect gene functions in *C. elegans*. RNAi reduces the expression of targeted genes (gene silencing) in a sequence-specific manner by translation inhibition or mRNA degradation, allowing researchers to investigate the phenotypic consequences of gene knockdown. The simplicity of delivering double-stranded RNA to *C. elegans* via feeding, soaking, or injection makes RNAi a versatile tool for genome-wide screens as well as targeted gene studies [12–15]. Many genes essential for neural structures or functions have been isolated via genetic screens and confirmed with cell-specific rescue or RNAi-directed knockdown for the dissection of genetic control in thermosensory behavioral circuits [16–21].

Subsequent to the isolation of genes through phenotype-based genetic screens, the function of the gene can be confirmed by the interrogation of specific gene functions by deliberate gene lesions facilitated by the CRISPR/Cas9 system—clustered regularly interspaced short palindromic repeats associated with Cas9 nuclease. This potent genome-editing technique empowers researchers to induce precise insertions, deletions, or substitutions at defined genomic loci, thereby enabling sequence-customized *C. elegans* strains for spatiotemporal gene manipulations [22–24].

The following section provides general guidelines for genome editing in *C. elegans*. Specifically, we describe nucleotides insertion for LoxP sequences (34 bp) or degron sequences (132 bp) on gene loci to enable tissue-specific knockout by the Cre–LoxP recombination, or temporal protein degradation by auxin-inducible degron (AID) protein degradation system, respectively. This protocol is based mainly on the detailed protocols by Dokshin et al. in 2018 and Paix et al. in 2015 and proved effective in our hands.

2 Materials

1. Cre/LoxP system: A 34-bp LoxP sequence (5-′-ataacttcgtataatgtatgctatacgaagttat-3′), Cre recombinase gene inserted in pPD95.75 vector (*see* **Note 1**).

2. AID system: Auxin (4 mM indole-3-acetic acid) for TIR1 activity, nematode growth media (NGM) agar or TTX assay plates (0.3% NaCl, 2% Bacto agar, and 25 mM potassium phosphate, pH 6.0), a 132-bp degron sequence derived from the *Arabidopsis thaliana* IAA17 protein.
3. CRISPR/Cas9 system: *Streptococcus pyogenes* Cas9 nuclease protein (10 μg/μL), tracrRNA, crRNA (2 nmol or 10 nmol), DNA repair template (standard desalting; 4 nmol), single-stranded oligo DNA donor (ssODN), co-injection plasmid, PCR machine, PCR primers for genotyping (*see* **Note 1**). *C. elegans*: adult hermaphrodites for micro-injection and nematode growth media (NGM) for maintenance and cultivation of the animals [10].

3 Methods

3.1 Insertion of LoxP Sequences for Tissue-Specific Knockout by the Cre-LoxP Recombination

The Cre/LoxP recombination system enables the design of site-specific DNA recombination in target tissues. In this system, the Cre recombinase enzyme catalyzes the recombination between two directional LoxP sites, leading to the excision, inversion, or translocation of the DNA segment flanked by the two LoxP sequences (Fig. 1). Expressing exogenous Cre recombinase using tissue-specific promoters to conditionally knock out genes in specific tissues or at developmental stages provides valuable insights into gene function in a spatial–temporal context in *C. elegans* [25]. The following protocol describes how to insert two LoxP sequences in the target gene by CRISPR/Cas9-based genome editing:

1. Use two ssODN donors for inserting LoxP sites (<130 bp) at the 5′-end and 3′-end of the desired insertion locus, respectively. Each insertion donor has 35 bp homology sequences from the insertion locus flanking the LoxP sequence (Fig. 2).
2. If necessary, mutate the PAM (protospacer adjacent motif) site or the guide-binding sequence to avoid re-editing on the repaired locus.
3. When a mutation is introduced for re-editing prevention, the 35 bp homology arm should start from the last mutation.

3.2 Insertion of Degron Sequences for Temporal Protein Degradation by Auxin-Inducible Degron Protein Degradation System

The auxin-inducible degron (AID) protein degradation system is a relatively recent addition to the *C. elegans* genetic toolkit [26, 27]. Based on a plant hormone signaling pathway modified from *Arabidopsis*, the AID system enables rapid and reversible degradation of degron-targeted proteins in response to the small molecule auxin [28]. By synergistically manipulating the tissue-specific expression of the auxin-binding receptor TIR1 and applying auxin within a precise temporal window, the AID system

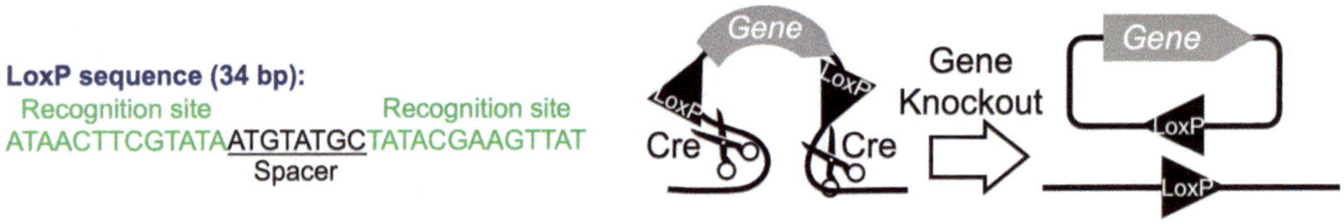

Fig. 1 Cre/LoxP recombination system. The Cre–Lox system operates through two key elements: a Cre recombinase and its target site, LoxP. The LoxP sequence comprises two Cre recognition sites with an eight-bp-long spacer sequence. When the Cre recombinase attaches to the LoxP sites flanking the targeted gene locus in the same direction, it forms a tetramer that permanently removes the gene of interest. Therefore, by driving the expression of Cre recombinase under tissue-specific promoters, we can achieve tissue-specific knockout of the desired gene

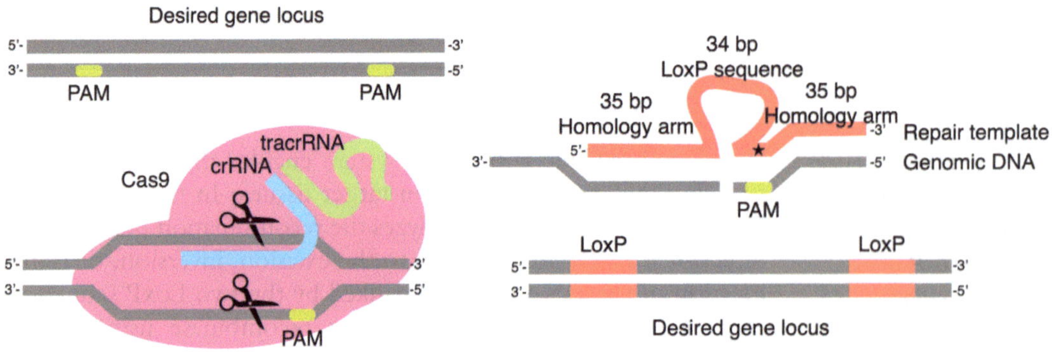

Fig. 2 Inserting two LoxP sites by CRISPR/Cas9 system. For short inserts less than 130 bp, flanking the insert with 35 bp homology arms is sufficient to give an efficient insertion rate in the progenies

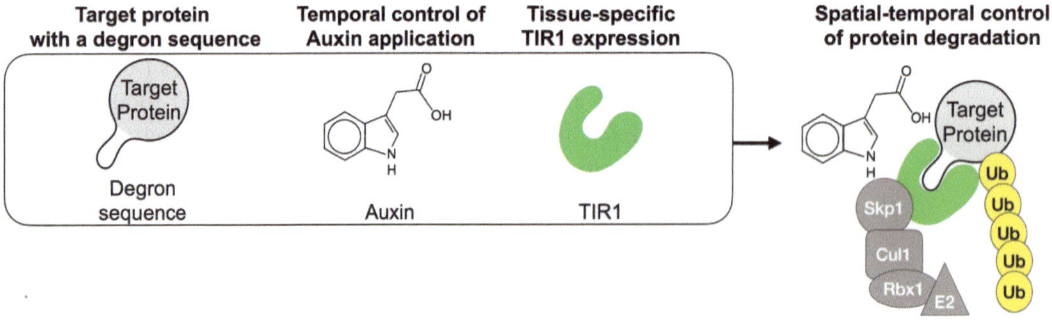

Fig. 3 Auxin-inducible degron protein degradation system. *Arabidopsis thaliana*-derived TIR1 F-box protein interacts with *C. elegans* endogenous components to form a functional E3 ubiquitin ligase. This E3 ubiquitin ligase forms and binds the degron-tagged protein only in the presence of the small molecule auxin. Upon temporal application of auxin and tissue-specific expression of TIR1, the degron-tagged proteins are poly-ubiquitinated and degraded by the proteasome in a spatial–temporal controlled manner

provides a mechanism of protein regulation at spatial and temporal precision (Fig. 3). Therefore, this method enables researchers to examine reversible and titratable phenotypic effects of protein depletion. The following protocol describes how to insert a degron

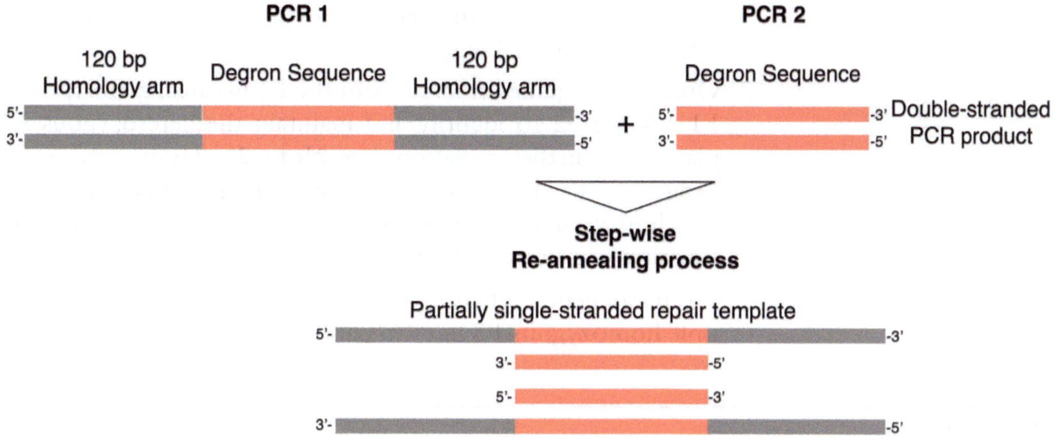

Fig. 4 Making a partially single-stranded repair template for degron insertion. For inserts longer than 130 bp, using a partially single-stranded repair template improves the insertion rates. Partially single-stranded repair templates can be generated by a stepwise cooling protocol

sequence upstream of the 5′-end or downstream of the 3′-end of the target gene by CRISPR/Cas9-based genome editing:

1. Order long oligonucleotide primers of 140 bp with 120 bp as homology arms and 20 bp complementary to the insert sequence. In addition, order standard primers complementary to the insert sequence without homology arms.
2. Set two PCR reactions: one with 120 bp homology arms and one only with insert sequence (Fig. 4), using a degron-containing plasmid as the PCR template.
3. Validate and purify the PCR products to achieve at least 300 ng/μL concentration for each.
4. Mix equal amounts of the purified PCR products from above and re-anneal the two PCR products to create a partially single-stranded DNA repair template via a stepwise cooling protocol detailed below in a regular PCR machine.

   Stepwise cooling protocol for re-annealing: 95 °C for 2: 00 min > > 85 °C for 10 sec > > 75 °C for 10 sec > > 65 °C for 10 sec > > 55 °C for 1:00 min > > 45 °C for 30 sec > > 35 °C for 10 sec > > 25 °C for 10 sec > > 4 °C on hold.
5. Add the donor mix and a co-injection plasmid as a selection marker for transgenic F1 progenies to the remaining injection mixture after pre-incubating Cas9, crRNA, and tracrRNA at 37 °C for 10 min.

3.3 Micro-injection and Screening

1. Inject the microinjection mixture pre-mixed in vitro, as described above, into approximately 20 adult hermaphrodites. After microinjection, transfer each P0 hermaphrodite onto a separate plate. After 3–3.5 days, identify and isolate F1 with the

visible marker from the co-injection plasmid onto individual NGM plates for F2 segregation.

2. Once F2 progenies emerge from the F1 isolates, genotype the F1 + some F2 to identify and maintain the heterozygous F2 plates for further isolation (*see* **Note 2**). Homozygous F2 progenies are likely to arise from F1 trans-heterozygotes, which may carry two independent edits. This step ensures the isolation of a homozygous allele from the edit.
3. From independent heterozygous F2, isolate and sequence multiple homozygous F3 lines to confirm correct edits.

4 Notes

1. Inserting two LoxP sequences can be made by adding the required and respective crRNAs and repair templates in one microinjection mixture.
2. Genotyping primers should be positioned outside the homology arms for later confirmation of the sequence by sequencing.

Acknowledgments

We thank Hironori J. Matsuyama and Moonsun Jang for their valuable input into this manuscript. This work was supported by KAKENHI grant number 19H05644 and Neuroscience Institute of the Graduate School of Science, Nagoya University, to I.M.

References

1. Brenner S (2003) Nobel lecture. Nature's gift to science. Biosci Rep 23:225
2. White JG, Southgate E, Thomson JN, Brenner S (1986) The structure of the nervous system of the nematode Caenorhabditis elegans. Philos Trans R Soc Lond B 314:1–340
3. C. Elegans Sequencing Consortium (1998) Genome sequence of the nematode C. elegans: a platform for investigating biology. Science (New York, N.Y.) 282:19981211
4. Cook SJ, Jarrell TA, Brittin CA et al (2019) Whole-animal connectomes of both Caenorhabditis elegans sexes. Nature 571:63
5. Witvliet D, Mulcahy B, Mitchell JK et al (2021) Connectomes across development reveal principles of brain maturation. Nature 596:257
6. Hedgecock EM, Russell RL (1975) Normal and mutant thermotaxis in the nematode Caenorhabditis elegans. Proc Natl Acad Sci USA 72:4061
7. Nakano S, Nakayama A, Kuroyanagi H, et al (2022) Genetic screens identified dual roles of microtubule-associated serine threonine kinase and CREB within a single thermosensory neuron in the regulation of Caenorhabditis elegans thermotaxis behavior 12
8. Mohri A, Kodama E, Kimura KD et al (2005) Genetic control of temperature preference in the nematode Caenorhabditis elegans. Genetics 169:1437
9. Perkins LA, Hedgecock EM, Thomson JN et al (1986) Mutant sensory cilia in the nematode Caenorhabditis elegans. Dev Biol 117:456
10. Brenner S (1974) The genetics of Caenorhabditis elegans. Genetics 77:71
11. Gengyo-Ando K, Mitani S (2000) Characterization of mutations induced by ethyl methanesulfonate, UV, and trimethylpsoralen in the nematode Caenorhabditis elegans. Biochem Biophys Res Commun 269:64

12. Tabara H, Grishok A, Mello CC (1998) RNAi in C. elegans: soaking in the genome sequence. Science 282:430–431
13. Fire A, Xu S, Montgomery MK et al (1998) Potent and specific genetic interference by double-stranded RNA in caenorhabditis elegans. Nature 391:806
14. Montgomery MK, Xu S, Fire A (1998) RNA as a target of double-stranded RNA-mediated genetic interference in Caenorhabditis elegans. Proc Natl Acad Sci U S A 95:15502
15. Kamath RS, Fraser AG, Dong Y et al (2003) Systematic functional analysis of the Caenorhabditis elegans genome using RNAi. Nature 421:231
16. Tanizawa Y, Kuhara A, Inada H et al (2006) Inositol monophosphatase regulates localization of synaptic components and behavior in the mature nervous system of C. elegans. Genes Dev 20:3296
17. Kuhara A, Inada H, Katsura I et al (2002) Negative regulation and gain control of sensory neurons by the C. elegans calcineurin TAX-6. Neuron 33:751
18. Satterlee JS, Sasakura H, Kuhara A et al (2001) Specification of thermosensory neuron fate in C. elegans requires ttx-1, a homolog of otd/Otx. Neuron 31:943
19. Hobert O, Mori I, Yamashita Y et al (1997) Regulation of interneuron function in the C. elegans thermoregulatory pathway by the ttx-3 LIM homeobox gene. Neuron 19:345
20. Okochi Y, Kimura KD, Ohta A et al (2005) Diverse regulation of sensory signaling by C. Elegans nPKC-epsilon/eta TTX-4. EMBO J 24:2127–2137
21. Chen YC, Chen HJ, Tseng WC et al (2016) A C. elegans thermosensory circuit regulates longevity through crh-1/CREB-dependent flp-6 neuropeptide signaling. Dev Cell 39:209
22. Dokshin GA, Ghanta KS, Piscopo KM et al (2018) Robust genome editing with short single-stranded and long, partially single-stranded DNA donors in caenorhabditis elegans. Genetics 210:781
23. Doudna JA, Charpentier E (2014) The new frontier of genome engineering with CRISPR-Cas9. Science 346:1258096
24. Paix A, Folkmann A, Rasoloson D et al (2015) High efficiency, homology-directed genome editing in Caenorhabditis elegans using CRISPR-Cas9ribonucleoprotein complexes. Genetics 201:47
25. Kage-Nakadai E, Imae R, Suehiro Y et al (2014) A conditional knockout toolkit for caenorhabditis elegans based on the Cre/loxP recombination. PLoS One 9:e114680
26. Negishi T, Kitagawa S, Horii N et al (2022) The auxin-inducible degron 2 (AID2) system enables controlled protein knockdown during embryogenesis and development in Caenorhabditis elegans. Genetics 220:iyab218
27. Zhang L, Ward JD, Cheng Z et al (2015) The auxin-inducible degradation (AID) system enables versatile conditional protein depletion in C. elegans. Development 142:4374
28. Nishimura K, Fukagawa T, Takisawa H et al (2009) An auxin-based degron system for the rapid depletion of proteins in nonplant cells. Nat Methods 6:917

Chapter 27

Analyses of Neural Circuits Governing Behavioral Plasticity in the Nematode *Caenorhabditis elegans*

Tzu-Ting Huang and Ikue Mori

Abstract

Behavioral plasticity is subjected to various sensory stimuli, experiences, and physiological states, representing the temporal and spatial patterns of neural circuit dynamics. Elucidation of how genes and neural circuits in our brain actuate behavioral plasticity requires functional imaging during behavioral assays to manifest temporal and spatial neural regulation in behaviors. The exploration of the nervous systems of *Caenorhabditis elegans* has catalyzed substantial scientific advancements in elucidating the mechanistic link between circuit dynamics and behavioral plasticity. The analyses of the nervous system of *C. elegans* have technologically flourished owing to the development of optogenetic instruments and fluorescent protein-based imaging compatible with its optically transparent body and the understanding of its completely revealed neural connectome and gene expression profiles at single-neuron resolution (The *C. elegans* Neuronal Gene Expression Map & Network, CeNGEN project). Using examples of the two temperature learning behaviors in *C. elegans*, this chapter delves into a selection of pivotal imaging tools, including genetically encoded calcium indicators, biosensors for second messenger imaging, and their usage in freely moving worms that have propelled our grasp of sensory representation in *C. elegans* neural circuits. To further connect the circuit dynamics to behavioral plasticity, this chapter will focus on technological advancements enabling simultaneous imaging and tracking system together with methodologies to quantify multiple behavioral elements of freely behaving *C. elegans* in a dynamic environment.

Key words *C. elegans*, Calcium imaging, Biosensors, Single-worm tracker, Multi-Worm Tracker (MWT)

1 Introduction

The translucent *Caenorhabditis elegans*, boasting comprehensively mapped nervous systems [1–3], is uniquely positioned for in vivo imaging for stimuli-evoked neural activation and artificial activation or inhibition by genetically encoded proteins that can be triggered by different wavelengths of light [4–7]. Genetically encoded calcium sensors such as GCaMP, yellow cameleon sensors, and their variants are indispensable imaging tools to monitor neural activity in vivo [8–11]. These fluorescent proteins change their fluorescent

Koh-ichi Nagata (ed.), *Cerebral Cortex Development: Methods and Protocols*, Methods in Molecular Biology, vol. 2794, https://doi.org/10.1007/978-1-0716-3810-1_27,

intensities proportionally to the dynamics of intracellular calcium concentration upon neural activation, providing a real-time indication of neuronal activity. GCaMP is especially notable due to its pronounced fluorescence alteration upon calcium binding. Conversely, the yellow cameleon proteins operate as a fluorescence resonance energy transfer (FRET)-based sensor, offering an alternative strategy for detecting calcium signals with dual-emission ratiometric imaging to correct artifacts from changes in focal planes and animals' movement.

Based on variant approaches for calcium indicators, several biosensors have been developed to gain insights into intracellular dynamics of second messengers in a range of sensory neurons in *C. elegans* and other organisms [12–14]. These genetically encoded sensors, frequently founded on the same fluorescent proteins as their calcium counterparts, modulate their fluorescence in response to cGMP or cAMP binding. Considering the pivotal role of second messengers in numerous sensory signaling and learning pathways [15–21], these biosensors offer a robust approach to the investigation of sensory transduction and associative learning in vivo.

Spatial and temporal dynamics of neural circuits manifest themselves in behavioral plasticity. One common strategy for unraveling the ultimate impact of neural dynamics on behaviors is to perturb neural activity by sensory stimuli or light-activating channels while simultaneously tracking the behavioral consequences. The small body size and compact neural circuit with a diverse range of behavioral repertoire of the nematode *C. elegans* empower it for the real-time correlation of neural dynamics with animal behaviors.

C. elegans exhibits two thermosensory learning behaviors: thermotaxis (TTX) and isothermal tracking (IT) behaviors, which are subjected to their feeding experience with a constant ambient temperature with food availability [22–24]. The association between food and temperature is reflected by the worm's tendency to migrate toward and keep tracking along the temperature corresponding to the cultivation temperature experienced with food availability (Fig. 1). Dissecting genetic pathways and circuit dynamics involved in thermotaxis and isothermal tracking have been instrumental in understanding the neural basis of sensory representation, sensorimotor integration, and the decision-making process for action selection [8, 10, 25, 26] and have led to significant advances to our understanding of how nervous systems perceive and transform environmental information into corresponding actions [27–33].

The quantitative analyses of thermosensory behaviors in *C. elegans* have been revolutionized by the development of the Multi-Worm Tracker (MWT) system, a technological advancement that enables the simultaneous tracking of multiple worms with a high temporal and spatial resolution [34]. The Multi-Worm Tracker, as detailed in the seminal work by Ikeda et al., automates the identification and trajectories of individual worms in a

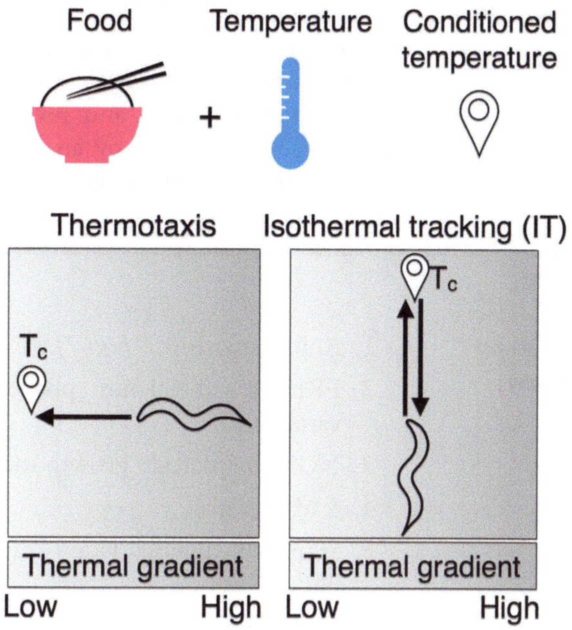

Fig. 1 Thermotaxis (TTX) and isothermal tracking (IT) behaviors in *C. elegans*. Thermotaxis and isothermal tracking behaviors in *C. elegans* are two experience-dependent thermosensory behaviors representing distinct behavioral features. TTX behavior represents learning and memory that changes the preference of an ambient temperature. IT behavior reflects the precision of the memory formation and retrieval on the tracking range in relation to the cultivation temperature

population and captures the dynamics of their thermotactic and isothermal tracking strategies on a temperature gradient over time [25, 29]. By correlating these behavioral strategies altered with specific genetic manipulations or acute neuronal activations/inhibition, the Multi-Worm Tracker provides a quantitative view of genetic and neural bases for the decision-making process in alternating behavioral strategies over time.

The subsequent section aims to present methodologies for recording the sensory activity from a single worm freely navigating on a temperature gradient based on the seminal works by Tsukada et al. in 2016 [10] and Nakano et al. in 2020 [29] and detail the quantitative methodologies employed in the Multi-Worm Tracker to dissect the temporal emergence of behavioral strategies of a worm population exploring on a temperature gradient [34].

The Multi-Worm Tracker (MWT) system detects and digitalizes the movement data from up to 150 worms at once (Fig. 4). It provides pivotal behavioral information about the worm population utilizing multiple parameters, including their centers of mass, body lengths, widths, outlines, and moving trajectories (Fig. 5). These data can be computed and analyzed into each behavioral strategy using a custom-built MATLAB script. Through digitized movement information, various behavioral parameters are obtained. These parameters can define the direction of movement, compute the curve, and identify behavioral elements such as omega turns, shallow turns, and reversals. Notably, when a reversal is immediately followed by an omega turn (within a 6 s interval), these two actions are grouped and classified as a reversal turn.

2 Materials

2.1 Single-Worm Tracking Assay

1. Animals with *lite-1(xu7)* background.
2. FRET-based calcium probe Yellow Cameleon 3.60 (or its variants).
3. NGM (nematode growth medium) plates.
4. 8 M glycerol.
5. Standard analog video camera with continuous halogen illumination.
6. Motorized stage.
7. Custom-built Galvano shutter system with a mercury lamp or SPECTRA light engine.
8. Macro microscope with high fluorescence efficiency.
9. Dual-view image splitter and EMCCD camera for fluorescence imaging.
10. MetaMorph software for calcium imaging regulation.

2.2 Multi-Worm Tracker Assay

1. Synchronized adult animals for the experiment.
2. NGM (nematode growth medium) plates.
3. NG buffer (0.3% NaCl, 1 mM CaCl2, 1 mM MgSO4, and 25 mM potassium phosphate, pH 6.0).
4. Assay plates (13.6 cm × 9.6 cm, 1.45 cm height).
5. 18 mL of assay medium (0.3% NaCl, 2% Bacto agar, and 25 mM potassium phosphate, pH 6.0) per assay plate.

2.3 Recording and Analysis in Multi-Worm Tracker

1. Camera.
2. PC installed with the Multi-Worm Tracker (MWT) program.
3. Worm-defining parameters file.
4. Black aluminum plate.
5. Ring LED light.
6. Kimwipes for excess buffer removal.
7. Cover glass.
8. MATLAB script for data processing and analysis.

3 Methods

3.1 Calcium Imaging and Single-Worm Tracking

1. Use the *lite-1(xu7)* mutant to minimize light-evoked behaviors from interfering with the tracking result (*see* **Note 1**).
2. Express the FRET-based calcium probe Yellow Cameleon 3.60 (or its variants) by neuron-specific promoters.
3. Pick L4 animals for a fresh NGM plate the day before imaging experiments and cultivate them at a stable temperature.
4. Cut a thin layer of 2% agar medium on the plastic dish into a 9 × 6 cm rectangle. Next, apply 8 M glycerol to the borders of the cut agar to keep the worms freely navigating within the agar field (*see* **Note 2**).
5. Acquire yellow fluorescent protein (YFP) and cyan fluorescent protein (CFP) fluorescence images using a dual-view image splitter and an EMCCD camera, taking images with 80 ms exposure time at 1 s intervals with 1 × 1 camera binning (Fig. 2).
6. Regulate and analyze calcium imaging using the software MetaMorph.
7. Simultaneously capture the freely moving animal at 30 frames per second (fps) using a standard analog video camera with continuous halogen illumination and a 600 nm long-pass filter (*see* **Note 3**).
8. Employ a high fluorescence efficiency macro microscope equipped with a device to provide sensory stimuli (for example, a temperature control device) and the motorized stage controlled by microscope-controlling software and MATLAB to track a single worm navigating on the imaging field.
9. Calculate the time course of sensory stimuli experienced by the moving animal based on the recorded x–y coordinates of the motorized stage and the stimuli map on the imaging field (for example, a temperature gradient) (Fig. 3).

3.2 Behavioral Analysis Using the Multi-Worm Tracker (MWT)

1. Activate the systems in the following sequence: Camera > > LED light > > PC (*see* **Note 4**).
2. Initiate the MWT program, selecting "Raw Image" as the format for the acquired image.
3. Load the worm-defining parameters file into the MWT program.
4. To avoid carrying over bacterial food, wash *C. elegans* twice with the NG buffer (containing 0.3% NaCl, 1 mM CaCl2,

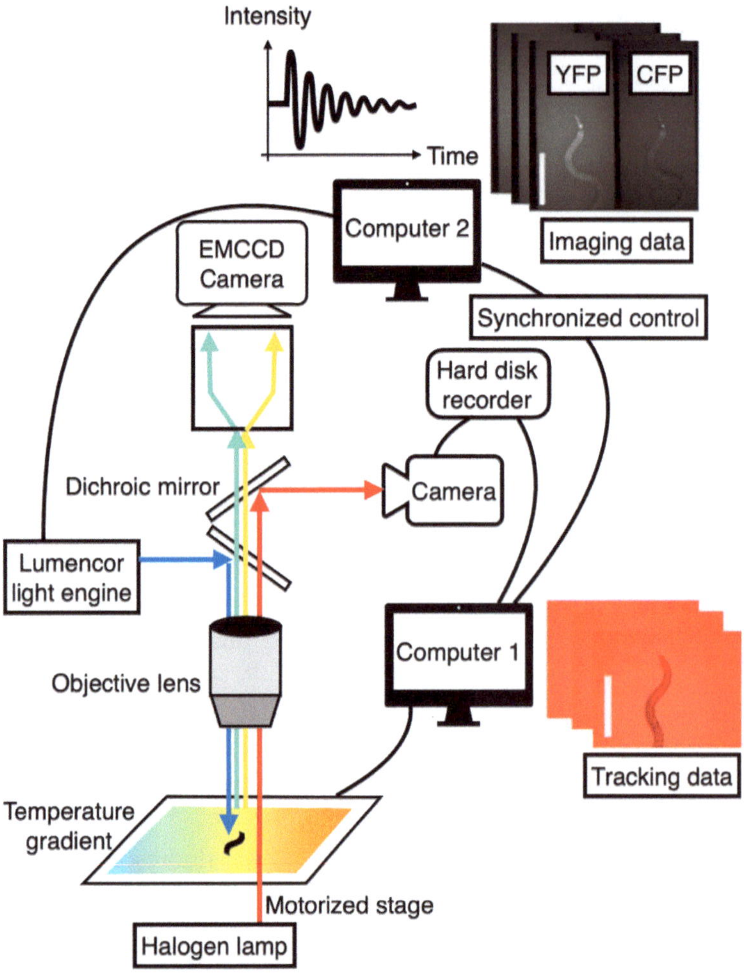

Fig. 2 Synchronized control for calcium imaging and tracking system. Synchronized control on calcium imaging and worm tracking systems ensures simultaneous recording of neural activity and worms' moving trajectory on a temperature gradient

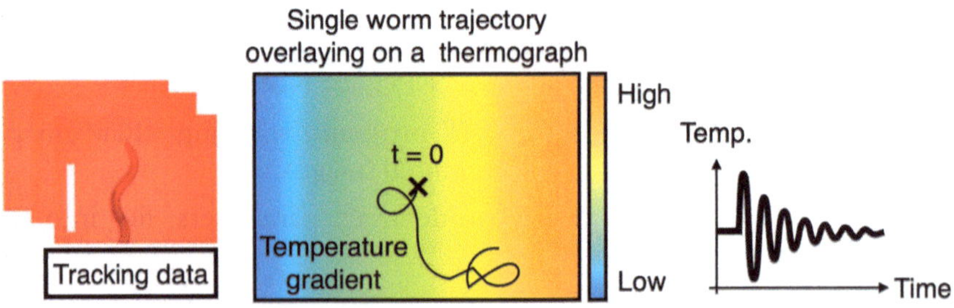

Fig. 3 Transforming tracking data to a temperature sequence on a thermograph. A motorized stage counteracts the worm's moving direction to keep the worm in the center of the imaging field. The x–y tracking information of the worm's trajectory is transformed offline into a temperature sequence experienced during navigation by overlaying the trajectory on a thermograph

1 mM MgSO4, and 25 mM potassium phosphate, pH 6.0). This process should take no more than 10 min.

5. Release the animals on the assay plate: Place the cleaned animals at the center of the assay plate and allow them to move freely for 60 min.
6. To enhance image contrast, illuminate the assay plate from the side with a ring LED light surrounding the assay plate (*see* **Note 5**).
7. Transfer animals from the NGM plates after washing them twice with the NG buffer. Release the animals on the central line of the assay plate and gently remove the excess buffer with Kimwipes tissues.
8. Cover the assay plate with a cover glass and start recording. Analyze the behavior after the recording (*see* **Note 6 and** 7).

3.3 Generating a Temperature Gradient

1. Prepare a custom thermal gradient device, which includes glass plates with indium tin oxide (ITO), an aluminum platform and aluminum plates, two water baths (one for warm water and the other for cool water), two dry baths for keeping the NG buffer and washing process, thermography and a thermometer for temperature measurement, a liquid pump, and a power supply.
2. To establish a thermal gradient on an assay plate in the Single-Worm Tracking system, use a thermal device consisting of glass plates with indium tin oxide (ITO) connected to a liquid pump and a power supply.
3. Place the assay plate, containing 18 ml of the medium with 2% agar, on the thermal device to establish a stable thermal gradient in 10 min.
4. Measure the thermal gradient at the start of each trial using thermography and a thermometer.
5. Create a stable temperature gradient for Multi-Worm Tracker: Place an aluminum platform over two water baths, one warm and one cool, at opposite ends. This process should create a temperature gradient of 0.5 °C/cm, with a preferred temperature at the center.
6. Prepare the TTX assay plate: Fill an assay plate (13.6 cm × 9.6 cm, 1.45 cm height) with 18 ml of assay medium containing 2% agar. Leave the plate to dry at room temperature (around 23 ° C), uncovered, for 30 min.
7. Set the assay plate temperature: Depending on the experiment, adjust the temperature at the center of the assay plate to 17 °C, 20 °C, or 23 °C, keeping the linear temperature gradient of roughly 0.5 °C/cm.

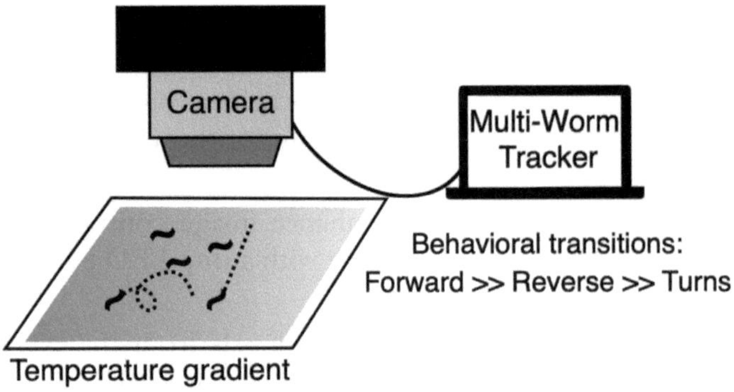

Fig. 4 Multi-Worm Tracker setup. The Multi-Worm Tracker can track up to 150 worms moving freely on a temperature gradient for an hour

8. Prepare the animals: Cultivate the animals at a constant temperature with food for at least 3 h to set the associative learning between temperature and food availability. Once the cultivation is complete, gently but promptly collect the animals from the NGM plates using the NG buffer kept at the same cultivation temperature to minimize the temperature changes before the assay.
9. Wait 10 min to stabilize the surface temperature of the assay plate, reflecting the temperature gradient on the aluminum platform connecting the warm and cool water baths mentioned in 3.3.5 (Fig. 4).

4 Notes

1. To record behaviors induced by neural activation instead of artifacts from light-induced repulsion by the fluorescent light illumination required for calcium imaging, use a light-insensitive *lite-1* mutant for the simultaneous calcium imaging and tracking experiments.
2. Worms are aversive to high-osmolarity stimuli. Therefore, 8 M glycerol is applied on the peripheral of the motorized imaging stage to contain worms freely moving within the imaging area.
3. In the single-worm tracking assay, we control high-speed illumination by a custom-built Galvano shutter system with a mercury lamp or SPECTRA light engine.
4. The camera used for the Multi-Worm Tracker was mounted at a distance above the assay plate resulting in an image resolution of 33.2 μm per pixel. The frame rate of recordings was approximately 13.5 frames per second (fps).

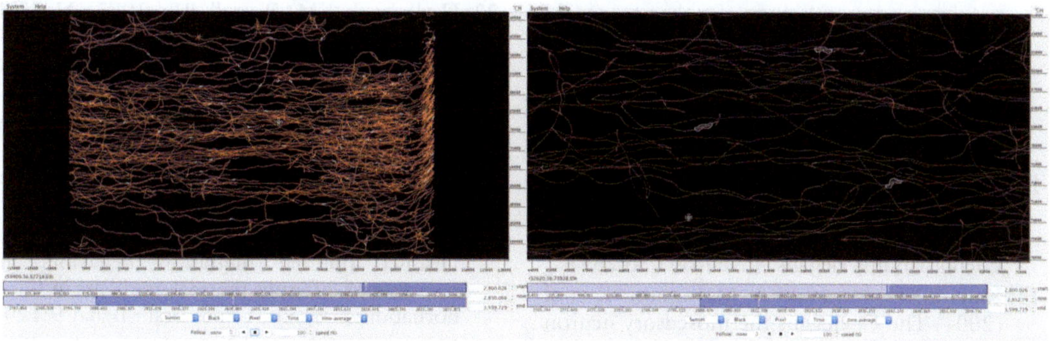

Fig. 5 Example data set acquired by the Multi-Worm Tracker. Multi-Worm Tracker records and digitalizes the time-course information about dozens of freely moving worms, including their centers of mass, body lengths, widths, outlines, and moving trajectories

5. Use a black aluminum plate under an assay plate to enhance image contrast and add water between the aluminum and assay plate to improve heat transmission on a temperature gradient.
6. Isothermal tracking is defined as a continuous movement within a range of ±25° of the isothermal direction for a duration exceeding 20 s. This isothermal tracking excludes any instances of turns or reversals.
7. When calculating the isothermal direction, slight distortions of the linear temperature gradient near the edge of the plates are adjusted to ensure accurate temperature information (Fig. 5).

Acknowledgments

We thank Hironori J. Matsuyama and Moonsun Jang for their valuable input into this manuscript. This work was supported by KAKENHI grant number 19H05644 and Neuroscience Institute of the Graduate School of Science, Nagoya University, to I.M.

References

1. White JG, Southgate E, Thomson JN, Brenner S (1986) The structure of the nervous system of the nematode Caenorhabditis elegans. Philos Trans R Soc Lond Ser B Biol Sci 314(1165):1–340
2. Cook SJ, Jarrell TA, Brittin CA et al (2019) Whole-animal connectomes of both Caenorhabditis elegans sexes. Nature 571:63
3. Witvliet D, Mulcahy B, Mitchell JK et al (2021) Connectomes across development reveal principles of brain maturation. Nature 596:257
4. Tsukada Y, Mori I (2021) Optogenetics in Caenorhabditis elegans. Adv Exp Med Biol 1293:321–334
5. Fang-Yen C, Alkema MJ, Samuel ADT (2015) Illuminating neural circuits and behaviour in Caenorhabditis elegans with optogenetics. Philos Trans R Soc B Biol Sci 370(1677):20140212
6. Kuhara A, Ohnishi N, Shimowada T et al (2011) Neural coding in a single sensory neuron controlling opposite seeking behaviours in Caenorhabditis elegans. Nat Commun 2:355
7. Chalfie M, Tu Y, Euskirchen G et al (1994) Green fluorescent protein as a marker for gene expression. Science 1979:263
8. Huang TT, Matsuyama HJ, Tsukada Y et al (2020) Age-dependent changes in response

property and morphology of a thermosensory neuron and thermotaxis behavior in Caenorhabditis elegans. Aging Cell 19:e13146
9. Kobayashi K, Nakano S, Amano M et al (2016) Single-cell memory regulates a neural circuit for sensory behavior. Cell Rep 14:11–21
10. Tsukada Y, Yamao M, Naoki H et al (2016) Reconstruction of spatial thermal gradient encoded in thermosensory neuron AFD in Caenorhabditis elegans. J Neurosci 36:2571
11. Kimura KD, Miyawaki A, Matsumoto K et al (2004) The C. elegans thermosensory neuron AFD responds to warming mechanism for sensation and memory of temperature information. Curr Biol 14:1291–1295
12. Aok I, Shiota M, Tsukada Y et al (2022) cGMP dynamics that underlies thermosensation in temperature-sensing neuron regulates thermotaxis behavior in C. elegans. PLoS One 17: e0278343
13. Russwurm M, Mullershausen F, Friebe A et al (2007) Design of fluorescence resonance energy transfer (FRET)-based cGMP indicators: a systematic approach. Biochem J 407(1):69–77
14. Wang L, Wu C, Peng W et al (2022) A high-performance genetically encoded fluorescent indicator for in vivo cAMP imaging. Nat Commun 13:5363
15. Komatsu H, Mori I, Rhee JS et al (1996) Mutations in a cyclic nucleotide-gated channel lead to abnormal thermosensation and chemosensation in C. elegans. Neuron 17:707
16. Coburn CM, Mori I, Ohshima Y et al (1998) A cyclic nucleotide-gated channel inhibits sensory axon outgrowth in larval and adult Caenorhabditis elegans: a distinct pathway for maintenance of sensory axon structure. Development 125(2):249–258
17. Takeishi A, Yu YV, Hapiak VM et al (2016) Receptor-type guanylyl cyclases confer thermosensory responses in C. elegans. Neuron 90(2): 235–244
18. Gray JM, Karow DS, Lu H et al (2004) Oxygen sensation and social feeding mediated by a C. elegans guanylate cyclase homologue. Nature 430:317
19. Zimmer M, Gray JM, Pokala N et al (2009) Neurons detect increases and decreases in oxygen levels using distinct guanylate cyclases. Neuron 61:865
20. Dudai Y, Jan YN, Byers D et al (1976) Dunce, a mutant of drosophila deficient in learning. Proc Natl Acad Sci USA 73:1684
21. Okochi Y, Kimura KD, Ohta A et al (2005) Diverse regulation of sensory signaling by C. elegans nPKC-epsilon/eta TTX-4. EMBO J 24(12):2127
22. Hedgecock EM, Russell RL (1975) Normal and mutant thermotaxis in the nematode Caenorhabditis elegans. Proc Natl Acad Sci USA 72:4061
23. Ito H, Inada H, Mori I (2006) Quantitative analysis of thermotaxis in the nematode Caenorhabditis elegans. J Neurosci Methods 154: 45
24. Jurado P, Kodama E, Tanizawa Y et al (2010) Distinct thermal migration behaviors in response to different thermal gradients in Caenorhabditis elegans. Genes Brain Behav 9:120
25. Matsuyama HJ, Mori I (2020) Neural coding of thermal preferences in the nematode Caenorhabditis elegans. eNeuro 7: ENEURO.0414
26. Aoki I, Jurado P, Nawa K et al (2022) OLA-1, an Obg-like ATPase, integrates hunger with temperature information in sensory neurons in C. elegans. PLoS Genet 18:e1010219
27. Mori I, Ohshima Y (1995) Neural regulation of thermotaxis in Caenorhabditis elegans. Nature 376:344
28. Ikeda M, Nakano S, Giles AC et al (2020) Context-dependent operation of neural circuits underlies a navigation behavior in Caenorhabditis elegans. Proc Natl Acad Sci USA 117: 6178
29. Nakano S, Ikeda M, Tsukada Y et al (2020) Presynaptic MAST kinase controls opposing postsynaptic responses to convey stimulus valence in Caenorhabditis elegans. Proc Natl Acad Sci USA 117:1638–1647
30. Yoshida A, Nakano S, Suzuki T et al (2016) A glial K+/Cl- cotransporter modifies temperature-evoked dynamics in Caenorhabditis elegans sensory neurons. Genes Brain Behav 15:429
31. Nakano S, Nakayama A, Kuroyanagi H et al (2022) Genetic screens identified dual roles of microtubule-associated serine threonine kinase and CREB within a single thermosensory neuron in the regulation of Caenorhabditis elegans thermotaxis behavior. G3 12:jkac248
32. Inada H, Ito H, Satterlee J et al (2006) Identification of guanylyl cyclases that function in thermosensory neurons of Caenorhabditis elegans. Genetics 172:2239
33. Tanizawa Y, Kuhara A, Inada H et al (2006) Inositol monophosphatase regulates localization of synaptic components and behavior in the mature nervous system of C. elegans. Genes Dev 20:3296
34. Swierczek NA, Giles AC, Rankin CH et al (2011) High-throughput behavioral analysis in C. elegans. Nat Methods 8:592

Chapter 28

Social Behavior in Animal Models of Autism Spectrum Disorder

Hitomi Kurahashi, Kazuo Kunisawa, and Akihiro Mouri

Abstract

Autism spectrum disorder (ASD) is a complex neurodevelopmental disorder characterized by social deficits and stereotyped, repetitive patterns of behaviors, limited interests, and cognitive impairment. Especially, social deficit has been considered a core feature of ASD. Because of the limitations of the experimental approach in humans, valid animal models are essential in an effort to identify novel therapeutics for social deficits in ASD. The genetic and environmental factors are clinically relevant to the pathophysiology of ASD. Epidemiological studies demonstrate environmental interventions such as prenatal exposure to valproic acid (VPA). Prenatal exposure to VPA represents a robust model of ASD exhibiting face, construct, and predictive validity. Here, we introduce protocols of the social interaction test and the three-chamber test for evaluating social deficits in mice prenatally exposed to VPA.

Key words Autism spectrum disorder, Valproic acid, Social interaction test, Three-chamber test

1 Introduction

Autism spectrum disorder (ASD) is a complex neurodevelopmental disorder characterized by social deficits and stereotyped, repetitive patterns of behaviors, limited interests, and cognitive impairment [1]. Especially, social deficit has been considered a core feature of ASD. Social deficits could appear as early as 6 months old. In early childhood, ASD patients often express lack of interest in other people and a strong interest in objects rather than people. From school age onward, it is characterized by difficulty in establishing close and equal relationships with friends. In rodents, when a pair of unfamiliar animals is placed in the familiar open field, normal pair of animals spend time for active social interaction. Here, we introduce the social interaction test and the three-chamber test that are often used to assess social deficit in ASD model mice. The genetic and environmental factors are clinically relevant to the pathophysiology of ASD. Since genetic mutation such as MECP2, SHANK3, and

Koh-ichi Nagata (ed.), *Cerebral Cortex Development: Methods and Protocols*, Methods in Molecular Biology, vol. 2794, https://doi.org/10.1007/978-1-0716-3810-1_28,

FMR1, and copy number variants such as 15q13.3 microdeletion and 22q11.2 deletion have been implicated in the development of ASD, various gene mutant mice are used as ASD model mice [2]. Epidemiological studies demonstrate that environmental interventions such as prenatal exposure to valproic acid (VPA), maternal autoantibodies, and maternal immune activation increase the prevalence of ASD [3]. Prenatal exposure to VPA represents a robust model of ASD exhibiting face, construct, and predictive validity [4]. In this protocol, mice prenatally exposed to VPA are used as ASD models.

2 Social Interaction Test

2.1 Materials

2.1.1 Animals

1. Test mice: VPA (500 mg/kg; Tokyo Chemical Industry, Tokyo, Japan) dissolved in saline (SAL) was administered intraperitoneally (i.p.) to pregnant C57BL/6 J mice on E12.5. As a control, saline (SAL) was administered to the pregnant mice. Eight-week-old male mice born from dams treated with VPA and SAL were used as model mice and control mice, respectively.
2. Partner mice: The mice used as the partner must be matched to the model mice in age, sex, and strain. The partner mice should be unfamiliar to model and control mice. Therefore, these mice should not be kept in the same cage.

2.1.2 Equipment (Fig. 1)

1. Social interaction test apparatus: Square open area with grey walls (25 cm × 25 cm × 30 cm).
2. Video camera.
3. Computer for video playback.
4. Brightness adjustable light.
5. Lux meter.
6. Timer.
7. Stopwatch.
8. Marker pen.
9. Alcohol wet tissue.

2.2 Methods (Fig. 2)

2.2.1 Preparation

1. Mark numbering on the tails of mice for individual identification. The color of the marker should be different for each group.
2. Place the apparatus in the experimental room and adjust the position of the apparatus to avoid shadows and equalize the brightness in the apparatus. Adjust the brightness of light to 10 lux in the apparatus. Set the video camera above the apparatus to record and monitor the social behavior via a monitor (Fig. 1).

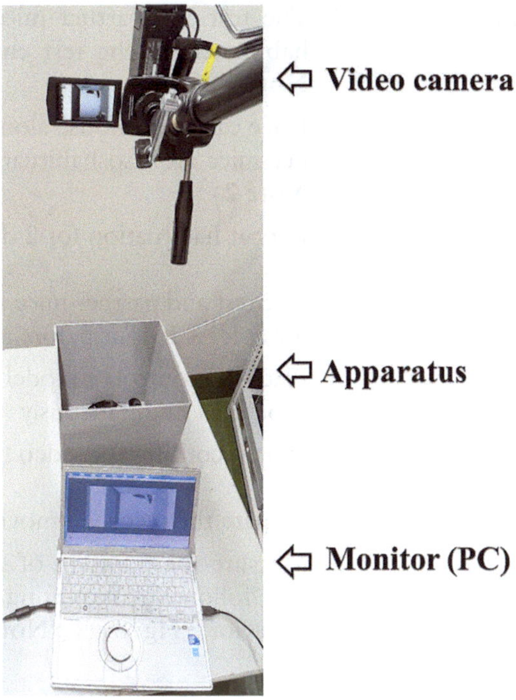

Fig. 1 Layout for social interaction test

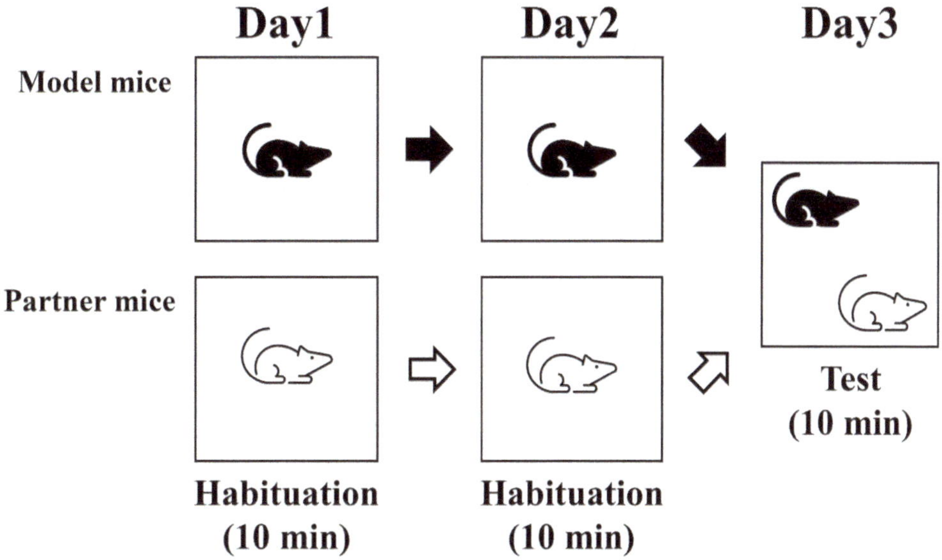

Fig. 2 Social interaction test Before the day of the social interaction test, each test and partner mouse is placed alone in the apparatus for 10 min for two consecutive days for habituation

2.2.2 Habituation

1. The test and partner mice are moved to the test room and habituated to the test environment for at least 1 hour (*see* **Note 1**).
2. Place each test mouse alone in the apparatus for 10 min. Partner mice are also habituated in the apparatus for 10 min (*see* **Note 2**).
3. Repeat habituation for 2 days.

2.2.3 Test

1. The test and partner mice are moved to the experimental room for at least 1 hour before the test.
2. Place the control or model mouse and the partner mouse in the apparatus simultaneously (*see* **Note 3**).
3. Start recording the video for 10 min (*see* **Notes 4** and **5**).

2.2.4 Analysis

1. Prepare the recorded movie data, stopwatch, and timer.
2. Measure the duration of social interactions (sniffing, grooming, following, mounting, and crawling, but not aggressive behavior) (Fig. 3) (*see* **Note 6**).

3 Three-Chamber Test

3.1 Materials

3.1.1 Animals

1. Test mice: Eight-week-old male mice born from dams treated with VPA and SAL were used as model mice and control mice, respectively. Refer to Subheading "2.1., Social interaction test" for further details.
2. Social stimulus mice: Social stimulus mice must be of the same age, sex, and strain. Furthermore, take care not to keep them in the same cage to avoid social contact.

3.1.2 Equipment (Fig. 4)

1. Three-chamber apparatus: Rectangular open area with grey walls (40 cm × 60 cm × 25 cm) divided into three chambers by transparent plexiglass. The apparatus chambers are numbered 1–3 from left to right (Fig. 5). The size of each chamber is 40 cm x 20 cm.
2. Two wire cages (8 cm diameter × 10 cm high) for housing the social stimulus mice. Plexiglass bottles should be placed on top of the wire cages to prevent the test mouse from climbing.
3. Video camera.
4. Computer-installed video tracking and analysis software such as ANY-maze (Stoelting Co, Wood Dale, IL) or EthoVision XT (Noldus, Wageningen, The Netherlands).
5. Brightness adjustable light.
6. Lux meter.

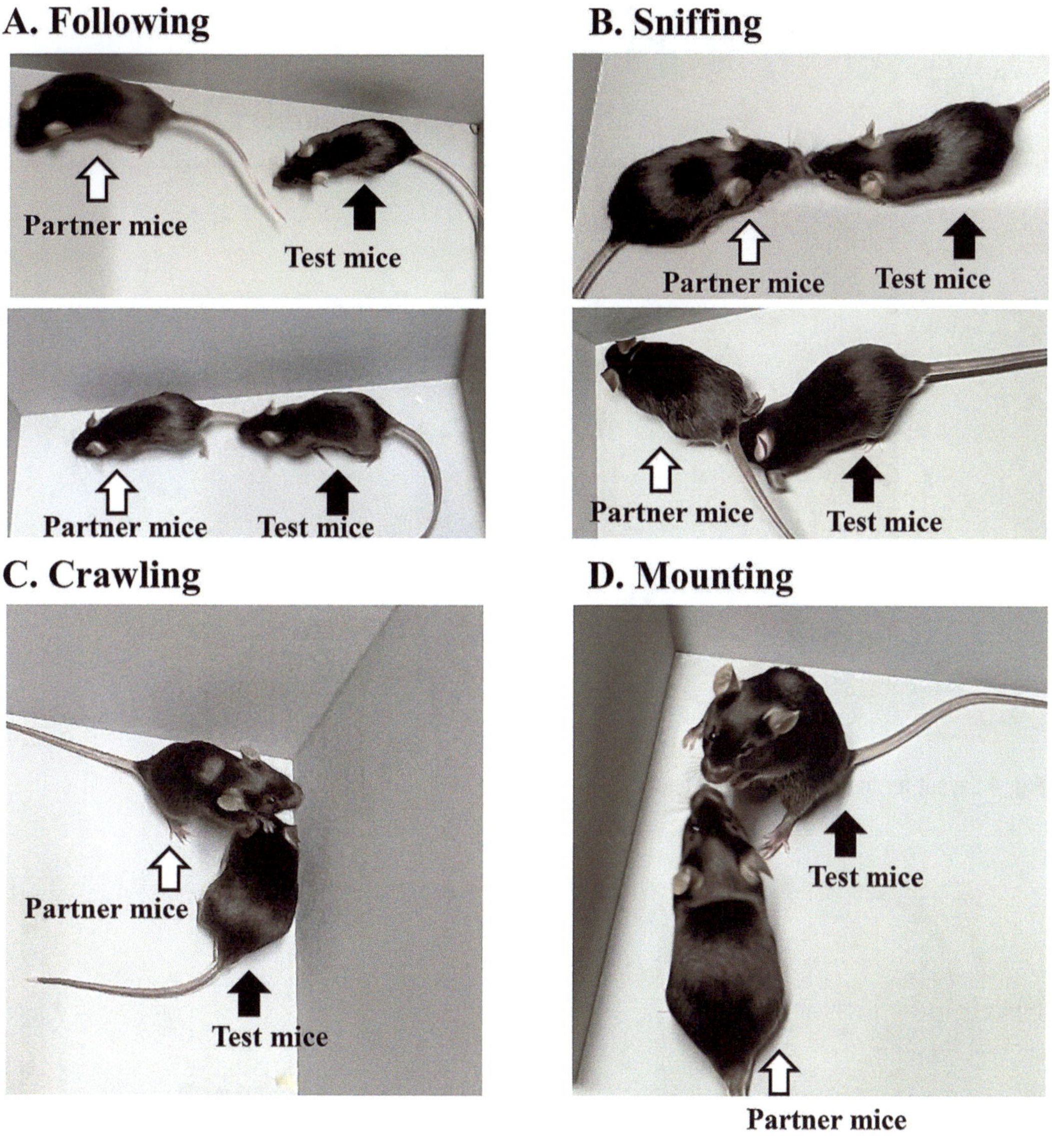

Fig. 3 Mouse-to-mouse social behaviors. The following (**a**) sniffing face and anus, (**b**) crawling, (**c**) and mounting the partner mouse are recognized as social behaviors of test mice

7. Timer.
8. Marker pen.
9. Alcohol wet tissue.

3.2 Methods

3.2.1 Preparation

1. Number the tails of mice with markers to identify individuals. The color of the marker should be different for each group.
2. Place the apparatus in the experimental room and adjust the position of the apparatus to avoid shadows and equalize the

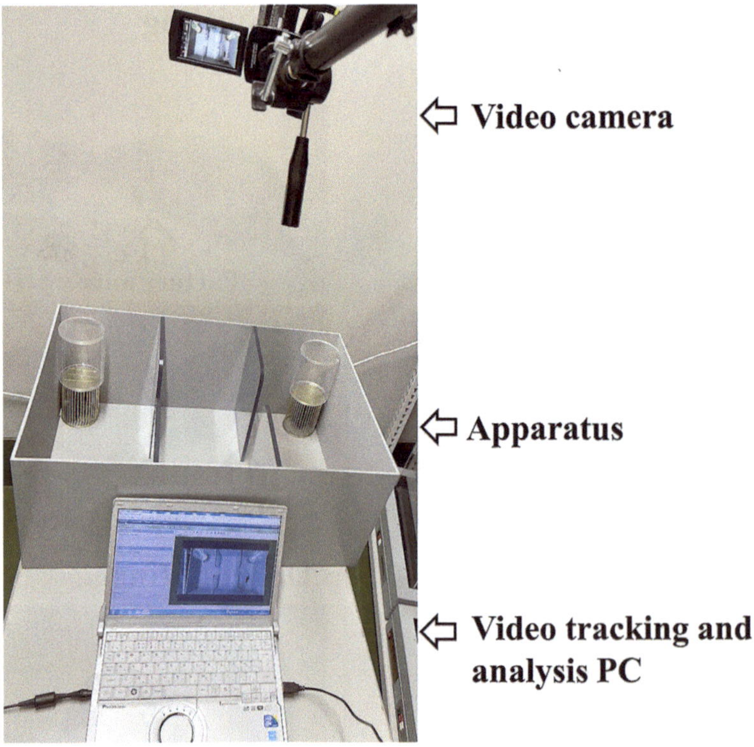

Fig. 4 Layout for the three-chamber test

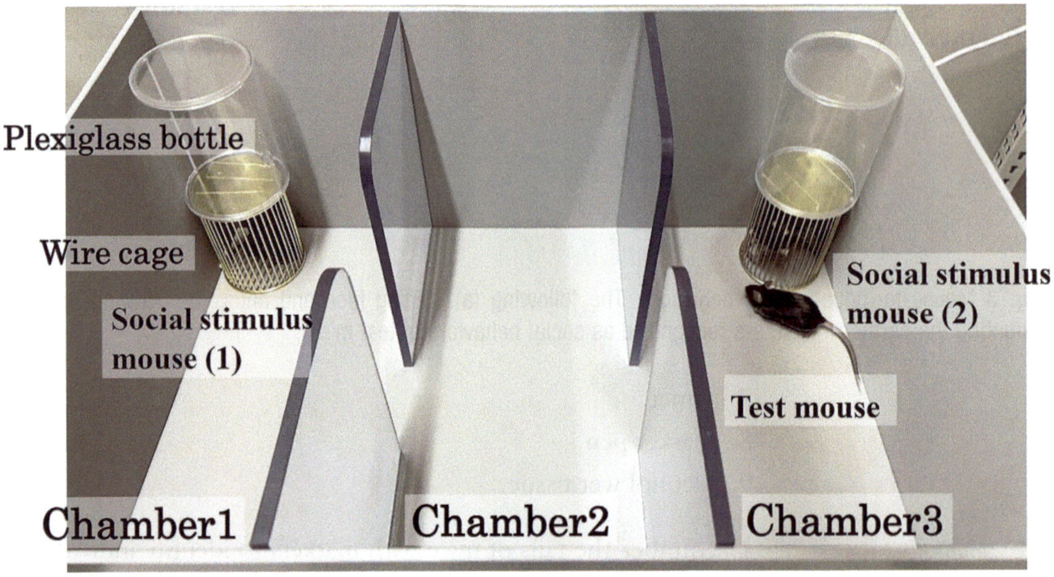

Fig. 5 Social interaction chamber with the test and partner mice

brightness in the apparatus. Adjust the brightness of light to 10 lux in the apparatus. Set the video camera above the apparatus and connect to video tracking and analysis via a PC (Fig. 4).

3. The wire cages are placed in chambers 1 and 3 (*see* **Notes 7** and **8**) (Fig. 5).

3.2.2 Habituation

1. The test and social stimulus mice are moved to the test room and habituated to the test environment for at least 1 hour (*see* **Note 9**).

3.2.3 Test (Fig. 6)

Phase 1 (Chamber 1: Empty; Chamber 3: Empty)

1. Start recording the video and place the test mouse into chamber 2 for the test mouse to explore freely while habituating to the apparatus and empty wire cages for 10 min.
2. Stop video recording and remove the test mouse from the apparatus and return to its home cage for a 5 min break. Clean the apparatus with an alcohol wet tissue to remove any residual odors that may affect subsequent tests.

Phase 2 (Chamber 1: Social Stimulus Mouse (1); Chamber 3: Empty)

1. Place social stimulus mice into a wire cage of chamber 1. Start recording the video and place the test mouse into chamber 2 for the test mouse to explore freely to the wire cages containing the social stimulus mice and empty wire cages for 10 min.
2. Stop video recording and remove the test mouse from the apparatus and return to its home cage for a 5 min break. Clean the apparatus with an alcohol wet tissue to remove any residual odors that may affect subsequent tests. Keep the social stimulus mouse in the wire cage.

Phase 3 (Chamber 1: Social Stimulus Mouse (1); Chamber 3: Social Stimulus Mouse (2))

1. Place "novel" social stimulus mice into the wire cage of chamber 3. Start recording the video and place the test mouse into chamber 2 for it to explore freely the wire cages containing the social stimulus mice and empty wire cages for 10 min.
2. Stop video recording and remove the test and social stimulus mice from the apparatus and return to their home cage for a 5 min break. Clean the apparatus with an alcohol wet tissue to remove any residual odors that may affect subsequent tests.
3. Perform phases 1–3 for each test mouse.

3.2.4 Analysis (Fig. 7)

1. Measure the interaction time taken from the amount of time spent in each chamber (A–C) or around each wire cage (A′ and C′) using video tracking and analysis software.

   Normal mice spend more time in cages with stimulate social mice (time spent in A or A′) than in empty cages (time spent in C or C′) in phase 2 as an index of sociality and spend more time in cages with stimulus social mice (2) (time spent in C or C′) than in stimulus social mice (1) (time spent in A or A′) in phase 3 as an index of social recognition memory.

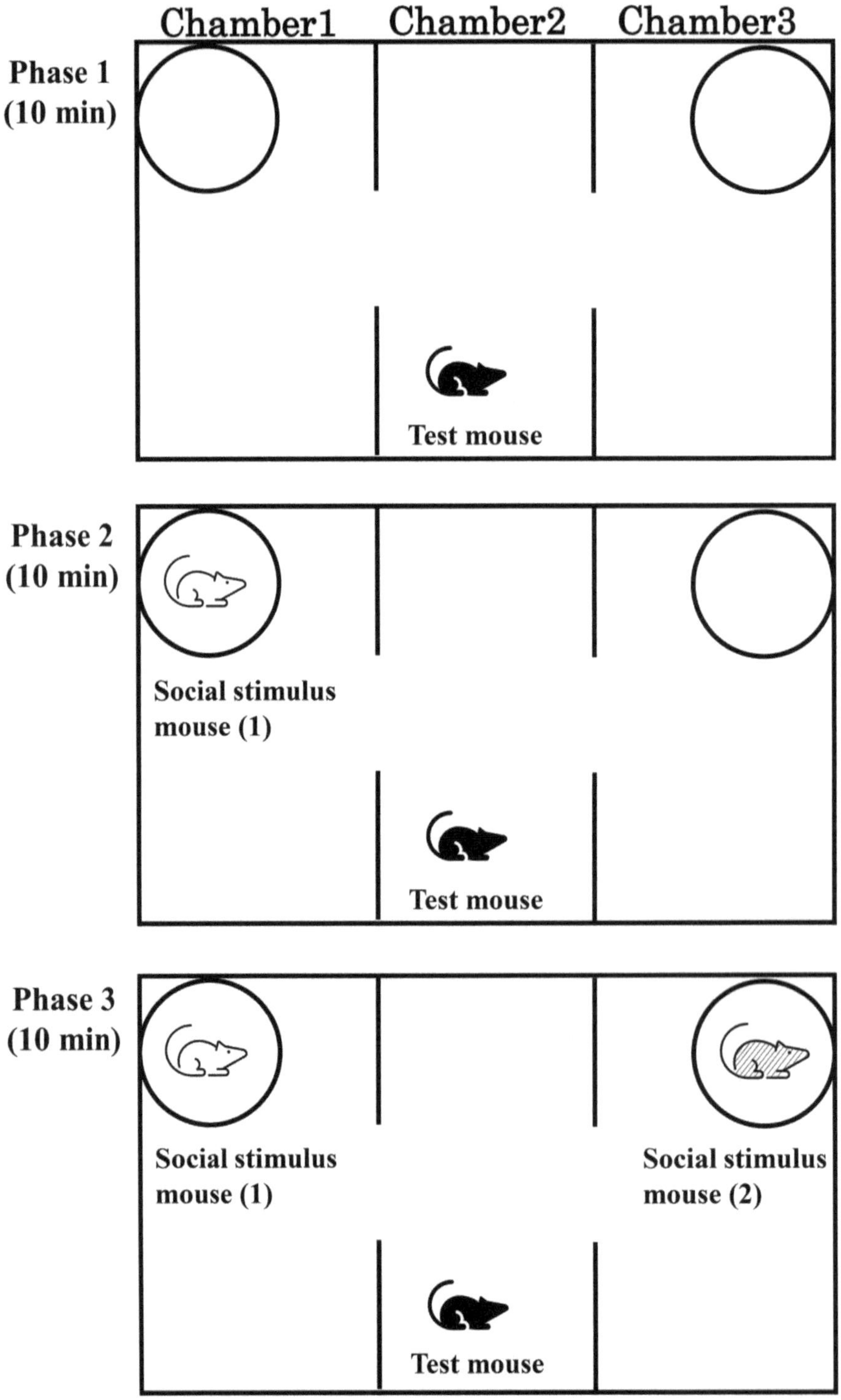

Fig. 6 Three-chamber test. After habituation to the apparatus and empty wire cages for 10 min (phase 1), a social stimulus mouse (1) is introduced into the wire cage in chamber 1 for evaluating sociality (phase 2). Thereafter, a social stimulus mouse (2) is introduced into the wire cage in chamber 3 for evaluating social recognition memory (phase 3)

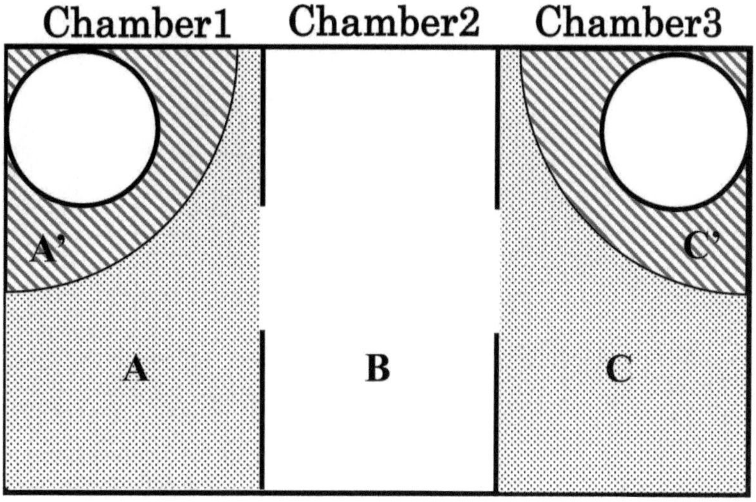

Fig. 7 Interaction area indicating each chamber (A–C) and around each wire cage (A′ and C′)

4 Notes

1. The cages of the test mice should be placed at a distance from that of the partner mice. When the cages are placed next to each other, mice may recognize the partner mice via the cage and not show social behavior during the test.
2. Clean the apparatus with alcohol wet tissue after each habituation.
3. When placing the mice, the mouse should face the corner of the apparatus, because the mouse should not recognize the partner mouse before the test.
4. Be careful not to make noises or look over the mice during the test. The noises or glances might disturb the social behavior.
5. Clean the apparatus after each test to remove any residual odors that may affect subsequent tests.
6. Passive social behavior by the partner mouse should not be counted.
7. The wire cages should be placed so that they are attached to the corner of the chamber. If there are gaps between the walls and the cages, the mice may enter them and social behavior cannot be assessed correctly.
8. It is likely that wire cages are moved by the social stimulus mice from the inside. If the wire cage is light, a weight should be placed on the plexiglass bottle on the wire cages.

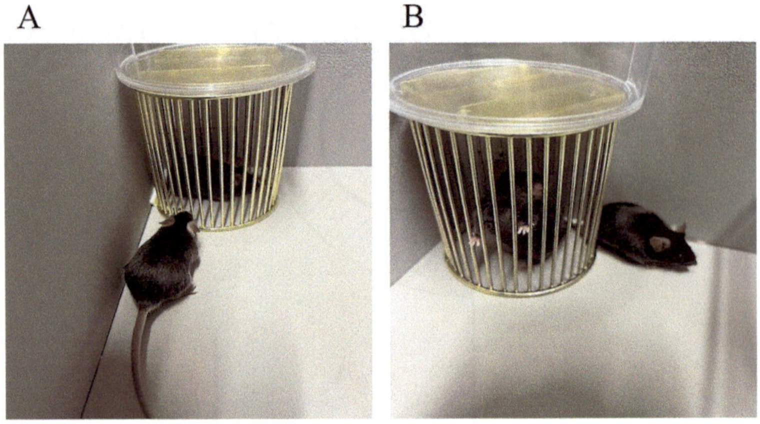

Fig. 8 Social behavior in the three-chamber test. Mice spend time in the interaction area to show social behavior (**a**). Mice spend time in the interaction area to stay in a corner without social behavior (**b**)

9. The cages of the control and model mice should be placed at a distance from that of the social stimulus mice. When the cages are placed next to each other, mice may recognize the partner mice via the cage and not show social behavior and/or social recognition behavior during the test.
10. When analyzing with video tracking and analysis software, it is recommended to visually confirm whether mice show social behavior in the interaction area (Fig. 8a), because some mice spend time in the interaction area to stay in a corner without social behavior (Fig. 8b).

References

1. American Psychiatric Association (2022) Diagnostic and statistical manual of mental disorders: DSM-5-TR
2. Varghese M, Keshav N, Jacot-Descombes S, Warda T, Wicinski B, Dickstein DL, Harony-Nicolas H, De Rubeis S, Drapeau E, Buxbaum JD, Hof PR (2017) Autism spectrum disorder: neuropathology and animal models. Acta Neuropathol 134(4):537–566. https://doi.org/10.1007/s00401-017-1736-4
3. Ornoy A, Weinstein-Fudim L, Ergaz Z (2015) Prenatal factors associated with autism spectrum disorder (ASD). Reprod Toxicol 56:155–169. https://doi.org/10.1016/j.reprotox.2015.05.007
4. Nicolini C, Fahnestock M (2018) The valproic acid-induced rodent model of autism. Exp Neurol 299(Pt A):217–227. https://doi.org/10.1016/j.expneurol.2017.04.017

Chapter 29

Preparation of Healthy Single Neuron or Astrocyte Suspension from Adult Mouse Brain for RNA-seq

Neng-Yuan Hu, Tian-Ming Gao, and Yi-Hua Chen

Abstract

Single-cell RNA sequencing (scRNA-seq) has been widely applied in neuroscience research, enabling the investigation of cellular heterogeneity at the transcriptional level, the characterization of rare cell types, and the detailed analysis of the stochastic nature of gene expression. Isolation of single nerve cells in good health, especially from the adult rodent brain, is the most difficult and critical process for scRNA-seq. Here, we describe methods to optimize protease digestion of brain slices, which enable yield of millions of cells in good health from the adult brain.

Key words Adult neurons, Adult astrocytes, Single-cell suspension, Flow cytometry, RNA-seq

1 Introduction

Neurons and astrocytes represent the most abundant cell types in the central nervous system (CNS). Despite the broad reach of neurons and astrocytes across numerous brain functions, the question of whether their diverse functions are executed by distinct subpopulations in the adult brain remains poorly defined. Recent advances in single-cell RNA sequencing (scRNA-seq) have offered unprecedented power to reveal fine-scale molecular differences that may underlie functional heterogeneity across cell types. The starting material for all scRNA-seq is a cell suspension.

To isolate neurons and astrocytes, it is necessary to target them selectively. A major advance in this regard came with the ability to fluorescently label cell types either with genetically fluorescent proteins [1, 2] or with antibodies to identify specific cell surface markers [3–5]. The fluorescent cells need to be physically isolated from their nonfluorescent neighbors using a cell sorting procedure. Although it is relatively easy to isolate embryonic or early postnatal neurons and astrocytes before they sprout, isolation of adult neurons and astrocytes with mature properties is problematic because

Koh-ichi Nagata (ed.), *Cerebral Cortex Development: Methods and Protocols*, Methods in Molecular Biology, vol. 2794,
https://doi.org/10.1007/978-1-0716-3810-1_29,

of tight adhesion of cell bodies and intermingled processes that mediate interactions with other cells in the CNS.

The extracellular matrix that holds cells together has to be processed to loosen this mesh and to induce the release of cells into suspension. Enzymes that target the composition of the extracellular matrices in terms of molecules such as collagen, fibronectin, and proteoglycan must be used. Papain is recommended for preparing single-cell suspensions in brain tissue because it efficiently preserves cell integrity and viability [6–9]. In addition, since most of the composition of the extracellular matrices is Ca^{2+}- or Mg^{2+}-binding protein, Ca^{2+}- and Mg^{2+}-free buffer is recommended for preparation of the digestion solution. Dissociation protocols use a combination of gentle mechanical means and enzymatic hydrolysis to decrease the overall processing time: a mechanical step breaks up the tissue, increasing the surface area, followed by enzymatic treatment. In addition, incubation with DNAse I is included in all dissociation conditions to eliminate DNA mucus that negatively affects cell viability [10].

Importantly, neurons and glial cells can be affected by the harsh conditions of tissue dissociation. Brain sections were blocked in an AP5, CNQX, and TTX cocktail to prevent excitotoxic cell death [11, 12], and neurotrophic factors (Glutamax and B27) were added to promote neuron survival [13]. In single-cell RNA sequencing data of other cell types, it has been shown that enzymatic digestions can induce gene expression changes of a set of immediate early genes, which then create artificial subpopulations in biologically homogeneous cell populations [14]. Decreasing the digestion time as much as possible is recommended to minimize these unwanted effects. In addition, there are still several key factors that need special attention: (1) the enzyme(s) used, (2) the oxygen saturation in the solution, (3) the osmotic pressure of the solution [15], (4) the pH of the solution, and (5) temperature changes.

In this chapter, we describe neuronal and astrocytic dissociation protocols that have been used to obtain single-cell suspensions that were then successfully used as input for single-cell protocols.

2 Materials

2.1 Reagents

1. Mice: Experiments using live rodents must conform to appropriate national and institutional regulations.
2. 0.1 μM Tetrodotoxin (TTX).
3. 50 μM 2-Amino-5-phosphonovaleric acid (DL-AP5).
4. 20 μM 6,7-Dinitroquinoxaline-2,3-dione (DNQX).
5. Compressed 95% O_2/5% CO_2.
6. Artificial cerebrospinal fluid (ACSF).

7. Papain: ≥100 units/tube, aliquoted and stored at −20 °C.
8. 100 mM L-Cysteine (100X): 6 mg/tube, stored in the dark at room temperature. Add 500 μL EBSS (or Hibernate A) to dissolve before use.
9. 50 mM EDTA (100X): Add 1 mL 0.5 M EDTA into 9 mL EBSS (or Hibernate A) and store at 4 °C.
10. Glutamax (100X).
11. Hibernate A.
12. Ca^{2+}- and Mg^{2+}-free EBSS (Phenol red).
13. DNAase (1000 units/tube).
14. 20 mL/L B27™ Plus Supplement (50X) (*see* **Note 1**).
15. DAPI (1 mg/tube).
16. DMEM/F12.
17. Fetal bovine serum (FBS).
18. 0.5 mg/mL TruStain FcX™ PLUS.

2.2 Equipment

1. Custom-made glass digestion bottle: To maintain proper temperature and oxygen saturation of the solution during tissue digestion, a custom-made glass digestion bottle was made (Fig. 1).
2. 0.6 mm, 0.3 mm, and 0.15 mm fire-polished Pasteur pipettes (Fig. 2).

2.3 Reagent Setup

1. Artificial cerebrospinal fluid (ACSF) [12]: 126 mM NaCl, 20 mM $NaHCO_3$, 20 mM D-glucose, 3 mM KCl, 1.25 mM NaH_2PO_4, 2 mM $CaCl_2$, 2 mM $MgCl_2$ (*see* **Note 2**). For

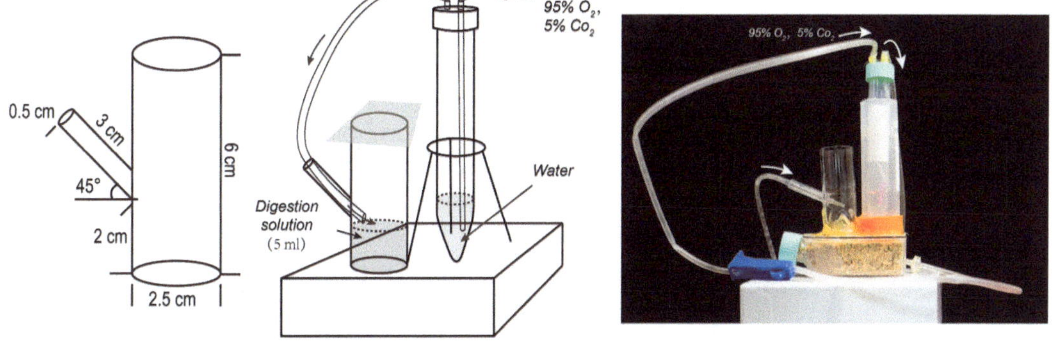

Fig. 1 Custom-made glass digestion bottle. (**a**) Installation diagram. (**b**) 2 mL water is added to the 15 mL centrifuge tube to moisten the gas and prevent the digestion solutions from evaporating when bubbling. During digestion, the digestion bottle was immersed in a 37 °C water bath to maintain its temperature, and the gas flowed above the digestion solution surface, causing the tissues to rotate

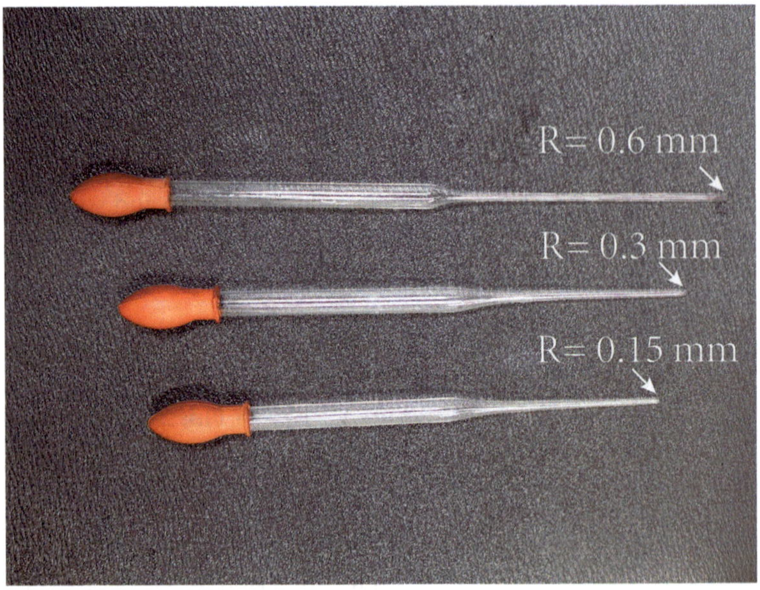

Fig. 2 Fire-polished Pasteur pipettes

neuron isolation, add 50 μM DL-AP5 sodium salt, 20 μM DNQX, and 0.1 μM TTX into the ACSF.

2. $Ca2^{+}$- and $Mg2^{+}$-free EBSS: equilibrated in 95% O2/5% CO_2 for 15 min before use.
3. Hibernate A: Equilibrated in 95% O2/5% CO_2 for 15 min before use.
4. DNAase (≥ 2000 units/mL): Add 500 μL $Ca2^{+}$- and $Mg2^{+}$-free EBSS to a tube containing DNAase (≥1000 units), temporarily store at −4 °C (*see* **Note 3**).
5. Digestion solution for astrocyte isolation: ≥ 20 units/mL papain (dissolved in Ca^{2+}- and Mg^{2+}-free EBSS), 1 mM L-cysteine, 0.5 mM EDTA, 100 units/mL DNase, temporarily stored at 4 °C (*see* **Note 4**).
6. Digestion solution for neuron isolation: ≥ 20 units/mL papain (dissolved in Hibernate A), 1 mM L-cysteine, 0.5 mM EDTA, 100 units/mL DNAase, 1X Glutamax 100X, 1X B_{27} 50X, temporarily stored at 4 °C (*see* **Note 5**).
7. Cell culture medium: 80% DMEM/F12 (equilibrated in 95% O2/5% CO_2 for 15 min before use) and 20% FBS.
8. DAPI: 1 mg/mL (dissolved in sterile double-distilled water), aliquoted and stored at −20 °C.

3 Methods

Using the example of isolating genetically labeled fluorescent cells, we describe the method for isolating single-cell suspensions. The different procedures between neuron isolation and astrocyte isolation will be indicated. The method of isolating antibody-labeled cells is also described, and the steps are the same as those for genetically labeled fluorescent cells unless otherwise specified. The streamline is shown in Fig. 3:

1. Solutions ready for **steps 2–6**: EBSS (7 mL) (to isolate neurons, Hibernate A (7 mL) instead), ice-cold ACSF (> 200 mL), and DMEM/F12 (10 mL) are equilibrated in 95% O2/5% CO_2 for 15–30 min (*see* **Note 6**).
2. Preparation of digestion solutions for astrocytes: Add 4.65 mL of EBSS from **step 1** to dissolve the papain, 50 μL of 100 mM L-cysteine, 50 μL of 50 mM EDTA, and 250 μL of DNase in the glass digestion bottle.
3. Preparation of digestion solutions for neurons: Add 4.5 mL Hibernate A from **step 1** to dissolve the papain, 50 μL 100 mM L-Cysteine, 50 μL 50 mM EDTA, 250 μL DNAase, 50 μL Glutamax 100X, and 100 μL B_{27} 50X in the glass digestion bottle.
4. Put the glass digestion bottle into a 37 °C water bath and flow the gas to rotate the liquid for 15 min (*see* **Note 7**).
5. Mice were anesthetized with 1% pentobarbital sodium and then transcardially perfused with ice-cold ACSF from **step 1**. The brain is removed gently and placed in the ice-cold ACSF from **step 1** (*see* **Note 8**).
6. Mount the brain on a Vibroslice microtome and 300–400 μm brain sections are cut at a rate of 0.2–0.3 μm/s of the desired plane (*see* **Note 9**).
7. Transfer the slice with a glass pipette to a dish on ice and dissect the desired region with microscope forceps (*see* **Note 10**).

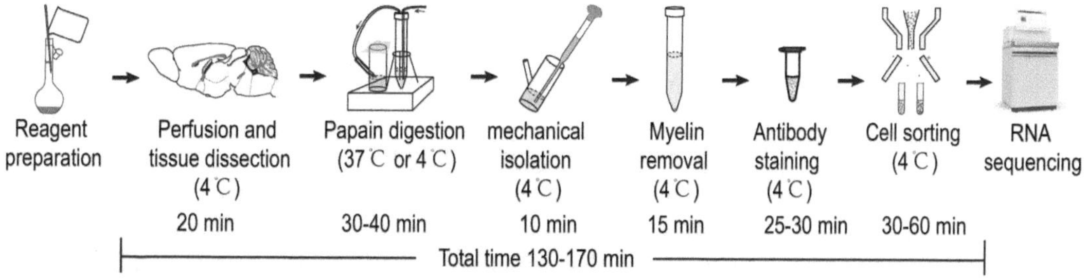

Fig. 3 Schematic diagram of the process of brain tissue dissociation

8. Wash the dissected pieces of brain tissue with EBSS (to isolate neurons, Hibernate A) from **step 1** twice at room temperature and add 1 mL digestion solution from **step 2** into the dish after discarding the EBSS (to isolate neurons, Hibernate A) (*see* **Note 11**).
9. Transfer the tissue and solution with a glass pipette to the digestion bottle, and the digestion solutions are immersed in a 37 °C warm water bath to maintain their temperature for 30–40 min, while the flowing gases caused the liquids and tissues to rotate.
10. Mechanical isolation:
    (a) Starting with the 0.6 mm fire-polished Pasteur pipette, gently triturate the sample 15–20 times, counting at least 1 sec in and 1 s out (*see* **Note 12**). The digestion solutions should turn noticeably cloudy, but large chunks should remain.
    (b) Allow the pieces to settle for 0.5 min and transfer 3 mL supernatant to a new 15 mL tube (mark tube-S).
    (c) Add 2 mL of cell culture medium into the digestion bottle, and then repeat the trituration of the 0.3 mm fire-polished Pasteur pipette.
    (d) Transfer 2 mL supernatant to tube S after settling for 0.5 min.
    (e) Add 1 mL cell culture medium into the digestion tube, and then repeat the trituration with the 0.15 mm fire-polished Pasteur pipette.
    (f) These "adding cell culture medium–trituration–transferring supernatant" procedures are repeated (*see* **Note 13**); at the end of the procedure, macroscopic pieces of tissue should be discarded (*see* **Notes 14** and **15**).
11. Centrifuge the cell suspension at 350 × *g* for 5 min at 4 °C. If the supernatant is cloudy, increase the centrifugation time.
12. When targeting cells are labeled with flow cytometry antibodies (*see* **Note 16**), mix the cell suspension with 26% Percoll (1560 μL Percoll and 4440 μL cell suspension from **step** 7) and perform gradient density centrifugation at 850 × *g* for 15 min at 4 °C to remove myelin sheath and cell debris (*see* **Note 17**):
    (a) Aspirate the supernatant and resuspend the sediment with an ice-cold cell culture medium, and then centrifuge at 350 × *g* for 5 min at 4 °C.
13. Aspirate the supernatant and resuspend the sediment gently with a 1 mL cell culture medium (add 50 μL DNAase to eliminate DNA mucus and 1 μL DAPI) using a 0.15 mm fire-polished Pasteur pipette.

14. When targeting cells are labeled with flow cytometry antibodies:
    (a) Aspirate the supernatant and resuspend the sediment gently with 100 μL EBSS (add 5 μL DNAase into it to eliminate DNA mucus) using the 0.15 mm fire-polished Pasteur pipette.
    (b) Incubate with 0.5 μL of TruStain FcX™ PLUS (0.5 mg/mL) on ice for 5–10 min, and then incubate the fluorescent antibody (such as APC-conjugated anti-ACSA-2 for astrocytes) on ice for 25–30 min in the dark.
    (c) Resuspend the cells with 1 mL cell culture medium and centrifuge the cell suspension at 350 × *g* for 5 min at 4 °C.
    (d) Aspirate the supernatant and resuspend the sediment gently with 1 mL cell culture medium (add 1 μL DAPI) using the 0.3 mm fire-polished Pasteur pipette.
15. Filter the cell suspension with a prewet 70 μm cell strainer.
16. The cells are immediately loaded into the Beckman MoFlo XDP Cell Sorter system for fluorescence-activated cell sorting (FACS). Large aperture and low-speed sorting mode (100 μm nozzle and the difference between sample PSI and sheath PSI is less than 0.5, 4 °C) is employed (*see* **Note 18**). Harvest the cells into a tube containing cell culture medium at 4 °C.
17. Centrifuge the sorting cells at 350 × *g* for 5 min at 4 °C. If the next step is scRNA-seq, aspirate most of the supernatant, leaving approximately 100 μL for resuspension. In general, the cell viability can reach more than 90% (Fig. 4) (*see* **Notes 19–21**). If the next step is bulk RNA-seq, discard the supernatant and quickly freeze the sediment in liquid nitrogen.

4 Notes

1. B27™ should be prepared fresh every week and stored at −20 °C.
2. The ACSF should be prepared on the day of tissue isolation. Ice-cold ACSF (e.g., stored at −80 °C for 30 min) is required and should be equilibrated in 95% O2/5% CO_2 for 15 min before use.
3. The DNAase solution should be prepared on the spot when in use.
4. The digestion solution for astrocyte isolation should be prepared on the spot when in use.
5. The digestion solution for neuron isolation should be prepared on the spot when in use.

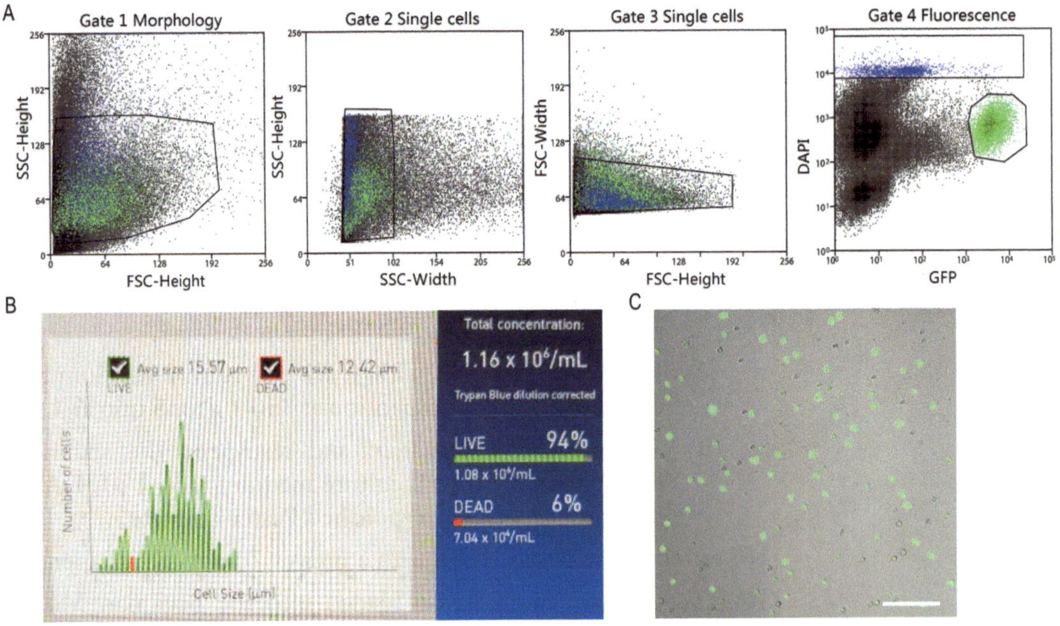

Fig. 4 Flow cytometry sorting of astrocytes from Aldh1l1-EGFP mice. (**a**) Gate setup for flow cytometry sorting. (**b**) Cell viability was detected by Countess™ 3 (>90% viability). (**c**) Cell purity was checked by fluorescence microscopy

6. Pay attention to the color of the EBSS: If the solution turns from dark red to pink, stop bubbling.
7. The time for water bath should not exceed 30 min; otherwise, the activity of the papain enzyme will decrease.
8. The time from anesthetizing the mouse to obtaining the target brain tissue (**steps 5–7**) is recommended to be no more than 20 minutes.
9. Bubbled ice-cold ACSF is needed throughout this process.
10. Do not squeeze to crush the tissue.
11. When cell surface markers are used for FACS, it is suggested to dissociate the tissues with mechanical means without enzymatic digestion, which may damage the membrane proteins, thereby reducing labeling.
12. Do not introduce bubbles when triturating and avoid aspirating tissue blocks directly.
13. Effective trituration requires a balance between being forceful enough to dissociate the tissue and gentle enough not to destroy all of the cells. The outcome should be monitored, and the procedure should be adjusted accordingly.
14. The concentration of the enzyme and the digestion time can be appropriately adjusted for optimal digestion, while it is important to note that the time from sacrificing the mouse to

finishing the FACS should not exceed 4 h to avoid poor cell viability and RNA degradation.

15. Except for **step 9**, the remaining steps were performed on ice or at 4 °C to reduce cell metabolism.
16. It is recommended to use transgenic mice expressing fluorescent protein to memorize the target cells (such as mice with astrocytes labeled by Aldh1l1-EGFP), which can reduce the stimulation of antibody staining, so that the obtained cells are more liable.
17. Myelin sheath and cell debris can be removed by several concentrations [10, 17, 18], and our results showed that 26% of Percoll [16] received higher cell production.
18. If the number of cells obtained by sorting can meet the requirements for subsequent experiments, it is recommended to use the single-cell sorting mode with higher purity; otherwise, the purification mode should be selected.
19. If the preparation contains enough cells but low viability, the most likely problem is (1) over-digestion and (2) triturating too vigorously. To prevent this, (1) reduce the concentration of enzyme and the digestion time; (2) triturate gently, reduce the triturating time, and avoid introducing bubbles.
20. If the preparation contains healthy cells but too few cells, this can be due to (1) under-digestion or (2) low density of labeled cells in the original tissue. Thus, (1) the concentration of the enzyme and the digestion time will have to increase, and (2) more tissue will have to be dissociated.
21. If the cell suspension is viscous, it can be due to (1) entanglement of DAN released by a large amount of cell lysis or (2) impurity interference in the tissue. To prevent this, (1) reduce the triturating time and increase the amount of DNAase I; (2) add hyaluronidase into the digestion solution.

Acknowledgments

This work was supported by grants from the STI2030-Major Projects (2022ZD0214300, to YHC; 2021ZD0202704 and 2022ZD0204702, to TMG); the National Natural Science Foundation of China (32271014, to YHC; 82090032 and 31830033, to TMG; 3220094 to NYH); the Key Area Research and Development Program of Guangdong Province (2018B030334001 and 2018B030340001, to TMG); and the Science and Technology Program of Guangzhou (202007030013, to TMG and SL2022A04J00182, to YHC).

References

1. Hu NY, Chen YT, Wang Q, Jie W, Liu YS, You QL, Li ZL, Li XW, Reibel S, Pfrieger FW, Yang JM, Gao TM (2020) Expression patterns of inducible Cre recombinase driven by differential astrocyte-specific promoters in transgenic mouse lines. Neurosci Bull 36(5):530–544. https://doi.org/10.1007/s12264-019-00451-z
2. Chen YH, Hu NY, Wu DY, Bi LL, Luo ZY, Huang L, Wu JL, Wang ML, Li JT, Song YL, Zhang SR, Jie W, Li XW, Zhang SZ, Yang JM, Gao TM (2022) PV network plasticity mediated by neuregulin1-ErbB4 signalling controls fear extinction. Mol Psychiatry 27(2):896–906. https://doi.org/10.1038/s41380-021-01355-z
3. Lin CCJ, Yu K, Hatcher A, Huang TW, Lee HK, Carlson J, Weston MC, Chen FJ, Zhang YQ, Zhu WY, Mohila CA, Ahnied N, Patel AJ, Arenkiel BR, Noebels JL, Creighton CJ, Deneen B (2017) Identification of diverse astrocyte populations and their malignant analogs. Nat Neurosci 20(3):396–405. https://doi.org/10.1038/nn.4493
4. Kantzer CG, Boutin C, Herzig ID, Wittwer C, Reiss S, Tiveron MC, Drewes J, Rockel TD, Ohlig S, Ninkovic J, Cremer H, Pennartz S, Jungblut M, Bosio A (2017) Anti-ACSA-2 defines a novel monoclonal antibody for prospective isolation of living neonatal and adult astrocytes. Glia 65(6):990–1004. https://doi.org/10.1002/glia.23140
5. Wang Q, Kong Y, Wu DY, Liu JH, Jie W, You QL, Huang L, Hu J, Chu HD, Gao F, Hu NY, Luo ZC, Li XW, Li SJ, Wu ZF, Li YL, Yang JM, Gao TM (2021) Impaired calcium signaling in astrocytes modulates autism spectrum disorder-like behaviors in mice. Nat Commun 12(1):ARTN 3321. https://doi.org/10.1038/s41467-021-23843-0
6. Joseph DJ, Von Deimling M, Hasegawa Y, Cristancho AG, Risbud R, McCoy AJ, Marsh ED (2021) Protocol for isolating young adult parvalbumin interneurons from the mouse brain for extraction of high-quality RNA. STAR Protoc 2(3):100714. https://doi.org/10.1016/j.xpro.2021.100714
7. Marques S, Zeisel A, Codeluppi S, van Bruggen D, Mendanha Falcao A, Xiao L, Li H, Haring M, Hochgerner H, Romanov RA, Gyllborg D, Munoz Manchado A, La Manno G, Lonnerberg P, Floriddia EM, Rezayee F, Ernfors P, Arenas E, Hjerling-Leffler J, Harkany T, Richardson WD, Linnarsson S, Castelo-Branco G (2016) Oligodendrocyte heterogeneity in the mouse juvenile and adult central nervous system. Science 352(6291):1326–1329. https://doi.org/10.1126/science.aaf6463
8. Zeisel A, Munoz-Manchado AB, Codeluppi S, Lonnerberg P, La Manno G, Jureus A, Marques S, Munguba H, He LQ, Betsholtz C, Rolny C, Castelo-Branco G, Hjerling-Leffler J, Linnarsson S (2015) Cell types in the mouse cortex and hippocampus revealed by single-cell RNA-seq. Science 347(6226):1138–1142. https://doi.org/10.1126/science.aaa1934
9. Zhang Y, Sloan SA, Clarke LE, Caneda C, Plaza CA, Blumenthal PD, Vogel H, Steinberg GK, Edwards MSB, Li G, Duncan JA, Cheshier SH, Shuer LM, Chang EF, Grant GA, Gephart MGH, Barres BA (2016) Purification and characterization of progenitor and mature human astrocytes reveals transcriptional and functional differences with mouse. Neuron 89(1):37–53. https://doi.org/10.1016/j.neuron.2015.11.013
10. Calvo B, Rubio F, Fernandez M, Tranque P (2020) Dissociation of neonatal and adult mice brain for simultaneous analysis of microglia, astrocytes and infiltrating lymphocytes by flow cytometry. IBRO Rep 8:36–47. https://doi.org/10.1016/j.ibror.2019.12.004
11. Paul A, Crow M, Raudales R, He M, Gillis J, Huang ZJ (2017) Transcriptional architecture of synaptic communication delineates GABAergic neuron identity. Cell 171(3):522–539 e520. https://doi.org/10.1016/j.cell.2017.08.032
12. Tasic B, Menon V, Nguyen TN, Kim TK, Jarsky T, Yao ZZ, Levi B, Gray LT, Sorensen SA, Dolbeare T, Bertagnolli D, Goldy J, Shapovalova N, Parry S, Lee C, Smith K, Bernard A, Madisen L, Sunkin SM, Hawrylycz M, Koch C, Zeng HK (2016) Adult mouse cortical cell taxonomy revealed by single cell transcriptomics. Nat Neurosci 19(2):335-+. https://doi.org/10.1038/nn.4216
13. Bhattacherjee A, Djekidel MN, Chen R, Chen W, Tuesta LM, Zhang Y (2019) Cell type-specific transcriptional programs in mouse prefrontal cortex during adolescence and addiction. Nat Commun 10(1):4169. https://doi.org/10.1038/s41467-019-12054-3
14. Proserpio V (2019) Single cell methods. Methods Mol Biol. https://doi.org/10.1007/978-1-4939-9240-9
15. Saunders A, Macosko EZ, Wysoker A, Goldman M, Krienen FM, de Rivera H,

Bien E, Baum M, Bortolin L, Wang SY, Goeva A, Nemesh J, Kamitaki N, Brumbaugh S, Kulp D, McCarroll SA (2018) Molecular diversity and specializations among the cells of the adult mouse brain. Cell 174(4): 1015–1030. https://doi.org/10.1016/j.cell.2018.07.028

16. Smith SM, Kimyon RS, Watters JJ (2014) Cell-type-specific Jumonji histone demethylase gene expression in the healthy rat CNS: detection by a novel flow cytometry method. ASN Neuro 6(3):193–207. https://doi.org/10.1042/AN20130050

17. Nikodemova M, Watters JJ (2012) Efficient isolation of live microglia with preserved phenotypes from adult mouse brain. J Neuroinflammation 9:147. https://doi.org/10.1186/1742-2094-9-147

18. Agalave NM, Lane BT, Mody PH, Szabo-Pardi TA, Burton MD (2020) Isolation, culture, and downstream characterization of primary microglia and astrocytes from adult rodent brain and spinal cord. J Neurosci Methods 340:108742. https://doi.org/10.1016/j.jneumeth.2020.108742

Index

A

Actin 34, 72, 95–99, 101, 102, 282
Action potential 229, 232–234, 241, 242, 245, 254
Adenosine 3′,5′-cyclic monophosphate (cAMP) 33–41, 144, 322
Adult astrocytes 341
Adult neurons 341
Alzheimer's disease 142, 158, 305
Arterial spin labeling (ASL) 282, 286, 287
Autism spectrum disorder (ASD) 3, 9, 331, 332
Automated immunostainer 22, 24, 27
Automatic analysis 305
Auxin-inducible degron (AID) 314–316

B

Behavior 2, 8, 178, 202, 246, 247, 305, 314, 322, 323, 325, 328, 331, 332, 334, 335, 339, 340
Biosensors 322
Biotin ligase 96, 97
Brain 1, 13, 21, 33, 50, 66, 73, 105, 142, 157, 169, 177, 187, 201, 212, 221, 245, 259, 271, 281, 305, 341
Brain dissection 170, 195
Brain interstitial fluid 259–261
Brain malformations 272, 277
Brain MRI 276–279, 282–286, 289
Brain organoids 6–7, 157–166
Brain slice 14, 22, 212–214, 217, 226

C

Caenorhabditis elegans 6, 305, 313–316, 322, 323, 325
Ca^{2+} imaging 211–218
Calcium imaging 324–326, 328
Cerebral cortex 1–9, 18, 170, 172, 173, 175, 179, 245, 283
Cerebrum 73, 76
Climbing assay 305, 306, 308, 310
c-Myc 121, 122
Computed tomography (CT) 271–279
Confocal laser microscopy 203, 205, 211, 213, 214
Cortical development 5, 221, 222, 231, 232, 236, 277
Crawling assay 305–307, 309
Cre/LoxP 314–316
CRISPR/Cas9 .... 188, 314–316
Critical period 237, 238, 242
Cryostat 59, 191, 196, 197
CT scan 272–275, 277–279
Cytoskeleton 95, 96, 99, 183, 282
Cytosolic soluble protein 178

D

Dendritic spine 33, 39, 75, 80, 177
Dental pulp (DP) 122, 124–126, 128–129, 132, 136, 137, 139
Development 1, 65, 72, 95, 142, 157, 170, 188, 201, 237, 246, 260, 271, 282, 313, 322, 332
Differentiation 138, 141–153, 158, 159, 162, 282
Diffusion weighted imaging (DWI) 281, 286
Digitonin 178–180, 183–185
Directed differentiation 142
Disease modeling 141, 142, 157
Dissociated cell culture 169–175
DNA copy numbers 294
DNA plasmid 38, 41, 152, 170, 172, 175, 188, 201, 216
Droplet digital PCR (ddPCR) 294–302
Drosophila larval and adult locomotor ability 305

E

Electrophysiology 248
Electroporator 189, 190, 192, 204, 213
Embryoid bodies (EBs) 158, 160, 161, 165, 188
Embryonic brain 170, 172, 173, 213, 216
EvaGreen 294–300, 302, 303

F

Flow cytometry 346–348
Fluorescence lifetime imaging microscopy (FLIM) 33–42
Fluorescence reporters 34, 38
Formalin-fixed paraffin-embedded (FFPE) 13–15, 18, 21–23, 28

Koh-ichi Nagata (ed.), *Cerebral Cortex Development: Methods and Protocols*, Methods in Molecular Biology, vol. 2794, https://doi.org/10.1007/978-1-0716-3810-1,

Förster resonance energy transfer (FRET) 33–42, 322
Freely moving animals 259, 260, 325
Freeze cracking 65
Freeze drying 67
Freeze-substitution 54, 55

G

Genetic screens 314
Genetically encoded Ca^{2+} indicator 211

I

Immature neurons 187, 214
Immunocytochemistry 132, 177, 178
Immunoelectron microscopy 45–47, 54
Immunohistochemistry (IHC) 21–31, 45, 46, 177, 189, 196, 197
Induced pluripotent stem cells (iPSCs) 121, 122, 128–129, 131, 132, 134–139, 141–153, 157–166, 233
In situ hybridization (ISH) 13–19
Intracellular recording 246–248
In utero electroporation (IUE) 187–190, 195, 196, 201–208, 211–214, 216, 217
In vitro reconstitution 81
In vitro single molecule motility assays 79
In vivo 169, 178, 188, 202, 221, 246, 247, 264, 282, 287, 321, 322
In vivo microdialysis 259–261, 263–267
Ion channel 54, 221, 232–236, 241, 247

K

Kinesins 79–81, 83, 85, 88–90, 93
Klf4 ...... 121, 122

L

Large biological molecules 260, 261
LifeAct (LA) 96–99, 101

M

Malformations of cortical development (MCD) 272, 273, 275, 277–279, 282, 283, 285, 286, 289–291
Mass spectrometry 103, 106
Membrane-associated protein 183
Membrane potential 230–232, 241, 245, 247
Membrane traffic 80
Microcapillary 189, 190, 192, 193, 198
Microtubule 79–82, 84, 85, 88–93, 183
Microtubule-associated septins 80
Mouse 19, 40, 47, 65, 72, 101, 106, 123, 171, 190, 203, 213, 247, 261, 334, 348
Mouse brain 19, 23, 24, 65, 69, 72, 106, 108, 109, 115, 158, 196, 260, 262–267
Multi-worm tracker (MWT) 322–325, 328, 329

N

Neurodevelopmental disorder 5, 141, 282, 286, 293, 331
Neurological diseases 157, 158
Neurons 21, 33, 71, 80, 115, 141, 169, 177, 187, 201, 211, 221, 245, 322, 341

O

Oct3/4 ...... 121, 122
Optogenetics 33, 255

P

Patch clamp 221, 222, 224–227, 229–233, 239–242, 246–248, 250, 254
Perfusion fixation 23, 29, 46, 49–50, 65
Photoactivatable adenylyl cyclase (PAC) 36, 37, 39–42
Postembedding method 45–61
Postsynaptic density (PSD) 71, 72, 74–76, 115
Preembedding method 45–61
Primary culture 170
Protein identification 106, 110, 113, 117
Proteomics 105–107, 109, 113–115, 117
Proximity ligation 96

Q

QuantaSoft software 295, 299, 303

R

Radiations 272, 273, 275, 279, 284, 287, 289
RNA-seq 347

S

Scanning electron microscopes 64
Sendai virus (SeV) 122, 128–129, 132, 135
Septins 80–83, 86–88, 90–93
Single-cell suspension 147, 172, 175, 342, 345
Single-color 294
Single-worm tracker 324, 325, 328
Slice culture 34, 36, 38, 41, 201–203, 211–214, 216, 217
Social interaction test 331–334
Sox2 ......121, 122, 158, 163
Stem cells 21, 124, 135, 187, 201
Subcellular fractionation 74
Synapse 55, 57, 60, 61, 71, 169, 233, 237
Synaptic plasticity 33, 34, 237, 238, 241

System....7, 8, 21, 22, 26, 27, 29, 31, 37, 39, 40, 55, 71, 84, 90, 95, 105, 113, 117, 121, 142, 164, 187, 203, 206, 212, 217, 224, 246, 249, 260, 261, 264, 267, 300, 301, 305, 306, 314–316, 321–326, 328, 341, 347

T

Three-chamber test....331, 334–340
Time-lapse imaging....170, 205, 207, 214
Total internal reflection fluorescence (TIRF) microscopy....79, 80, 84, 88, 89, 91, 92
Triton X-100....72, 73, 75, 77, 101, 164, 171, 178–180, 182–185, 191
T1-weighted sequences....285
T2-weighted sequences....286
Two-photon microscope....37

U

Ultracentrifugation....143, 153

V

Valproic acid (VPA)....332, 334

W

Whole-cell recording....231, 247, 255

MIX
Papier aus verantwortungsvollen Quellen
Paper from responsible sources
FSC® C105338

If you have any concerns about our products,
you can contact us on
ProductSafety@springernature.com

In case Publisher is established outside the EU,
the EU authorized representative is:
Springer Nature Customer Service Center GmbH
Europaplatz 3, 69115 Heidelberg, Germany

Printed by Libri Plureos GmbH
in Hamburg, Germany